Lecture Notes in Mathematics

Edited by A. Dold and B. Eckmann

Subseries: Department of Mathematics, University of Maryland
Adviser: J. Alexander

1024

Lie Group Representations I

Proceedings of the Special Year
held at the University of Maryland, College Park, 1982–1983

Edited by R. Herb, R. Lipsman and J. Rosenberg

Springer-Verlag
Berlin Heidelberg New York Tokyo 1983

Editors

Rebecca Herb
Ronald Lipsman
Jonathan Rosenberg
Department of Mathematics, University of Maryland
College Park, MD 20742, USA

AMS Subject Classifications (1980): 22E47, 22E40, 17B35

ISBN 3-540-12725-9 Springer-Verlag Berlin Heidelberg New York Tokyo
ISBN 0-387-12725-9 Springer-Verlag New York Heidelberg Berlin Tokyo

Library of Congress Cataloging in Publication Data Main entry under title: Lie group representations. (Lecture notes in mathematics; 1024) Sponsored by the Dept. of Mathematics, University of Maryland, College Park. 1. Lie groups–Congresses. 2. Representations of groups–Congresses. I. Herb, R. (Rebecca), 1948-. II. Lipsman, Ronald L. III. Rosenberg, J. (Jonathan), 1951-. IV. University of Maryland, College Park. Dept. of Mathematics. V. Series: Lecture notes in mathematics (Springer-Verlag); 1024. QA3.L28 no. 1024 510s [512'.55] 83-16871 [QA387]
ISBN 0-387-12725-9 (U.S. : v. 1)

Printed in Germany

Printing and binding: Beltz Offsetdruck, Hemsbach/Bergstr.
2146/3140-543210

PREFACE

The Department of Mathematics of the University of Maryland conducted a Special Year in Lie Group Representations during the academic year 1982-1983. This volume is the first (of three) containing articles submitted by the main speakers during the Special Year. Most of the invited speakers submitted articles, and virtually all of those appearing here deal with the subject matter on which the authors lectured during their visits to Maryland.

The Special Year program at Maryland represents a thriving departmental tradition—this being the fourteenth consecutive year in which such an event has taken place. As usual, the subject matter was chosen on the basis of active current research and the interests of departmental members. The modern theory of Lie Group Representations is a vast subject. In order to keep the program within bounds, the Special Year was planned around five distinct intensive periods of activity—each one (of three weeks duration) devoted to one of the main branches of current research in the subject. During those periods (approximately) eight distinguished researchers were invited to present lecture series on areas of current interest. Each visitor spent 1-3 weeks in the department and gave 2-5 lectures. In addition, during each period approximately 8-10 other visitors received financial support in order to attend and participate in the Special Year activities. Thus each period had to some extent the flavor of a mini-conference; but the length of the periods, the fact that visitors were provided with office space and the (relatively) low number of lectures per day also left ample time for private discussion and created the atmosphere of "departmental visitor" rather than "conference participant." Furthermore, as part of the Special Year the department was fortunate to have in residence D. Barbasch, J.

Bernstein and J.-L. Brylinski for the Fall 1982 semester, and B. Blank for the Spring 1983 semester. These visitors ran semester-long seminars in Group Representations. All of the activities of the Special Year were enthusiastically supported by the department, its faculty and graduate students.

Although most of the cost of the Special Year was borne by the department, the NSF did provide a generous amount of supplementary support. In particular, the contributions to the additional visitors were entirely funded by NSF. The Mathematics Department is grateful to the Foundation for its support of the Special Year. The Organizing Committee would also like to express its gratitude to the Department for its support. In particular the splendid efforts of Professors W. Kirwan, J. Osborn, G. Lehner, as well as of N. Lindley, D. Kennedy, D. Forbes, M. Keimig, and J. Cooper were vital to the success of the Special Year. The outstanding job of preparation of manuscripts by Berta Casanova and her staff June Slack, Anne Eberly and Linda Fiori, was of immense help in producing this volume so quickly. Also we are grateful to Springer-Verlag for its cooperation. Finally we are very pleased that so many of our participants provided us with high quality manuscripts, neatly prepared and submitted on time. It is our conviction that the theory of Group Representations has profited greatly from the efforts of all the above people towards the Special Year.

The Editors
April 1983

INTRODUCTION

We have made a serious attempt to group the papers (within the three volumes) according to the Periods in which they were presented and according to subject matter. However we were also influenced by the time at which manuscripts became available, and by a desire to equalize the size of the volumes. This (first) volume contains papers from Periods I and III of the Special Year. The programs for those periods were as follows:

PERIOD I. Algebraic Aspects of Semisimple Theory — Harish-Chandra Modules, Verma Modules, Kazhdan-Lusztig Conjectures, Methods of Homological Algebra, D-Modules, Primitive Ideals of Enveloping Algebras

T. Enright	--	Unitary representations
A. Joseph	--	Primitive ideals in the enveloping algebra of a semisimple Lie algebra
B. Kostant	--	Remarkable elements of finite order in semisimple Lie groups
G. Lusztig	--	Left cells in Weyl groups
W. Schmid	--	Asymptotics and intertwining operators
D. Vogan	--	Organizing the unitary dual
N. Wallach	--	Asymptotic expansions of generalized matrix entries

PERIOD III. Analytic Aspects of Semisimple Theory—Invariant Eigendistributions, L^p-Analysis, Schwartz Space, Irreducibility Criteria, Inversion Theorems, Semisimple Symmetric Spaces, Geometric Realization of Unitary Representations

M. Flensted-Jensen	--	Harmonic analysis on semisimple symmetric spaces - a method of duality
Sigurdur Helgason	--	Wave equations on homogeneous spaces
Anthony Knapp	--	Unitary representations and basic cases
Paul Sally	--	Tempered spectrum of SL(n) over a p-adic field
V. S. Varadarajan	--	Eigenvalues and eigenfunctions on homogeneous spaces
Garth Warner	--	Toward the trace formula
Gregg Zuckerman	--	Quantum physics and semisimple symmetric spaces

The additional participants during these periods of the Special Year were:

I.
- R. Gupta
- D. King
- A. Rocha
- P. Sally

III.
- D. Barbasch
- D. Collingwood
- J. Kolk
- R. Kunze
- B. Ørsted
- R. Stanton
- E. van den Ban
- J. Wolf

SPECIAL YEAR DATA

A. The five periods of activity of the Special Year were as follows:

I. Algebraic Aspects of Semisimple Theory -- Sept. 7, 1982 - Oct. 1, 1982
II. The Langlands Program -- Nov. 1, 1982 - Nov. 19, 1982
III. Analytic Aspects of Semisimple Theory -- Jan. 24, 1983 - Feb. 11, 1983
IV. The Orbit Method -- Feb. 28, 1983 - March 18, 1983
V. Applications -- April 18, 1983 - May 6, 1983

B. The speakers and the dates of their visits were:

Period I

Thomas Enright, UCSD (9/7 -9/22)
Anthony Joseph, Weizmann Institute (9/21 - 9/25)
Bertram Kostant, MIT (9/7 - 9/14)
George Lusztig, MIT (9/7 - 9/11)
Wilfried Schmid, Harvard (9/13 - 9/18)
David Vogan, MIT (9/27 - 10/1)
Nolan Wallach, Rutgers (9/20 - 10/1)

Period II

James Arthur, Toronto (11/1 - 11/19)
William Casselman, British Columbia (11/3 - 11/12)
Stephen Gelbart, Cornell (11/1 - 11/12)
Roger Howe, Yale (11/8 - 11/12)
Hervé Jacquet, Columbia (11/1 - 11/12)
David Kazhdan, Harvard (11/1 - 11/12)
Robert Langlands, IAS (11/1 - 11/12)
Ilya Piatetski -Shapiro, Yale (11/1 - 11/12)

Period III

Mogens Flensted-Jensen, Copenhagen (1/24 - 2/11)
Sigurdur Helgason, MIT (1/24 - 1/28)
Anthony Knapp, Cornell (2/2 - 2/4)
Paul Sally, Chicago (1/24 - 2/11)
V. S. Varadarajan, UCLA (1/24 - 2/11)
Garth Warner, Washington (2/7 - 2/8)
Gregg Zuckerman, Yale (1/24 - 2/4)

Period IV

Lawrence Corwin, Rutgers (3/7 - 3/11)
Michael Cowling, Genova (3/2 - 3/4)
Michel Duflo, Paris (2/28 - 3/11)
Roger Howe, Yale (3/7 - 3/11)
Henri Moscovici, Ohio State (3/7 - 3/18)
Richard Penney, Purdue (3/7 - 3/11)
Lajos Pukanszky, Penn (3/7, 3/11 - 3/18)
Wulf Rossmann, Ottawa (2/28 - 3/4)
Michèle Vergne, MIT (3/3 - 3/15)

Period V

Lawrence Corwin, Rutgers (4/18 - 4/29)
Bernard Helffer, Nantes (4/18 - 5/6)
Sigurdur Helgason, MIT (4/18 - 4/22)
Roger Howe, Yale (4/18 - 4/22)
Adam Koranyi, Washington Univ. (4/18 - 5/6)

Henri Moscovici, Ohio State (4/25 - 4/30)
Richard Penney, Purdue (4/25 - 5/6)
Linda Rothschild, Wisconsin (4/18 - 4/22)

C. The Organizing Committee for the 1982-1983 Special Year in Lie Group Representations is:

Rebecca Herb
Raymond Johnson
Stephen Kudla
Ronald Lipsman (Chairman)
Jonathan Rosenberg

TABLE OF CONTENTS

*For papers with more than one author, an asterisk indicates the author who delivered the lectures.

UNITARY REPRESENTATIONS FOR TWO REAL FORMS OF A SEMISIMPLE LIE ALGEBRA: A THEORY OF COMPARISON

Thomas J. Enright*
Department of Mathematics
University of California, San Diego
La Jolla, California 92093

§1. INTRODUCTION AND SUMMARY OF RESULTS

Unitary representations for semisimple Lie groups have been constructed by three main techniques. The first is the construction of the discrete series representations. The second is the general technique of unitary induction and unitary induction followed by analytic continuation of the invariant Hermitian form. The third technique is based on the Weil representation, the theory of dual pairs and the resulting decomposition of the Weil representation when restricted to certain subgroups of the symplectic group. This article concerns another general method for constructing unitary representations. This method is a theory for the comparison of admissible modules for two different real forms of a complex semisimple Lie algebra.

In this article we consider an example of this comparison theory. Let $\mathfrak{g}_{o,\mathbb{R}}$ be a real semisimple Lie algebra. Let G_o be the corresponding simply connected, connected Lie group with Lie algebra $\mathfrak{g}_{o,\mathbb{R}}$ and let K_o denote a maximal compactly embedded subgroup of G_o. Let $\mathfrak{k}_{o,\mathbb{R}}$ be the subalgebra of $\mathfrak{g}_{o,\mathbb{R}}$ corresponding to K_o. Delete the subscript $\mathbb{R}$ to denote the complexified Lie algebras. Assume (G_o,K_o) is an irreducible Hermitian symmetric pair. Let G be the simply connected, connected complex simple Lie group with Lie algebra $\mathfrak{g}_o$ and let $\mathfrak{g} = \mathfrak{g}_o \times \mathfrak{g}_o$. Then $\mathfrak{g}$ is the abstract complexification of $\mathfrak{g}_o$

---

*The author has been supported in part by NSF grant MCS-7802896.

when $\mathfrak{g}_o$ is considered as a Lie algebra over $\mathbb{R}$. Both $\mathfrak{g}_o$ and $\mathfrak{g}_{o,\mathbb{R}} \times \mathfrak{g}_{o,\mathbb{R}}$ are real forms of $\mathfrak{g}$ with corresponding Lie groups G and $G_o \times G_o$. The main results of this article describe a correspondence between special unitary representations of these two groups. In this example, the representations of G will be components of degenerate series; the representations of $G_o \times G_o$ will be the representations in the analytic continuation of the holomorphic discrete series.

In the remainder of this section, we introduce the basic notation kept throughout the article and describe in some detail the main results.

Fix a Cartan subalgebra (CSA) $\mathfrak{h}_o$ of $\mathfrak{k}_o$. Then $\mathfrak{h}_o$ is also a CSA of $\mathfrak{g}_o$. Since (G_o,K_o) is a Hermitian symmetric pair, $\mathfrak{k}_o$ is the reductive part of a maximal parabolic subalgebra. Write $\mathfrak{g}_o = \mathfrak{p}^- \oplus \mathfrak{k}_o \oplus \mathfrak{p}^+$ with $\mathfrak{p}^+$ the nilradical of the parabolic subalgebra $\mathfrak{k}_o \oplus \mathfrak{p}^+$. Let $\mathfrak{h}_o^*$ be the algebraic dual of $\mathfrak{h}_o$. Let Δ_o denote the roots of $(\mathfrak{g}_o,\mathfrak{h}_o)$, $\Delta_{o,c}$ the roots of $(\mathfrak{k}_o,\mathfrak{h}_o)$ and $\Delta_{o,n}$ the complement of $\Delta_{o,c}$ in Δ_o. We call $\Delta_{o,c}$ the compact roots and $\Delta_{o,n}$ the noncompact roots. Let Δ_o^+ be a positive system of roots for Δ_o and put $\Delta_{o,c}^+ = \Delta_o^+ \cap \Delta_{o,c}$, $\Delta_{o,n}^+ = \Delta_o^+ \cap \Delta_{o,n}$. <u>We may and do assume</u> Δ_o^+ is chosen so that $\mathfrak{p}^+$ is the span of the noncompact positive root spaces. We now define generalized Verma modules. For $\Delta_{o,c}$- integral $\lambda \in \mathfrak{h}_o^*$, let $F(\lambda)$ be the irreducible finite dimensional $\mathfrak{k}_o$-module with extreme weight λ. Now define

$$(1.1) \qquad N(\lambda) = U(\mathfrak{g}_o) \otimes_{U(\mathfrak{k}_o \oplus \mathfrak{p}^+)} F(\lambda).$$

Let $L(\lambda)$ denote the unique irreducible quotient of $N(\lambda)$. Let C denote the (closed) Weyl chamber corresponding to the positive system of roots $\Delta_{o,c}^+ \cup -\Delta_{o,n}^+$. Write ρ (resp. ρ_c, ρ_n) for half the sum of the elements in Δ_o^+ (resp. $\Delta_{o,c}^+$, $\Delta_{o,n}^+$). Then $\rho = \rho_c + \rho_n$. For $\lambda + \rho \in C$, $N(\lambda)$ is irreducible and is infinitesimally equivalent to a holomorphic discrete series or limit of holomorphic discrete series representation [12]. For $\Delta_{o,c}$ - integral λ, but with $\lambda + \rho$ not necessarily in the chamber C, we refer to the $\mathfrak{g}_o$-modules $N(\lambda)$ as the analytic continuation of the holomorphic discrete series. The reason for this terminology will become apparent in section three.

To define a map from admissible representations of $G_o \times G_o$ to admissible representations of G we use the right derived functors introduced by Zuckerman [20], [10].

Let $\mathfrak{k}$ be the diagonal subalgebra of $\mathfrak{g}_o \times \mathfrak{g}_o$ and let $\mathfrak{m}$ equal the diagonal in $\mathfrak{k}_o \times \mathfrak{k}_o$. Then $\mathfrak{k}$ is isomorphic to $\mathfrak{g}_o$ and this isomorphism carries $\mathfrak{m}$ onto $\mathfrak{k}_o$. Let $\mathfrak{n}^+$ (resp. $\mathfrak{n}^-$) denote the subalgebra corresponding to $\mathfrak{p}^+$ (resp. $\mathfrak{p}^-$). We have: $\mathfrak{k} = \mathfrak{n}^- \oplus \mathfrak{m} \oplus \mathfrak{n}^+$. We will carry over from $\mathfrak{g}_o$ to $\mathfrak{k}$ all the notation as above regarding positive roots, Weyl chambers and generalized Verma modules. It will be clear from the context whether we are considering roots, chambers or modules for $\mathfrak{g}_o$ or for $\mathfrak{k}$.

Let $C(\mathfrak{g},\mathfrak{m})$ be the category of $\mathfrak{g}$-modules which as $\mathfrak{m}$-modules are $U(\mathfrak{m})$ - locally finite and completely reducible. For a $\mathfrak{g}$-module $A \in C(\mathfrak{g},\mathfrak{m})$ define ΓA to be the subspace of $U(\mathfrak{k})$ - locally finite vectors. We call Γ the $\mathfrak{k}$-finite submodule functor. The category $C(\mathfrak{g},\mathfrak{m})$ has enough injective objects to construct injective resolutions. If $0 \to A \to I^*$ is an injective resolution of A then define $\Gamma^i A$ to be the ith homology group of the complex $0 \to \Gamma I^0 \to \Gamma I^1 \to \cdots$. The Γ^i are the right derived functors of Γ on $C(\mathfrak{g},\mathfrak{m})$. For all i, $\Gamma^i A$ is a $U(\mathfrak{k})$-locally finite $\mathfrak{g}$-module. If it is finitely generated it is infinitesimally equivalent to an admissible representation of G.

Next we consider some special degenerate series representations of G. Put $\mathfrak{q}_o = \mathfrak{k}_o \oplus \mathfrak{p}^+$. Then $\mathfrak{q}_o$ is a maximal complex parabolic subalgebra of $\mathfrak{g}_o$. In fact, in this way, we obtain all maximal parabolic complex subalgebras of $\mathfrak{g}_o$ with abelian nilradical. Let Q be the corresponding parabolic subgroup of G and let $Q = MAN$ be the Langlands decomposition. Since G is complex, both M and Q are connected. Also A is one dimensional (over $\mathbb{R}$) and Q is a maximal parabolic subgroup of G. Now define induced representations using the standard (shifted) unitary induction. Let ξ be a finite dimensional irreducible representation of MA and let H_ξ denote the $\mathfrak{g}$-module of $U(\mathfrak{k})$-locally finite vectors in the unitarily induced representation of ξ from Q to G (cf. (6.1)).

We can now state the main results. Put $L = L(\lambda)$, $N = N(\lambda)$, $L' = L(-\lambda-2\rho_n)$ and $N' = N(-\lambda-2\rho_n)$. Assume $F(\lambda)$ is one dimensional. We may and do (by symmetry of the parameters) assume $L' = N'$. Put $s = \frac{1}{2}\dim(\mathfrak{k}/\mathfrak{m})$.

Theorem 1.2. (i) $\Gamma^i(L \otimes N') = \Gamma^i(N \otimes N') = 0$ if $i \neq s$.

(ii) Γ^s carries any Jordan-Hölder series for $N \otimes N'$ to one for $\Gamma^s(N \otimes N')$.

(iii) The complexified Lie algebra of MA is $\mathfrak{k}_o \times \mathfrak{k}_o$. Let ξ be the one dimensional representation of MA corresponding to $F(\lambda+\rho_n) \otimes F(-\lambda-\rho_n)$. Then $\Gamma^s(N \otimes N')$ has the same distribution character as the degenerate series representation H_ξ.

(iv) Assume λ is real and $\Gamma^s(L \otimes N')$ admits an invariant Hermitian form. Then $\Gamma^s(L \otimes N')$ is unitarizable for G if and only if $L \otimes N'$ is unitarizable for $G_o \times G_o$.

The theorem is proved in several parts. Parts (i) and (ii) are given as (5.4) and (5.6). The connection with degenerate series is established in (6.7), and the most interesting result, part (iv), is the content of (8.13).

From the work of Wallach [19], it is known precisely when $L \otimes N'$ is unitarizable for $G_o \times G_o$. The result is quite intricate; and so, from (1.2(iv)) we obtain some interesting unitarizable class one components of the degenerate series H_ξ. These results are summarized in case by case form in section nine. Also, from the work of Jantzen [14], it is easy to determine precisely when N is reducible. Thus, by (1.2(ii) and (iii)), we obtain necessary and sufficient conditions for the reducibility of the degenerate series H_ξ. We include these results in the form of a table in section nine.

§2. NOTATION

We include here some standard notation, definitions and facts used throughout the article. In addition to the Lie algebras defined in section one, put $\mathfrak{h} = \mathfrak{h}_o \times \mathfrak{h}_o$. Then $\mathfrak{h}$ is a CSA of $\mathfrak{g}$. Let Δ denote the roots of $(\mathfrak{g},\mathfrak{h})$ and let $\mathcal{W}$ (resp. $\mathcal{W}_o$) be the Weyl group of Δ (resp. Δ_o). Let $\mathcal{W}_{o,c}$ be the Weyl group of $\Delta_{o,c}$. Each Weyl group contains a unique element of maximal length. Let w_o (resp. u_o) be the element of maximal length in $\mathcal{W}_o$ (resp. $\mathcal{W}_{o,c}$). Put $r_o = w_o u_o = u_o w_o$. The element r_o maps the chamber C to the dominant chamber for Δ_o^+. The Hermitian symmetric pair $(\mathfrak{g}_o,\mathfrak{k}_o)$ will break into two types depending as $r_o\rho_n = -\rho_n$ or not. The cases for which $r_o\rho_n \neq -\rho_n$ are: $su(p,q)$, $p \neq q$; $so^*(2n)$, n odd and the pair (E_6,D_5).

For any Lie algebra $\mathfrak{a}$, let $U(\mathfrak{a})$ denote the universal enveloping algebra and $Z(\mathfrak{a})$ the center of $U(\mathfrak{a})$. For an $\mathfrak{a}$-module A and a homomorphism γ of $Z(\mathfrak{a})$ let the generalized eigenspace for γ be the maximal submodule of A where the operators $z - \gamma(z)$ are locally nilpotent for all $z \in Z(\mathfrak{a})$. If $Z(\mathfrak{a})$ acts by γ on A we say A has infinitesimal character γ.

For any Lie algebras $\mathfrak{a}$ and $\mathfrak{b}$ with $\mathfrak{b} \subset \mathfrak{a}$, let $C(\mathfrak{a},\mathfrak{b})$ denote the category of $\mathfrak{a}$-modules which are $U(\mathfrak{b})$ - locally finite and completely reducible as $\mathfrak{b}$-modules. If τ is any anti-involution of $U(\mathfrak{a})$ and A is an object in $C(\mathfrak{a},\mathfrak{b})$ then define the τ-dual to A, denoted by A^τ, to be the $\mathfrak{a}$-module whose vector space is the $U(\mathfrak{b})$ - locally finite vectors in the algebraic dual to A and whose action is given by:

$$(x\cdot f)(a) = f(\tau(x)\cdot a), \qquad x \in U(\mathfrak{a}), \quad a \in A, \quad f \in A^\tau.$$

Let $\mathcal{A}(\mathfrak{a},\mathfrak{b})$ be the subcategory of $C(\mathfrak{a},\mathfrak{b})$ of modules which have finite dimensional isotypic subspaces as $\mathfrak{b}$-modules.

For any root α and $\lambda \in \mathfrak{h}_o^*$ define the coroot $\alpha^\vee$ by $(\lambda,\alpha^\vee) = \frac{2(\alpha,\lambda)}{(\alpha,\alpha)}$.

Let $\mathbb{N}$ be the natural numbers $0,1,2,\ldots$ and $\mathbb{Z}$ the integers. Let $\mathbb{N}^*$ and $\mathbb{Z}^*$ denote the sets obtained by deleting zero.

§3. HIGHEST WEIGHT MODULES FOR HERMITIAN SYMMETRIC PAIRS

In this section we summarize the results of Wallach on the analytic continuation of the discrete series [19]. In addition to the standard notations and definitions we need here the strongly orthogonal noncompact roots as introduced by Harish-Chandra [12]. Let r equal the split rank of $\mathfrak{g}_{o,\mathbb{R}}$. Let γ_1 be the unique simple noncompact root in Δ_o^+. Assume $\gamma_1,\ldots,\gamma_j$ have been defined and define γ_{j+1} to be a minimal (with respect to Δ_o^+) element in $\Delta_{o,n}^+$ which is orthogonal to γ_i, $1 \le i \le j$, if such an element exists. The set $\{\gamma_i\}$ contains r elements and has the properties: (1) $\gamma_1 < \cdots < \gamma_r$ and $\gamma_i \pm \gamma_j \notin \Delta_o$ for all $1 \le i < j \le r$, (2) let H_{γ_i} be dual to γ_i in $\mathfrak{h}_o$ and let $\mathfrak{h}_o^-$ be the span of H_{γ_i}, $1 \le i \le r$. If $\alpha \in \Delta_{o,n}^+$, then the restriction of α to $\mathfrak{h}_o^-$ has one of the forms:

a) $\alpha|_{\mathfrak{h}_o^-} = \frac{1}{2}\gamma_i$, for some i, $1 \le i \le r$,

b) $\alpha|_{\mathfrak{h}_o^-} = \frac{1}{2}(\gamma_i+\gamma_j)$ for some i,j, $1 \le i \le j \le r$.

If $\alpha \in \Delta_{o,c}^+$ then the restriction to $\mathfrak{h}_o^-$ has one of the forms:

a) $\alpha|_{\mathfrak{h}_o^-} = -\frac{1}{2}\gamma_j$ for some j, $1 \le j \le r$,

b) $\alpha|_{\mathfrak{h}_o^-} = \frac{1}{2}(\gamma_i-\gamma_j)$ for some i,j, $1 \le j < i \le r$,

c) $\alpha|_{\mathfrak{h}_o^-} = 0$.

Recall from section one, the decomposition $\mathfrak{g}_o = \mathfrak{p}^- \oplus \mathfrak{k}_o \oplus \mathfrak{p}^+$ and the definitions of the generalized Verma modules $N(\lambda)$ and their irreducible quotients $L(\lambda)$. Since we have the $\mathfrak{k}_o$-module isomorphism:

$$N(\lambda) \simeq S(\mathfrak{p}^-) \otimes F(\lambda),$$

the decomposition of $S(\mathfrak{p}^-)$ as a $\mathfrak{k}_o$-module will be especially useful. Schmid has analyzed this decomposition [17].

<u>Proposition 3.1</u>. $S(\mathfrak{p}^-)$ is a multiplicity free $\mathfrak{k}_o$-module and $F(\mu)$ occurs in $S(\mathfrak{p}^-)$ if and only if $\mu = -m_1\gamma_1 - \cdots -m_r\gamma_r$, $m_i \in \mathbb{N}$, $m_1 \geq \cdots \geq m_r$.

We now specialize to the case where $F(\lambda)$ is one dimensional. $\mathfrak{k}_o$ has a one dimensional center. Therefore define $\zeta \in \mathfrak{h}_o^*$ by the conditions

$$(3.2)\qquad \begin{cases} \text{i)} & \zeta \text{ is orthogonal to } \Delta_{o,c} \\ \text{ii)} & 2\langle\zeta,\gamma_1\rangle/\langle\gamma_1,\gamma_1\rangle = 1. \end{cases}$$

Let c_j equal the number of positive compact roots α with $\alpha|_{\mathfrak{h}^-} = \frac{1}{2}(\gamma_j-\gamma_i)$ with $i < j$. Note that $c_1 = 0$. A case by case check shows that $c_j = 2(j-1)c$ with c given by the table:

(3.3)

real form	su(p,q)	sp(n,ℝ)	so*(2n)	so(2,2n-2)	so(2,2n-1)	E III	E VII
c	1	1/2	2	n-2	$n-\frac{3}{2}$	3	4

The analytic continuation of the discrete series as described by Wallach [19] is given by:

<u>Theorem 3.4</u>. $L(z\zeta)$ is unitarizable if and only if $z < -(r-1)c$ or $z = -(j-1)c$ for $j \in \mathbb{N}$, $1 \leq j \leq r$.

For $z < -(r-1)c$, $L(z\zeta) = N(z\zeta)$; and so, the $\mathfrak{k}_o$-structure is given by Schmid's theorem. At the points $z = -(j-1)c$, $j \in \mathbb{N}$, $1 \leq j \leq r$, $N(z\zeta)$ is reducible. However, at these points Wallach has determined the $\mathfrak{k}_o$-structure of $L(z\zeta)$. This result is included in remarks ending the proof of (5.10) [19].

<u>Theorem 3.5</u>. Let $z = -(j-1)c$, $j \in \mathbb{N}$, $1 \leq j \leq r$. Then $L(z\zeta)$ is a multiplicity free $\mathfrak{k}_o$-module and $F(\mu)$ occurs in $L(z\zeta)$ if and only if μ has the form $z\zeta - m_1\gamma_1 - \cdots - m_{j-1}\gamma_{j-1}$, $m_i \in \mathbb{N}$, $m_1 \geq \cdots \geq m_{j-1}$.

The following lemma will be used in later sections.

Lemma 3.6. Let C be the positive chamber associated to the positive system of roots $\Delta^+_{o,c} \cup -\Delta^+_{o,n}$. Then the elements $-m_1\gamma_1 - \cdots - m_r\gamma_r$, $m_i \in \mathbb{N}$, $m_1 \geq \cdots \geq m_r$, are integral elements in C.

Proof. Let β be the maximal root in Δ^+_o. Then the simple roots for C are $-\beta$ and the simple roots of $\Delta^+_{o,c}$. From Schmid's result (3.1), our elements are dominant for $\Delta^+_{o,c}$; and so, we need only check the inner product with $-\beta$. It is sufficient to compute the inner product with the restriction of $-\beta$ to $\mathfrak{h}^-_o$. From the properties listed above, $-\beta$ restricts to either $-\frac{1}{2}\gamma_i$, $1 \leq i \leq r$, or $-\frac{1}{2}(\gamma_i+\gamma_j)$, $1 \leq i \leq j \leq r$. Now since the γ_i are orthogonal, we conclude that the inner product of $-\beta$ and $-m_1\gamma_1 - \cdots - m_r\gamma_r$ is ≥ 0. This proves the lemma.

The elements ρ_n and ζ are linearly dependent. A case by case check (cf. [19]) shows that $-\rho_n$ corresponds to a point on the line $z\zeta$ with $z < -(r-1)c$. This proves

Lemma 3.7. $N(-\rho_n - z\zeta)$ is irreducible and unitarizable for all $z \geq 0$.

Remark 3.8. For all $\lambda \in \mathfrak{h}^*_o$ which are $\Delta_{o,c}$ - integral it is known when $L(\lambda)$ is unitarizable [9]. Since we do not use these more general results we do not take the time to describe them here.

§4. THE DERIVED FUNCTORS OF THE $\mathfrak{k}$-FINITE SUBMODULE FUNCTOR

Here we summarize the basic results for the Zuckerman derived functors. For more details the reader should consult [10] or [18].

Let notation be as in sections one and two. Put $\mathfrak{t}$ equal to the diagonal in $\mathfrak{h} = \mathfrak{h}_o \times \mathfrak{h}_o$. Then $\mathfrak{t}$ is a CSA of both $\mathfrak{m}$ and $\mathfrak{k}$. For $\lambda \in \mathfrak{t}^*$ and integral for $\mathfrak{m}$ define as in (1.1), $N(\lambda) = U(\mathfrak{k}) \otimes_{U(\mathfrak{m}+\mathfrak{n}^+)} F(\lambda)$. As noted in section one the isomorphism of $\mathfrak{k}$ and $\mathfrak{g}_o$ carries $\mathfrak{m}$ to $\mathfrak{k}_o$, $\mathfrak{n}^+$ to $\mathfrak{p}^+$ and $\mathfrak{t}$ to $\mathfrak{h}_o$. Under these isomorphisms we write Δ_o (resp. $\Delta_{o,c}$, Δ^+_o, etc.) for the roots of $\mathfrak{k}$ (resp. roots of $\mathfrak{m}$, positive roots of $\mathfrak{k}$, etc). Let Γ^i be defined as in section one. For our purposes the usefulness of Γ^i comes from its computability. Although the Γ^i above are defined on the category $C(\mathfrak{g},\mathfrak{m})$, for A an object in $C(\mathfrak{g},\mathfrak{m})$ the $\mathfrak{k}$-structure of $\Gamma^i A$ depends only on the $\mathfrak{k}$-structure of A. The basic computational result for Γ^i is

Proposition 4.1. Let $\lambda \in \mathfrak{t}^*$ be Δ_o^+-dominant integral and regular and let $\omega \in W_o$ satisfy: $\omega\Delta_o^+ \supset \Delta_{o,c}^+$. Then, for $i \in \mathbb{N}$, $2s = \dim(\mathfrak{k}/\mathfrak{m})$,

$$\Gamma^{2s-i}(N(\omega\lambda-\rho)) = \begin{cases} L(\lambda-\rho) & \text{if } i = \text{length of } \omega \\ 0 & \text{otherwise.} \end{cases}$$

For a proof of 4.1 see [10] or [18].

§5. STRUCTURE OF THE MODULE $N \otimes N'$

In this section we describe the structure of the modules $N \otimes N'$ which enter in the main theorem. We continue with the notation of sections one and two. For $\mathfrak{g}_o$ - modules A and B, $A \otimes B$ is a $\mathfrak{g}$-module and the restriction of the action to the diagonal subalgebra $\mathfrak{k}$ is the standard tensor product action of $\mathfrak{g}_o$ on $A \otimes B$. Therefore, to analyze the $\mathfrak{k}$-module structure of $A \otimes B$ we must decompose the tensor product action.

Fix $\lambda \in \mathfrak{h}_o^*$ and assume $F(\lambda)$ is one dimensional. Put

(5.1) $N = N(\lambda)$, $L = L(\lambda)$, $N' = N(-\lambda-2\rho_n)$ and $L' = L(-\lambda-2\rho_n)$.

Since $\mathfrak{g}_o$ is simple, $\mathfrak{k}_o$ has a one dimensional center and so $\lambda = z\rho_n$, $z \in \mathbb{C}$. By symmetry of our parameter we may assume, interchanging N and N' if necessary, $\mathrm{Re}\, z \geq -1$. Then by (3.7) we find N' is irreducible.

Proposition 5.2. As a $\mathfrak{k}$-module, $N \otimes N'$ is a direct sum of irreducible generalized Verma modules $N(\mu - 2\rho_n)$. The multiplicity of $N(\mu-2\rho_n)$ is either one or zero depending as μ has the form

$-m_1\gamma_1 - \cdots - m_r\gamma_r$, $m_i \in \mathbb{N}$, $m_1 \geq \cdots \geq m_r$, or not. Moreover, for each summand the highest weight plus ρ lies in C.

Note: The γ_i are as in (3.1) and, by our convention, we use the notation of $(\mathfrak{g}_o, \mathfrak{k}_o)$ since $\mathfrak{k} \simeq \mathfrak{g}_o$ and $\mathfrak{m} \simeq \mathfrak{k}_o$.

Proof. As a $\mathfrak{k}_o$-module, $N \simeq S(\mathfrak{p}^-) \otimes F(\lambda)$. Let $S(\mathfrak{p}^-) = \sum F_i$ be the decomposition of $S(\mathfrak{p}^-)$ into irreducible $\mathfrak{k}_o$-modules. Then

$$N \otimes N' = N \otimes U(\mathfrak{g}_o) \otimes_{U(\mathfrak{k}_o \oplus \mathfrak{p}^+)} F(-\lambda-2\rho_n)$$

(5.3)

$$\simeq U(\mathfrak{g}_o) \otimes_{U(\mathfrak{k}_o \oplus \mathfrak{p}^+)} (N \otimes F(-\lambda-2\rho_n)).$$

Now combining (5.3) and $N \simeq \sum F_i \otimes F(\lambda)$, we obtain a filtration A_i of $N \otimes N'$ where A_i/A_{i+1} is isomorphic to $U(\mathfrak{g}_o) \otimes_{U(\mathfrak{k}_o + \mathfrak{p}^+)} (F_i \otimes F(-2\rho_n))$. So by (3.1) we obtain a filtration with the correct multiplicities. To prove that this filtration splits we observe that the highest weights plus ρ all have the form $-m_1\gamma_1 - \cdots - m_r\gamma_r - \rho_n + \rho_c$. But $-\rho_n + \rho_c$ is half the sum of the positive roots for C and thus (3.6) implies that the infinitesimal characters of A_i/A_{i+1} are all different for different i. So the decomposition into generalized eigenspaces for $Z(\mathfrak{k})$, the center of $U(\mathfrak{k})$, gives the direct sum decomposition. Since the highest weight plus ρ for each summand lies in C, the generalized Verma module is irreducible. This proves (5.2).

Using the derived functor from section four, we now define some special admissible $(\mathfrak{g},\mathfrak{k})$-modules. Let $s = \frac{1}{2} \dim \mathfrak{k}/\mathfrak{m}$ and let w_o (resp. u_o) be the maximal element of W_o (resp. $W_{o,c}$).

Proposition 5.4. i) For any $\mathfrak{g}$-module subquotient A of $N \otimes N'$, $\Gamma^i A = 0$ if $i \neq s$.

ii) $\Gamma^s(N \otimes N')$ is an admissible $(\mathfrak{g},\mathfrak{k})$-module. As a $\mathfrak{k}$-module it is multiplicity free and the finite dimensional $\mathfrak{k}$-module $L(\mu)$ occurs if and only if μ has the form: $\mu = w_o u_o(-m_1\gamma_1 - \cdots - m_r\gamma_r)$, $m_i \in \mathbb{N}$, $m_1 \geq \cdots \geq m_r$.

Proof. Since $N \otimes N'$ is the direct sum of irreducible generalized Verma modules, the same is true for any subquotient. But i) and ii) now follow from the direct sum decomposition (5.2) and the computation of Γ^i in (4.1).

For convenience we put:

(5.5) $$M = \Gamma^s(N \otimes N').$$

Proposition 5.6. i) Let $A \subset N \otimes N'$ be a $\mathfrak{k}$-submodule. Then A is a $\mathfrak{g}$-submodule if and only if $\Gamma^s A$ is a $\mathfrak{g}$-submodule of M.

ii) Γ^s maps a Jordan-Hölder series for $N \otimes N'$ to one for M. In particular, M is irreducible if and only if $N \otimes N'$ is irreducible.

Note: From the work of Jantzen on generalized Verma modules, it is known precisely when $N \otimes N'$ is irreducible. We shall return to this in detail in section eight.

To prove (5.6) we will need several technical results given now as lemmas.

Let C be a summand in the decomposition of $N \otimes N'$; i.e., $C = N(\mu-2\rho_n)$ with $\mu = -m_1\gamma_1 - \cdots - m_r\gamma_r$, $m_i \in \mathbb{N}$, $m_1 \geq \cdots \geq m_r$. Let χ_o denote the infinitesimal character of C. As above, let β denote the maximal root in Δ_o^+. Let $E = \mathfrak{g}_o \otimes C$ and consider E as a $\mathfrak{k}$-module by the tensor product action of ad on $\mathfrak{g}_o$ and the given action on C.

Lemma 5.7. i) If $\langle\mu,\beta\rangle < 0$ then E is the direct sum of irreducible generalized Verma modules $N(\xi)$ with $\xi + \rho \in \mathcal{C}$.

ii) If $\langle\mu,\beta\rangle = 0$ then for every infinitesimal character χ different from χ_o, the generalized eigenspace for χ in E is isomorphic to an irreducible generalized Verma module $N(\xi)$ with $\xi + \rho \in \mathcal{C}$.

Proof. Since $E \simeq U(\mathfrak{g}_o) \otimes_{U(\mathfrak{k}_o+\mathfrak{p}^+)} (\mathfrak{g}_o \otimes F(\mu-2\rho_n))$, to prove i) we need only show:

(5.8) if $F(\nu)$ occurs in $\mathfrak{g}_o \otimes F(\mu-2\rho_n)$ and ν is $\Delta_{o,c}^+$-dominant, then $\nu + \rho \in \mathcal{C}$.

We now prove (5.8). Since ν is a $\Delta^+_{o,c}$-highest weight, $\nu = \alpha + \mu - 2\rho_n$ for some root $\alpha \in \Delta$ or $\alpha = 0$. Since $-\beta$ is a simple root for C, $\nu + \rho \in C$ if and only if $(\nu+\rho)_\beta \leq 0$. Also since $-\beta$ is simple for C, $(-2\rho_n+\rho)_\beta = -1$. Therefore, $(\nu+\rho)_\beta = \alpha_\beta + \mu_\beta - 1$. From the form of μ and since $\beta|_{\mathfrak{h}_o^-} = \frac{1}{2}\gamma_i$ or $\frac{1}{2}(\gamma_i+\gamma_j)$ for some $1 \leq i$, $j \leq r$, we have $\mu_\beta \geq 0$. So $\nu + \rho \in C$ unless $\mu_\beta = 0$ and $\alpha_\beta = 2$. This proves i).

For ii), observe that if $\nu + \rho \notin C$ then $\mu_\beta = 0$ and $\alpha_\beta = 2$. But β is the maximal root; and so, is long. This means $\alpha = \beta$ and then $\nu + \rho = s_\beta(\mu-2\rho_n+\rho)$. This proves the lemma.

Next we consider the $\mathfrak{k}$-module map induced by the action of $\mathfrak{g}$ on $N \otimes N'$. Define a map:

(5.9) $\qquad \theta: E \to N \otimes N'; \quad y \otimes c \mapsto (y,-y)\cdot c \qquad$ for $y \in \mathfrak{g}_o$, $c \in C$.

<u>Lemma 5.10</u>. $\Gamma^s C + \text{Image } \Gamma^s\theta = \Gamma^s(\text{Image } \theta) + \Gamma^s C$.

<u>Proof</u>. Let χ be the infinitesimal character of a $\mathfrak{k}$-module. For a $\mathfrak{k}$-module A, let A_χ denote the generalized eigenspace in A for χ. By (5.7), if χ is not the infinitesimal character of C, then Γ^s is an exact functor on $E_\chi \to \theta E_\chi$. This gives $\Gamma^s\theta(\Gamma^s E_\chi) = \Gamma^s(\theta E_\chi)$. Since $\Gamma^s C$ is the full generalized eigenspace for the infinitesimal character of C by (5.2), this equality proves the lemma.

We now prove the proposition. Let A be a $\mathfrak{k}$-submodule of a $\mathfrak{g}$-module. Then A is a $\mathfrak{g}$-submodule if and only if $(y,-y)\cdot A \subset A$ for all $y \in \mathfrak{g}_o$. Equivalently, with notation as above, A is a $\mathfrak{g}$-submodule of $N \otimes N'$ if and only if for each submodule C of A, $\theta(\mathfrak{g}_o \otimes C) \subsetneq A$. From the basic properties of Γ^i, $\Gamma^s(\mathfrak{g}_o \otimes C) \simeq \mathfrak{g}_o \otimes \Gamma^s C$ and, under this isomorphism, $\Gamma^s\theta(y \otimes c) = (y,-y)\cdot c$, for $c \in \Gamma^s C$. Therefore, from (5.10) and (5.2), A is a $\mathfrak{g}$-submodule of $N \otimes N'$ if and only if $\Gamma^s A$ is a $\mathfrak{g}$-submodule of M. This completes the proof of (5.6)(i); and in fact gives a bijection between $\mathfrak{g}$-submodules of $N \otimes N'$ and M. Finally (5.6)(ii) follows from this bijection.

For later reference, we give:

Lemma 5.11. i) The infinitesimal character of $N \otimes N'$ is parameterized by the $W_o \times W_o$ orbit of $(\lambda+\rho,-\lambda-2\rho_n+\rho)$.

ii) If $\dim F(\lambda)$ equals one then the orbit in i) is the orbit of $(\lambda+\rho,-\lambda-\rho)$.

Proof. i) is clear. For ii), let u_o be the maximal element of $W_{o,c}$. Then $u_o\rho_c = -\rho_c$, $u_o\rho_n = \rho_n$, $u_o\lambda = \lambda$. Now apply $(1,u_o)$ to $(\lambda+\rho,-\lambda-2\rho_n+\rho)$. This gives ii).

§6. DEGENERATE SERIES FOR G

In this section we describe the connection between $\Gamma^s(N \otimes N')$ and the degenerate series for the complex Lie group G. The main result here is (6.7).

We begin by viewing $\mathfrak{g}_o$ from a point of view somewhat different from the preceeding sections. Here we consider $\mathfrak{g}_o$ as the real Lie algebra of G. Then $\mathfrak{q}_o = \mathfrak{k}_o \oplus \mathfrak{p}^+$ is a parabolic subalgebra with nil radical $\mathfrak{p}^+$. The center of $\mathfrak{k}_o$ is one dimensional (over $\mathbb{C}$); and so, we have a Langlands decomposition for $\mathfrak{q}_o$: $\mathfrak{q}_o = \mathfrak{m}_o \oplus \mathfrak{a}_o \oplus \mathfrak{n}_o$ where $\mathfrak{n}_o = \mathfrak{p}^+$, $\mathfrak{m}_o \oplus \mathfrak{a}_o = \mathfrak{k}_o$ and $\dim_{\mathbb{R}} \mathfrak{a}_o = 1$. Let $Q = MAN$ be the corresponding parabolic subgroup of G. Since G is complex, both M and Q are connected subgroups of G.

We now define induced representations from Q to G. Let $\rho_{G/Q}$ be the functional on $\mathfrak{a}$ equal to $\frac{1}{2}$ trace ad(h) restricted to $\mathfrak{p}^+$, $h \in \mathfrak{a}$. Let ξ be an irreducible finite dimensional representation of Q on a vector space E. Let H_ξ^∞ be the set of C^∞-functions f on G with values in E satisfying:

$$(6.1) \qquad f(g\,man) = (e^{\rho_{G/Q}(a^{-1})} \otimes \xi(m^{-1}a^{-1}))f(g),$$
$$g \in G,\ m \in M,\ a \in A,\ n \in N.$$

Now G acts on H_ξ^∞ by left translation. Let K be the maximal compact subgroup of G whose complexified Lie algebra is $\mathfrak{k} \hookrightarrow \mathfrak{g}$. The dense subspace of H_ξ^∞ of K-finite vectors is a $U(\mathfrak{g})$-module which we denote by H_ξ. We call this admissible $(\mathfrak{g},\mathfrak{k})$-module, the <u>degenerate series</u> module induced from (Q,ξ) to G.

These degenerate series modules are in fact coinduced modules. We now give the correspondence. Let $\mathfrak{q}_o^- = \mathfrak{k}_o \oplus \mathfrak{p}^-$. Then the complexification of $\mathfrak{q}_o$ in $\mathfrak{g}$ is given by $\mathfrak{q} = \mathfrak{q}_o \times \mathfrak{q}_o^-$. For any $\mathfrak{q}$-module W, let $\mathrm{Hom}_{U(\mathfrak{q})}(U(\mathfrak{g}),W)$ be the space of linear maps $\varphi\colon U(\mathfrak{g}) \to W$ such that

$$(6.2) \qquad \varphi(yx) = y\cdot\varphi(x); \qquad x \in U(\mathfrak{g}),\ y \in U(\mathfrak{q}).$$

$U(\mathfrak{g})$ acts on this space, by the action induced by right multiplication on $U(\mathfrak{g})$. We have:

$$(6.3) \qquad (x\cdot\varphi)(z) = \varphi(zx), \qquad x,z \in U(\mathfrak{g}).$$

This $U(\mathfrak{g})$-module is called the module coinduced from the pair $(\mathfrak{q},W)$. Let $H(W)$ denote the $U(\mathfrak{q})$-submodule of locally $U(\mathfrak{k})$-finite vectors. For $f \in H_\xi^\infty$, let Tf be the linear map of $U(\mathfrak{g})$ into E defined by:

$$Tf(x) = (x\cdot f)(1), \qquad x \in U(\mathfrak{g}).$$

<u>Lemma 6.4</u>. The map T is a $\mathfrak{g}$-module isomorphism of H_ξ onto $H(E \otimes \rho_{G/Q})$.

This is a standard result. For a proof see (4.2) [11] or [5].

Coinduced modules are dual to induced modules. We now describe the degenerate series as the $\mathfrak{k}$-finite dual of certain generalized Verma modules.

<u>Lemma 6.5</u>. The degenerate series module H_ξ is isomorphic to the $\mathfrak{g}$-module of $\mathfrak{k}$-finite vectors in the algebraic dual of $U(\mathfrak{g}) \otimes_{U(\mathfrak{q})} (E \otimes \rho_{G/Q})^*$

This follows from (6.4) and the natural duality between $\mathrm{Hom}_{U(\mathfrak{q})}(U(\mathfrak{q}),A^*)$ and $U(\mathfrak{g}) \otimes_{U(\mathfrak{q})} A$. For details see [5].

From this last isomorphism we read off easily the infinitesimal character of H_ξ.

<u>Lemma 6.6</u>. If $\mathfrak{r}$ is the reductive part of $\mathfrak{q}$, then $\mathfrak{r} = \mathfrak{k}_o \times \mathfrak{k}_o$ and

$\rho_{\mathfrak{r}} = (\rho_c, \rho_c)$ is half the sum of positive roots for $\mathfrak{r}$. Assume E is an irreducible finite dimensional $\mathfrak{q}$-module with lowest weight μ. Then, the infinitesimal character of H_ξ is parameterized by the Weyl group orbit of $\mu - \rho_{\mathfrak{r}}$.

<u>Proof</u>. If μ is a lowest weight of E, then $-\mu - \rho_{G/Q}$ is the highest weight of $U(\mathfrak{g}) \otimes_{U(\mathfrak{q})} (E \otimes \rho_{G/Q})^*$. The infinitesimal character of this module is given by the orbit of the highest weight plus ρ which equals $-\mu + \rho_{\mathfrak{r}}$. So, being dual, H_ξ has infinitesimal character given by the orbit of $\mu - \rho_{\mathfrak{r}}$.

From the work of Casselman [3], we know that any admissible $(\mathfrak{g},\mathfrak{k})$-module X has a (distribution) character which we denote by $\Theta(X)$.

<u>Proposition 6.7</u>. Let notation be as in section five and assume that $F(\lambda)$ is one dimensional. Let (ξ,E) be the one dimensional $\mathfrak{q}$-module $F(\lambda+\rho_n) \otimes F(-\lambda-\rho_n)$. Then $\Gamma^s(N \otimes N')$ has the same distribution character as the degenerate series module H_ξ.

<u>Proof</u>. The set of λ with $F(\lambda)$ one dimensional is a complex line. For a Zariski open set of this line both N and N' are irreducible. Therefore, by (5.6), $M = \Gamma^s(N \otimes N')$ is irreducible. For complex groups, if two irreducible representations have a $\mathfrak{k}$-fixed vector and the same infinitesimal character they are isomorphic. Both M and H_ξ have $\mathfrak{k}$-fixed vectors and the same infinitesimal character by (5.11) and (6.6). Therefore, on the Zariski open set, M will occur as an irreducible component of H_ξ. But results of Berline and Duflo [4] show that H_ξ is irreducible for a Zariski open set of parameters. Thus on some open set, M and H_ξ are both irreducible and are isomorphic. The characters of M and H_ξ vary continuously in the parameter λ. Since they are equal on a Zariski open set, they are equal for all λ. This proves (6.7).

§7. INVARIANT BILINEAR AND HERMITIAN FORMS

In this section we study the canonical invariant bilinear and

Hermitian forms on $N \otimes N'$ as well as on $\Gamma^s(N \otimes N')$. We continue with the notation of sections one and two. Let σ_o denote the standard involutive antiautomorphism of $U(\mathfrak{g}_o)$ which equals the identity on $\mathfrak{h}_o$. From 25.2 [13], we may choose a Chevalley basis of $\mathfrak{g}_o$, $\{H_i\}_{1\le i\le r} \cup \{X_\alpha\}_{\alpha\in\Delta_o}$ such that i) if $\{\alpha_1,\ldots,\alpha_r\}$ are the simple roots of Δ_o^+ then $H_i = H_{\alpha_i} = [X_{\alpha_i}, X_{-\alpha_i}]$, ii) $\{H_i, X_{\pm\alpha_i}\}$ is a standard basis for $sl(2)$, and iii) $\sigma_o H_i = H_i$ and $\sigma_o X_\alpha = X_{-\alpha}$, $\alpha \in \Delta$. Let $\bar{\mathfrak{g}}_o$ denote the normal real form of $\mathfrak{g}_o$ spanned by this Chevalley basis. Put $\bar{\mathfrak{g}} = \bar{\mathfrak{g}}_o \times \bar{\mathfrak{g}}_o$ and, in general, let a superscript bar denote the real form of a Lie subalgebra determined by intersection with $\bar{\mathfrak{g}}_o$ or $\bar{\mathfrak{g}}$, when this intersection is a real form. For example, $\bar{\mathfrak{h}}_o = \mathfrak{h}_o \cap \bar{\mathfrak{g}}_o$ is a real form of $\mathfrak{h}_o$. Let σ denote the standard involutive antiautomorphism on $U(\mathfrak{g})$ which equals 1 on $\mathfrak{h}$. Then, since $U(\mathfrak{g}) \simeq U(\mathfrak{g}_o) \otimes U(\mathfrak{g}_o)$, $\sigma = \sigma_o \otimes \sigma_o$.

Let L be any real Lie group whose complexified Lie algebra is $\mathfrak{L}$. Letting $\mathfrak{L}(L)$ denote the Lie algebra of L, then $\mathfrak{L}(L)$ is a real form of $\mathfrak{L}$.

<u>Definition 7.1</u>. Let A be an $\mathfrak{L}$-module and $\{\cdot,\cdot\}$ a Hermitian form on A. We call this form invariant (for L or $\mathfrak{L}(L)$) if $\{X\cdot a, a'\} = \{a, -Xa'\}$ for all $X \in \mathfrak{L}(L)$, $a, a' \in A$.

<u>Definition 7.2</u>. Let τ be some fixed involutive antiautomorphism of $U(\mathfrak{L})$. For any $\mathfrak{L}$-module A and bilinear form $\langle\cdot,\cdot\rangle$ on A, we say the form is invariant if

$$\langle x\cdot a, a'\rangle = \langle a, \tau(x)\cdot a'\rangle \quad \text{for all } x \in U(\mathfrak{L}),\ a, a' \in A.$$

Throughout this section we shall always have either $\tau = \sigma_o$ or $\tau = \sigma$. From the context, the choice will be obvious. By convention, we use $\{\cdot,\cdot\}$ to denote invariant Hermitian forms and $\langle\cdot,\cdot\rangle$ to denote invariant bilinear forms.

<u>Lemma 7.3</u>. i) The modules N, N', L and L' all admit nonzero canonical invariant bilinear forms.

ii) These modules admit nonzero Hermitian forms if and only if λ is real (here Hermitian is defined with respect to $\mathfrak{L}(G_o)$).

iii) In all cases above, when the forms exist, they are unique up to scalar multiple.

Proof. The forms in i) are called contravariant by Jantzen. A proof of i) is given in [15] and in [7]. The proofs of ii) and iii) are essentially the same and are given in [8].

For each $\lambda \in \mathfrak{h}_o^*$ which is $\Delta_{o,c}^+$-dominant integral, fix a highest weight vector 1_λ in $F(\lambda)$ and then let $1 \otimes 1_\lambda$ denote the canonical cyclic vector in $N(\lambda)$. Let φ_λ be the invariant bilinear form on $N(\lambda)$ with $\varphi_\lambda(1 \otimes 1_\lambda, 1 \otimes 1_\lambda) = 1$. We call φ_λ the canonical bilinear form on $N(\lambda)$.

For any $\mathfrak{g}_o$-module A in $C(\mathfrak{g}_o, \mathfrak{k}_o)$ recall from section two the notion of σ_o-dual to A; as well as, the definitions of the categories $C(\mathfrak{a},\mathfrak{b})$ and $\mathcal{A}(\mathfrak{a},\mathfrak{b})$. It is an easy exercise in linear algebra to show that if A has finite $\mathfrak{k}_o$-multiplicities, invariant forms on A correspond to $\mathfrak{g}_o$-module maps of A to A^{σ_o}. Our next goal will be to apply Γ^s to such maps and then interpret the result as invariant forms on $\Gamma^s A$. To do this we shall need a duality theorem for Γ^*.

Recall our convention which identifies $\mathfrak{g}_o$ and $\mathfrak{k}$; and in turn, $\mathfrak{k}_o$ and $\mathfrak{m}$. For any $\mathfrak{k}$-module A, A^{σ_o} is the σ_o-dual $\mathfrak{k}$-module.

Proposition 7.4. Let A be an object in $\mathcal{A}(\mathfrak{k},\mathfrak{m})$ (resp. $\mathcal{A}(\mathfrak{g},\mathfrak{m})$). Then, for all i, $\Gamma^i(A)$ and $\Gamma^{2s-i}(A^{\sigma_o})$ are dual objects in $\mathcal{A}(\mathfrak{k},\mathfrak{k})$ (resp. $\mathcal{A}(\mathfrak{g},\mathfrak{k})$). Stated more precisely, let $\sim$ denote the $U(\mathfrak{k})$-locally finite vectors in the σ_o-dual (resp. σ-dual) module. Then, $A \mapsto \Gamma^i(A)^\sim$ and $A \mapsto (\Gamma^{2s-i}(A^{\sigma_o}))$ are naturally equivalent functors on $\mathcal{A}(\mathfrak{k},\mathfrak{m})$ (resp. $\mathcal{A}(\mathfrak{g},\mathfrak{m})$).

This is a restatement of a result of Zuckerman. In this form it is given as (4.2) and (4.4) in [10]; but the reader must replace the usual module action on the dual by the action above. This means the antiautomorphism induced by $X \mapsto -X$ on $\mathfrak{g}_o$ is replaced by σ_o (resp. σ).

Using the natural equivalence we can now define the image of an invariant bilinear form under Γ^S. Let φ be an invariant form on a module A in $\mathcal{A}(\mathfrak{k},\mathfrak{m})$ (resp. $\mathcal{A}(\mathfrak{g},\mathfrak{m})$). Let θ denote the natural equivalence given in (7.4). Let $\varphi^\vee$ be the $\mathfrak{k}$ (resp. $\mathfrak{g}$)-module map $\varphi^\vee: A \to A^{\sigma_o}$ defined by $(\varphi^\vee(a))(a') = \varphi(a,a')$, $a,a' \in A$. Now applying Γ^S we have: $\Gamma^S\varphi^\vee: \Gamma^S A \to \Gamma^S (A^{\sigma_o})$. Finally composing with θ, we have:

$$\theta \circ \Gamma^S\varphi^\vee: \Gamma^S A \to (\Gamma^S A)^\sim. \tag{7.5}$$

But this map represents an invariant bilinear form on $\Gamma^S A$. We call this form the image of φ under Γ^S and denote it by $\Gamma^S\varphi$.

As in section five, put $M = \Gamma^S(N \otimes N')$ and assume $F(\lambda)$ is one dimensional.

<u>Lemma 7.6</u>. i) The modules M and $\Gamma^S(L \otimes N')$ admit nonzero invariant forms.

ii) Since $F(\lambda)$ is one dimensional, we may write $\lambda = (z-1)\rho_n$ for some $z \in \mathbb{C}$. There are three cases where M and $\Gamma^S(L \otimes N')$ admit nonzero invariant Hermitian forms (w.r.t. $\mathcal{L}(G)$): the first is for pure imaginary values of z; the second case occurs when $w_o\rho_n = -\rho_n$ and z is real; the third case occurs when $w_o\rho_n \neq -\rho_n$, z is real and for some $r \in W_o$, $r(2\rho_n+\rho_c) = -(2\rho_n+\rho_c)$.

iii) In all cases above, when forms exist they are unique up to scalar multiple.

<u>Proof</u>. $N \otimes N'$ is a generalized Verma module with unique irreducible quotient $L \otimes N'$. By (5.6), M has a unique irreducible quotient $M_1 = \Gamma^S(L \otimes N')$. Now exactly as with highest weight modules, this implies that every invariant form is a pull back of a form on M_1. Since M_1 is irreducible, the space of such forms is at most one dimensional. This proves iii). Let φ be the canonical invariant bilinear form on

the highest weight module $L \otimes N'$. By the remarks following (7.5), $\Gamma^s\varphi$ is an invariant form on M_1. Since $(L \otimes N')^\sigma$ is isomorphic to $L \otimes N'$ the induced map $\varphi^\vee: L \otimes N' \to (L \otimes N')^\sigma$ is a nonzero multiple of the identity. Thus by functoriality, $\Gamma^s\varphi$ is nondegenerate and so nonzero. This proves i).

We now prove ii). Arguing as above it is sufficient to prove ii) for M_1. The $\mathfrak{g}$-module M_1 has a $\mathfrak{k}$-fixed vector and infinitesimal character given by the $W_o \times W_o$ orbit of $(\lambda+\rho, -\lambda-\rho)$. M_1 is an irreducible representation infinitesimally equivalent to a representation of G. So by [6], M_1 admits a nonzero invariant Hermitian form if and only if, for some $s,t \in W_o$, $(s,t)(\lambda+\rho,-\lambda-\rho) = \overline{(-\lambda-\rho,\lambda+\rho)}$. Splitting this into real and imaginary parts, both s and t must satisfy the two identities:

(7.7) $$r\,(\mathrm{Im}\, z)\rho_n = (\mathrm{Im}\, z)\rho_n$$

(7.8) $$r\,(\mathrm{Re}\, z\rho_n+\rho_c) = -(\mathrm{Re}\, z\rho_n+\rho_c).$$

The roots orthogonal to ρ_n are precisely the compact roots. So, if $\mathrm{Im}\, z \neq 0$ then r is a product of compact root reflections. But compact root reflections fix ρ_n and map $[\mathfrak{k}_o,\mathfrak{k}_o] \cap \mathfrak{h}_o$ into itself. Therefore, by (7.8), $\mathrm{Re}\, z = 0$ and $r\rho_c = -\rho_c$. This means r is the maximal element of $W_{o,c}$ and z is pure imaginary, which is the first case. Now assume $\mathrm{Im}\, z = 0$. If $w_o\rho_n = -\rho_n$ then $w_o\rho_c = -\rho_c$ and w_o is a solution to (7.7) and (7.8). This is case two. If $w_o\rho_n \neq -\rho_n$ then (7.8) is equivalent to case three. This completes the proof of (7.6).

<u>Remark 7.9</u>. For case one above, by (6.7) and (5.6), M is irreducible and isomorphic to the unitary degenerate series representation H_ξ. So, in this case, the invariant Hermitian form is positive definite and M is unitarizable.

§8. COMPARISON OF SIGNATURE

We now come to the main results of this article, the comparison of unitary representations of G and $G_o \times G_o$. Keeping the notation of sections one and two, we assume $F(\lambda)$ is one dimensional. Then $\lambda = z\rho_n$ and, recalling (5.1), (7.6(ii)) and (7.9), <u>we assume</u> z is real, $z \geq -1$ and either $w_o\rho_n = -\rho_n$ or $r(z\rho_n+\rho) = -(z\rho_n+\rho)$ for some $r \in \mathcal{W}_o$.

In order to compare invariant Hermitian and invariant bilinear forms we introduce real forms for both Lie algebras and modules. Recall from section seven the normal real forms $\bar{\mathfrak{g}}_o$ and $\bar{\mathfrak{g}}$. Since λ is real, let $\mathbb{R}_\lambda$ denote the one dimensional $(\bar{\mathfrak{k}}_o \oplus \bar{\mathfrak{p}}^+)$-module over $\mathbb{R}$ given by λ. Then the module

$$\overline{N(\lambda)} = U(\bar{\mathfrak{g}}_o) \otimes_{U(\bar{\mathfrak{k}}_o+\bar{\mathfrak{p}}^+)} \mathbb{R}_\lambda \tag{8.1}$$

is a real form of $N(\lambda)$. By restricting the canonical invariant Hermitian form $\{\cdot,\cdot\}$ and the invariant bilinear form $\langle\cdot,\cdot\rangle$ from $N(\lambda)$ to $\overline{N(\lambda)}$, we obtain two real bilinear forms on $\bar{N}(\lambda)$. In this way, we can compare Hermitian and bilinear forms.

Let $\bar{\sigma}_o$ (resp. $\bar{\sigma}$) denote the conjugate linear involutive anti-automorphism of $U(\mathfrak{g}_o)$ (resp. $U(\mathfrak{g})$) induced by the map $X \mapsto -X$ on $\mathcal{L}(G_o)$ (resp. $\mathcal{L}(G)$).

<u>Lemma 8.2</u>. i) $U(\bar{\mathfrak{g}}_o)$ and $U(\bar{\mathfrak{g}})$ are invariant under $\bar{\sigma}_o$ and $\bar{\sigma}$ respectively.

ii) $\bar{\sigma}_o = \sigma_o$ on $\bar{\mathfrak{k}}_o$ and $\bar{\sigma}_o = -\sigma_o$ on $\bar{\mathfrak{p}}^+ \oplus \bar{\mathfrak{p}}^-$.

iii) Let $\bar{\mathfrak{k}}$ (resp. $\bar{\mathfrak{p}}$) be the diagonal subalgebra (resp. skew diagonal subspace) of $\bar{\mathfrak{g}}_o \times \bar{\mathfrak{g}}_o$. Then $\bar{\sigma} = \sigma$ on $\bar{\mathfrak{k}}$ and $\bar{\sigma} = -\sigma$ on $\bar{\mathfrak{p}}$.

The proof of (8.2) is a short calculation which we omit.

We now decompose N, L and N' into even and odd parts. Since $\mathfrak{k}_o$ has a one dimensional center, let H_o be the unique central element of $\mathfrak{k}_o$ with $\alpha(H_o) = 1$ for all $\alpha \in \Delta^+_{o,n}$. For any $\mathfrak{g}_o$-module A, the eigenspaces for H_o are $\mathfrak{k}_o$-modules. Assume A has highest weight ξ. Then let A^{even} (resp. A^{odd}) be the sum of the eigenspaces with H_o-eigenvalue minus $\xi(H_o)$ an even (resp. odd) integer. This gives decompositions:

(8.3) $\quad N = N^{even} \oplus N^{odd}, \quad N' = N'^{even} \oplus N'^{odd}, \quad L = L^{even} \oplus L^{odd}$

as well as decompositions for underlying real forms.

Lemma 8.4. i) On the even part of $\bar{N}$, the canonical Hermitian and bilinear forms are equal.

ii) On the odd part of $\bar{N}$, the Hermitian and bilinear forms differ only in sign; i.e., $\{\cdot,\cdot\} = -\langle\cdot,\cdot\rangle$.

Proof. Let $x,y \in S(\overline{\mathfrak{p}})$ be homogeneous of degrees d and e respectively. Consider $x \otimes 1$ and $y \otimes 1$ as elements of N. If $d \neq e$, then these elements are orthogonal for either form. If $d = e$, then

$$\begin{aligned} \langle x \otimes 1, y \otimes 1\rangle &= \langle 1,\sigma_o(x)y \otimes 1\rangle \\ &= (-1)^d \langle 1,\bar{\sigma}_o(x)y \otimes 1\rangle \quad \text{by (8.2)} \\ &= (-1)^d \{x \otimes 1, y \otimes 1\}. \end{aligned}$$

This proves (8.4).

For a tensor product of $\mathfrak{g}_o$-modules, use (H_o,H_o) in place of H_o to define the even and odd parts. Then $(N \otimes N')^{even}$ is the sum of $N^{even} \otimes N'^{even}$ and $N^{odd} \otimes N'^{odd}$; $(N \otimes N')^{odd}$ is the sum of $N^{even} \otimes N'^{odd}$ and $N^{odd} \otimes N'^{even}$. Note that the even and odd parts of $N \otimes N'$ are <u>not</u> stable under $\mathfrak{k}$.

Definition 8.5. i) Let $N \otimes N' = \sum_i N_i$ be the decomposition of $N \otimes N'$ as a sum of irreducible generalized Verma modules for $\mathfrak{k}$ as in (5.2). We call the summand N_i of even type (resp. odd type) if its highest weight lies in the even (resp. odd) part of $N \otimes N'$. Each N_i is either of odd or even type.

ii) Since Γ^s gives a bijection from irreducible $\mathfrak{k}$-summands of $N \otimes N'$ to those of M, we call an irreducible $\mathfrak{k}$-submodule of M of even or odd type if the corresponding generalized Verma module is of even or odd type respectively.

Lemma 8.6. Let χ_i be the restriction of the canonical $\mathfrak{g}$-invariant bilinear form on $N \otimes N'$ to the $\mathfrak{k}$-submodule N_i and let φ_i be the canonical bilinear form on this generalized Verma module. Then the following are equivalent.

i) For all i, χ_i is a constant multiple of φ_i and the constant is nonnegative (resp. nonpositive) if N_i is of even (resp. odd) type.

ii) Both L and N' are unitarizable representations of G_o.

Proof. Assume (ii). Then by (8.4), the canonical bilinear forms on L and N' are positive (resp. negative) definite on the even (resp. odd) parts of L and N'. So the tensor product form has the same property on the even and odd parts of $L \otimes N'$. By the multiplicity free property of (5.2), χ_i is a constant multiple of φ_i; and so, the constant is positive (resp. negative) if the highest weight of N_i is in the even (resp. odd) part of $L \otimes N'$. Pulling back to $N \otimes N'$, this gives i).

By our conventions, $\lambda = z\rho_n$ with $z \geq -1$. Thus by (3.7), N' is irreducible and unitarizable. If L is not unitarizable, then by (8.4) the canonical bilinear form on $L \otimes N'$ will not be positive definite on the even part and negative definite on the odd part. By (5.2) each N_i which occurs in $N \otimes N'$ is infinitesimally equivalent to a holomorphic discrete series for G_o and so is unitarizable. Therefore, by (8.4) the canonical bilinear form on N_i is positive definite on N_i^{even} and negative definite on N_i^{odd}. (Here we view this $\mathfrak{k}$-module as a $\mathfrak{g}_o$-module). This fact and i) imply that the canonical form is positive definite (resp. negative definite) on the even (resp. odd) part of $L \otimes N'$. This contradiction shows not ii) implies not i) and completes the proof.

Next we consider the effect of the functor Γ^S on signature.

Lemma 8.7. For Δ_o^+-dominant integral λ in $\mathfrak{g}_o^*$ define constants a_λ by

$$\Gamma^s \varphi_{r_o(\lambda+\rho)-\rho} = a_\lambda \cdot \varphi_\lambda .$$

Then the constants a_λ are real, nonzero and all of the same sign.

Proof. Let K be the connected simply connected compact Lie group with complexified Lie algebra $\mathfrak{g}_o$. Then, the finite dimensional modules $L(\lambda)$ are representations of K; and so, admit invariant Hermitian forms with respect to $\mathfrak{L}(K)$. Since K is compact these forms are positive definite. Let τ be the conjugate linear involutive antiautomorphism induced by $X \mapsto -X$ on $\mathfrak{L}(K)$. Then on $U(\bar{\mathfrak{g}}_o)$, τ and τ_o are equal; and so, the Hermitian forms for $\mathfrak{L}(K)$ pull back and restrict to the invariant bilinear forms on $\overline{N(\lambda)}$. This proves that the canonical form φ_λ is positive semidefinite on $\overline{N(\lambda)}$.

We can replace σ_o by τ in section seven. The effect of this substitution is that (7.5) defines a map of invariant Hermitian forms on $\mathfrak{g}_o$-modules A to invariant Hermitian forms on $\Gamma^s A$. For irreducible A the space of such invariant forms is one dimensional over $\mathbb{R}$. The canonical bilinear forms are equal to canonical invariant Hermitian forms on $\overline{N(\xi)}$, for all real $\xi \in \mathfrak{h}_o^*$, by the remark in the preceeding paragraph. So in (8.7) we could assume the forms were invariant Hermitian for $\mathfrak{L}(K)$; and thus, the constant is real.

To see that $a_\lambda \neq 0$ note that $N(r_o(\lambda+\rho)-\rho)$ is irreducible; and thus $\varphi^\vee_{r_o(\lambda+\rho)-\rho}$ is a nonzero multiple of the identity map (cf. (7.5)). So Γ^s maps this to a nonzero multiple of the identity. Thus $\Gamma^s\varphi \neq 0$; and so, $a_\lambda \neq 0$.

Let λ and μ be dominant integral. We now prove:

(8.8) $\qquad a_\lambda$ and $a_{\lambda+\mu}$ have the same sign.

Write $F = L(-w_o\mu)$, $\varphi = \varphi_{-w_o\mu}$, $N = N(r_o(\lambda+\mu+\rho)-\rho)$ and $L = L(\lambda+\mu)$. Let φ_1 (resp. φ_2) denote the canonical form on N (resp. L). Let ψ denote the Zuckerman translation which carries $L(\lambda+\mu)$ to $L(\lambda)$.

The functors $A \mapsto \Gamma^i(F \otimes A)$ and $A \mapsto F \otimes \Gamma^i A$ are naturally equivalent; and so, it is a short exercise (which we omit) to show:

$$\Gamma^s(\varphi \otimes \varphi_1) = \varphi \otimes \Gamma^s\varphi_1. \tag{8.9}$$

If a (resp. b) is a highest weight vector of F (resp. N), then $a \otimes b$ is a highest weight vector for $N(r_o(\lambda+\rho)-\rho)$. Since $\varphi \otimes \varphi_1(a \otimes b, a \otimes b) = \varphi(a,a)\varphi_1(b,b) > 0$, the restriction of $\varphi \otimes \varphi_1$ to $N(r_o(\lambda+\rho)-\rho)$ equals $a \cdot \varphi_{r_o(\lambda+\rho)-\rho}$, $a > 0$. The form $\varphi \otimes \Gamma^s\varphi_1$ is either positive or negative definite depending on whether $a_{\lambda+\mu}$ is positive or negative. Therefore, $\Gamma^s a \cdot \varphi_{r_o(\lambda+\rho)-\rho} = a \cdot a_\lambda \varphi_\lambda$ is positive (resp. negative) definite if $a_{\lambda+\mu}$ is positive (resp. negative). This proves (8.8).

Now to prove the lemma observe that for any two λ, λ' integral elements, we can choose μ, μ' dominant integral so that $\lambda + \mu = \lambda' + \mu'$. Then by (8.8), a_λ and $a_{\lambda+\mu}$ and $a_{\lambda'}$ have the same sign. This proves (8.7).

Lemma 8.10. Let $\bar{M} = \Gamma^s(\bar{N} \otimes \bar{N}')$ and let $v \in \bar{M}$ be a nonzero $\mathfrak{k}$-fixed vector. Choose an invariant bilinear (resp. Hermitian) form φ (resp. ψ) so that $\varphi(v,v) = 1$ (resp. $\psi(v,v) = 1$). Then φ equals ψ on the even $\mathfrak{k}$-submodules of M and φ equals $-\psi$ on the odd $\mathfrak{k}$-submodules.

Proof. We first recall a natural grading on $L \otimes N'$ and then proceed by induction on the level in the grading.

Recall the element H_o in the center of $\mathfrak{k}_o$ from the paragraph above (8.4). Let N^i equal the $\mathfrak{k}_o$-submodule of N where H_o acts by eigenvalue $-i + \lambda(H_o)$. Let A^i be the $\bar{\mathfrak{k}}$-submodule of $\bar{N} \otimes \bar{N}'$ isomorphic to $U(\bar{\mathfrak{g}}_o) \otimes_{U(\mathfrak{k}_o \oplus \bar{\mathfrak{p}}^+)} (N^i \otimes F(-\lambda-2\rho_n))$. The submodules A^i are sums of $\mathfrak{k}$-modules of even (resp. odd) type if i is even (resp. odd). Since $\mathfrak{p}^-$ is abelian, $U(\mathfrak{p}^-)$ is isomorphic to the symmetric algebra of $\mathfrak{p}^-$; and so, N^{i+1} equals $\bar{\mathfrak{p}}^- \cdot N^i$. This implies easily that A^i is contained in $\bar{\mathfrak{g}} \cdot (\sum_{1 \le j < i} A^j)$. However, since the H_o-eigenvalue changes by 0, ± 1 by the action of $\mathfrak{g}$ and all $\mathfrak{k}$-summands of A^ℓ

have highest weights with H_o-eigenvalue $-\ell - 2\rho_n(H_o)$, we have:

(8.11) $$A^i \subset \bar{\mathfrak{g}}\cdot A^{i-1}.$$

Since $\mathfrak{g} = \mathfrak{k} \oplus \mathfrak{p}$ and A^{i-1} is $\mathfrak{k}$- stable, (8.11) becomes: $A^i \subset \bar{\mathfrak{p}}\cdot A^{i-1}$. Now applying Γ^s and (5.10), we obtain:

(8.12) $$\Gamma^s A^i \subset \bar{\mathfrak{p}}\cdot\Gamma^s A^{i-1}.$$

Note that by normalization $\varphi = \psi$ on $\Gamma^s A^o$. Now let $a \in \Gamma^s A^i$, for some $i \geq 1$, and assume $\psi = (-1)^{i-1}\varphi$ on $\Gamma^s A^{i-1}$. By (8.12) we can write $a = \sum X_j\cdot a_j$ with $X_j \in \bar{\mathfrak{p}}$ and $a_j \in \Gamma^s A^{i-1}$. Then

$$\begin{aligned}\psi(a,a) &= \sum_{j,k} \psi(a_j,\bar{\sigma}(X_j)X_k a_k)\\ &= -\sum \psi(a_j,\sigma(X_j)X_k a_k) \quad \text{by } (8.2\,(iii))\\ &= -(-1)^{i-1}\sum \varphi(a_j,\sigma(X_j)X_k a_k) \quad \text{by assumption}\\ &= (-1)^i\varphi(a,a).\end{aligned}$$

This proves the lemma.

We can now prove the main result of this article.

<u>Theorem 8.13</u>. $\Gamma^s(L \otimes N')$ is unitarizable for G if and only if both L and N' are unitarizable for G_o.

<u>Proof</u>. Let $M' = \Gamma^s(L \otimes N')$ and let $\varphi = \Gamma^s(\varphi_\lambda \otimes \varphi_{-\lambda-2\rho_n})$. M' is the direct sum of irreducible $\mathfrak{k}$-modules of even and odd type (cf. (8.5)). Let $a = 1$ or -1 depending as the constants in (8.7) are all positive or all negative. Now combining (8.6) and (8.7), we have:

(8.14) L and N' are unitarizable if and only if $a\cdot\Gamma^s\psi_i$ is positive (resp. negative) definite on the $\bar{\mathfrak{k}}$-submodules of even (resp. odd) type.

By naturality of the functor Γ^s, $\Gamma^s\psi_i$ is the restriction of φ to $\Gamma^s N_i$. So (8.14) becomes:

(8.15) L and N' are unitarizable if and only if $a\cdot\varphi$ is positive (resp. negative) definite on the $\overline{k}$-submodules of M' of even (resp. odd) type.

To complete the proof we merely combine (8.10) and (8.15).

§9. SUMMARY OF RESULTS IN TABLE FORM

In this section we bring together the main results of earlier sections. We summarize the results on reducibility and unitarizability of certain degenerate series for complex groups. The reducibility results are obtained by combining (5.6), (6.7) and Jantzen's necessary and sufficient conditions for the reducibility of $N \otimes N'$ [14]. The unitarizability results all follow by combining (8.13) and Wallach's description of the unitary $L(\lambda)$ (cf. (3.4)).

Recall the notation of section six and let ζ be given by (3.2). For $z \in \mathbb{C}$, let $H(z\zeta)$ be the degenerate series module H_ξ where ξ is the one dimensional $\mathfrak{m}$-module with weight $(z\zeta,-z\zeta)$. For all z, $H(z\zeta)$ contains a $\mathfrak{k}$-fixed vector. By (6.7), $H(z\zeta)$ has the same distribution character as $\Gamma^s(N(z\zeta-\rho_n) \otimes N(-z\zeta-\rho_n))$. If $\mathrm{Im}\, z \neq 0$ then the highest weight plus ρ is not integer valued at $\alpha^\vee$, for any $\alpha \in \Delta_{o,n}$. Therefore by Korollar 4 [14], both of these generalized Verma modules are irreducible. So, in this case, $H(z\zeta)$ is always irreducible, and, by (7.6) and (7.9), $H(z\zeta)$ is unitary if and only if z is pure imaginary.

For the remainder of this section <u>we assume</u> z is real. Note that by (7.9), $H(0)$ is unitary. For results which we state case by case, we label the types by the root systems for the pair $(\mathfrak{g}_o,\mathfrak{k}_o)$. The first root system is the root system of $\mathfrak{g}_o$ and the second is the root system for $\mathfrak{k}_o$, the reductive part of the maximal parabolic $\mathfrak{q}_o = \mathfrak{k}_o \oplus \mathfrak{p}^+$.

<u>Proposition 9.1</u>. $H(z\zeta)$ is reducible only for certain real values of z. These values are given in the following table:

$(\mathfrak{g}_o, \mathfrak{k}_o)$	Values of z
$(A_{n-1}, A_{p-1} \times A_{q-1})$ $n = p + q, \quad p \le q$	$z \in \mathbb{Z}$ and $\|z\| \ge \frac{q-p}{2} + 1$
(B_n, B_{n-1})	$z \in \mathbb{Z}^*$ <u>or</u> z an odd half integer and $\|z\| \ge n - \frac{1}{2}$
(C_n, A_{n-1})	$2z \in \mathbb{Z}$ and $\|z\| \ge 1$
(D_n, A_{n-1}) n even	$z \in \mathbb{Z}^*$
(D_n, A_{n-1}) n odd	$z \in \mathbb{Z}$ and $\|z\| \ge 2$
(D_n, D_{n-1})	$z \in \mathbb{Z}^*$
(E_6, D_5)	$z \in \mathbb{Z}$ and $\|z\| \ge 3$
(E_7, E_6)	$z \in \mathbb{Z}^*$

<u>Note</u>: The reducibility result for (E_7, E_6) is a corollary of a result of B. Boe about generalized Verma modules (cf. Theorem 7.3 [1]).

<u>Proposition 9.12</u>. Assume z is real. The values of z for the class one component of $H(z\zeta)$ to be unitarizable are given in the following table.

$(\mathfrak{g}_o, \mathfrak{k}_o)$	Values of z
$(A_{n-1}, A_{p-1} \times A_{q-1})$ $n = p + q, \quad p < q$	$z = 0$ <u>or</u> $\pm \frac{n}{2}$
$(A_{2p-1}, A_{p-1} \times A_{p-1})$	$\|z\| \le 1$ <u>or</u> $z \in \mathbb{Z}$ and $\|z\| \le p$
(B_n, B_{n-1})	$\|z\| \le 1$ <u>or</u> $\|z\| = n - \frac{1}{2}$

$(\mathfrak{g}_o,\mathfrak{k}_o)$	Values of z
(C_n,A_{n-1})	$\lvert z\rvert \le 1$ or $2z \in \mathbb{Z}$ and $\lvert 2z\rvert \le n+1$
(D_n,A_{n-1}) n even	$\lvert z\rvert \le 1$ or z an odd integer and $\lvert z\rvert \le n-1$
(D_n,A_{n-1}) n odd	z an even integer and $\lvert z\rvert \le n-1$
(D_n,D_{n-1})	$\lvert z\rvert \le 1$ or $\lvert z\rvert = n-1$
(E_6,D_5)	$z=0$ or $\lvert z\rvert = 6$ and possibly* $\lvert z\rvert = 1, 2,$ or 3
(E_7,E_6)	$\lvert z\rvert \le 1$ or $\lvert z\rvert = 5$ or $\lvert z\rvert = 9$

*For (E_6,D_5), each point z with $|z| = 1, 2$ or 3 is unitary if and only if $H(z\zeta)$ admits an invariant Hermitian form. This is equivalent to checking that $-(z\zeta+\rho_c)$ is in the Weyl group orbit of $z\zeta+\rho_c$. For E_6 this computation is not easy to check and we have not done it.

REFERENCES

[1] B. Boe, Homomorphisms between generalized Verma modules, Ph.D. Thesis, Yale University, 1982.

[2] N. Bourbaki, Groupes et algèbres de Lie, IV-VI. Hermann, Paris, 1968.

[3] W. Casselman and D. Miličic, Asymptotic behavior of matrix coefficients of admissible representations. Preprint.

[4] N. Conze-Berline and M. Duflo, Sur les représentations induites des groupes semi-simples complexes, Compositio Math. 34 (1977), 307-336.

[5] J. Dixmier, Algèbres Enveloppantes, Gauthier-Villars, Paris, 1974.

[6] M. Duflo, Représentations irreductibles des groupes semi-simples complexes, Lecture Notes 497 (1975), 26-88.

[7] T. Enright, On the fundamental series of a real semisimple Lie algebra: their irreducibility, resolutions and multiplicity formulae, Ann. of Math., 110 (1979), 1-82.

[8] ______, Lectures on Representations of Complex Semi-simple Lie Groups, Tata Institute Lecture Notes, Springer-Verlag, Berlin, 1981.

[9] T. Enright, R. Howe and N. Wallach, A classification of unitary highest weight modules, to appear in Proceeding of Park-City conference on Representations of Reductive Groups, March 1982.

[10] T. Enright and N. Wallach, Notes on homological algebra and representations of Lie algebras, Duke Math. J. 47 (1980), 1-15.

[11] ______, The fundamental series of semisimple Lie algebras and semisimple Lie groups, manuscript.

[12] Harish-Chandra, Representations of semisimple Lie groups VI, Amer. J. Math., 78 (1956), 564-628.

[13] J. Humphreys, Introduction to Lie Algebras and Representation Theory, Springer-Verlag, 1972.

[14] J. Jantzen, Kontravariante Formen auf induzierten Darstellungen halbeinfacher Lie-Algebren, Math. Ann. 226 (1977), 53-65.

[15] ______, Moduln mit einen höchsten Gewicht, Lecture Notes 750, Springer-Verlag, Berlin, 1979.

[16] A. Knapp, Investigations of unitary representations of semisimple Lie groups, preprint of article submitted to Torino-Milano Conference, 1982.

[17] W. Schmid, Die Randwerte holomorpher Functionen auf hermitesch symmetrischen Räumen, Inv. Math. 9 (1969), 61-80.

[18] D. Vogan, Representations of real reductive Lie groups, Birkhäuser, 1981.

[19] N. Wallach, The analytic continuation of the discrete series I, II, T.A.M.S., 251 (1979), 1-17.

[20] G. Zuckerman, Construction of some modules via derived functors, Lectures at I.A.S., 1977.

ON THE CLASSIFICATION OF PRIMITIVE IDEALS IN THE ENVELOPING ALGEBRA OF A SEMISIMPLE LIE ALGEBRA

A. Joseph
Department of Theoretical Mathematics
The Weizmann Institute of Science
Rehovot 76100, Israel

and

Laboratoire de mathématiques fondamentales
(Equipe de recherche associée au CNRS)
Université de Pierre et Marie Curie, France

1. INTRODUCTION

1.1. The classification of the primitive ideals in the enveloping algebra of a complex semisimple Lie algebra—a major goal in the theory of enveloping algebras—is now nearly completed. Nevertheless had it not been for the necessity of providing the Maryland Special Year in Group Representations with a manuscript, the author would probably have still further delayed any attempt to review this subject. Thus whereas the main lines of the classification are now firmly established there are still many fascinating questions whose solution can at least in some cases be shortly anticipated. Furthermore, the dazzling success of algebraic geometry in determining the multiplicities of the simple factors in Verma modules has reduced many an expert in enveloping algebras to the role of a mere bystander and has so much reorientated the field that one particular protagonist has suggested, no doubt with much justification, that enveloping algebras should now be relegated to a subdivision of the theory of rings of differential operators.

1.2. Let us first just attempt an informal historical sketch of the main steps in the classification of primitive ideals.

The first major breakthrough was undoubtedly Duflo's theorem (3.5) which showed it was enough to consider annihilators of simple highest weight modules and in more detail asserts that for a given central character these annihilators are indexed by a subset of the involutions of the Weyl group—the still mysterious Duflo set (3.5). About the same time Borho and Jantzen established a translation principle (3.6) which related annihilators having different central character, grouping them into infinite subsets which we shall call t-clans (4.1).

The notion of the characteristic variety (2.7) of a primitive ideal gave information on the inclusion relations between primitive ideals and via Knuth's combinatorial theorem (see [20]) provided an explanation to Jantzen's remarkable observation (in type A_n) that the primitive ideals having a given central character appear to be grouped into sets which we shall call r-clans (4.3) whose cardinality is the dimension of an irreducible representation of the Weyl group. Following this Spaltenstein showed how Knuth's theorem could be used to describe the Steinberg correspondence (8.7) between the Weyl group and nilpotent orbits and brought to the attention of the author, Springer's correspondence (8.7) between such orbits and Weyl group representations. This was of particular interest as Borho and Kraft [9] had pointed out that the associated variety (2.1) of a primitive ideal was indeed a union of nilpotent orbits and conjecturally the closure of just one orbit.

A powerful technique for separating primitive ideals was provided by the Goldie rank (Sect. 4) of the corresponding quotient algebras. Berline and Duflo [11] had already given some indication how this invariant could be calculated; but it was the establishment of an additivity principle (4.2) which under suitable refinement led to the following two major results. One, that Goldie rank on t-clans is a polynomial and those for distinct t-clans are linearly independent. Two, that the set of polynomials defined with respect to an r-clan form a basis of an irreducible representation of the Weyl group. Shortly afterwards it was established that the associated variety was constant on clans and the obvious link with Springer's correspondence was conjectured.

It was now becoming clear that much precise information on primitive ideals could be expressed in terms of the multiplicities of the simple highest weight modules in the Jordan-Hölder series of Verma modules and which we refer to as forming the entries of the Jantzen matrix (Jantzen having made great progress in computing these coefficients). In particular the Goldie rank polynomials discussed above could be expressed in terms of this matrix (at least up to a scalar). Again following a suggestion of the author, Vogan showed that this was also true of the inclusion relations (Sect. 5) between primitive ideals and thereby implicitly solved a problem which earlier had needed laborious calculations to describe even low rank cases.

Partly motivated by this work and partly by the Springer correspondence, Kazhdan and Lusztig formulated their now famous conjecture (Sect. 6) for the Jantzen matrix. An important aspect of their conjecture was to bring into the arena an entirely new participant; namely the intersection cohomology theory of Deligne-Goretsky-MacPherson (DGM) thereby establishing a remarkable link between the singularities of Schubert cells in the flag variety and the Jantzen matrix.

Before it was realized that the Kazhdan-Lusztig conjecture was too hard for algebraists, Vogan established an equivalence of this conjecture with his semisimplicity hypothesis (6.5), and shortly afterwards, it was shown that an earlier conjecture of Jantzen concerning filtrations of Verma modules, implied the Kazhdan-Lusztig conjecture and even a significant refinement of it providing the composition factors in each gradation step - information which we refer to as the JKL data (6.3).

Brylinski and Kashiwara established the Kazhdan-Lusztig conjecture through the study of differential operators over the flag variety and the use of deep results on DGM sheaves. Beilinson and Bernstein simultaneously obtained similar results and further realized as also did Vogan that this new viewpoint could even encompass and go beyond the

Langlands classification of simple Harish-Chandra modules. Bernstein also reported a proof of the Jantzen conjecture based on a significant refinement due to Gabber of a main step in Deligne's proof of the Weil conjectures.

Returning to primitive ideals per se, it was shown that the Duflo set was implicitly determined by the JKL data and furthermore this and the Vogan semisimplicity hypothesis also gave some refined information on what we shall call the Kostant ring (Sect. 7) of a simple highest weight module. This is turn gave information on the calculation of the scale factors in the Goldie rank polynomials.

Barbasch and Vogan developing combinatorial results analogous to the Knuth theorem for the symmetric group S_n extended to the remaining Lie algebras results concerning the cellular decomposition of the Weyl group (Sect. 5) defined by the primitive ideals via Duflo's theorem. From these results Brylinski and Borho through a theorem on induced ideals established in the so-called integral case the Borho-Kraft conjecture described above and linked the representations of the Weyl group defined by Goldie rank polynomials to those defined by the Springer correspondence.

1.3. Let us now indicate briefly some of the main techniques in the proof of the above results.

Dixmier first pointed out that primitive quotients of the enveloping algebra could be viewed as Harish-Chandra modules for the corresponding complex Lie group. Using classical results of Kostant, it was further shown by Duflo that for minimal primitives the corresponding quotient algebra was in fact a so-called principal series module, an observation which eventually led to the Duflo theorem. For this early work see ([13], Chap. 8). A more precise formulation established an equivalence of categories (3.4) of such Harish-Chandra modules with the so-called $\underline{\mathcal{O}}$ category (roughly speaking the category of highest weight modules). This result is used in refining the additivity principle for

Goldie rank based on the use of Small's theorem and in the proof of Vogan's theorem on inclusion relations between primitive ideals.

The translation principle which goes back to Harish-Chandra is one of the main techniques in studying primitive ideals. Here a module in say the $\underline{\mathcal{O}}$ category is tensored by some finite dimensional module, and this allows one to translate results from one central character to another thus giving the Borho-Jantzen translation principle and together with the additivity principle, the polynomial behaviour of the Goldie ranks on t-clans. Moreover, by "reflecting off walls" in the Weyl chamber one is able to implement an action of the Weyl group and it is this which leads essentially to the representations of the Weyl group defined on r-clans. Moreover, by replacing the base field with a discrete valuation ring use of the translation principle gives the JKL data as a consequence of the Jantzen conjecture. Finally it is again a version of the translation principle which allows one to study the enveloping algebra through algebraic geometry via sheaves of differential operators on the flag variety.

1.4. To close this introduction we indicate some open problems and related questions.

Rather little is known about the Jordan-Hölder series for primitive quotients, though, of course, the general philosophy is that these should be describable in terms of the Jantzen matrix. A conjecture is indicated along these lines (7.6).

The scale factors occuring in the Goldie rank polynomials have not yet been completely determined. This is a rather delicate problem. One step involves the construction of sufficiently many completely prime primitive ideals. Although this latter problem was one of the first to be considered, practically no progress on it has been reported.

The use of differential operators over the flag variety allows one to introduce an additional invariant associated to a module (over the enveloping algebra), namely its singular support (8.9). This may be used, for example, in computing the Krull dimension of primitive quotients (as yet only for minimal primitive ideals). It turns out that there are some rather natural conjectures (8.14) for the singular support of a simple highest weight module and of its annihilator. As pointed out by Borho and Brylinski (8.10), these invariants are closely related to, but more refined than, associated varieties; moreover, their computation elucidates the relation between the classification of primitive ideals through Goldie rank polynomials and the Springer correspondence alluded to previously.

One may also associate to a primitive ideal its Goldie skew-field (2.2). This is known to be constant on t-clans; but it is not yet known if this is true on r-clans. Conjecturally it is always isomorphic to a Weyl skew-field (4.5).

The Kostant ring (Sect. 7) of a simple module is always a Harish-Chandra module (for the corresponding complex group) and has furthermore been shown to be a maximal order. In analogy with highest weight modules one expects that as a module, the Kostant ring should have a large socle (7.2) and to be a direct sum of modules having a simple socle. A rather more daring hypothesis is that the ring is completely determined by the module structure of the socle.

Primitive ideals corresponding to the same central character can be viewed as prime ideals of the quotient of the enveloping algebra by a minimal primitive ideal. From this latter viewpoint one may ask if all such ideals are idempotent. This is false for singular central characters; but there is some evidence that it holds for regular characters. Again for singular characters the quotient algebra generally has infinite global dimension; but one expects to find finite global dimension at regular characters. One may also give a conjecture (9.4) for the Grothendieck group generated by representatives of the projective modules.

There seems to be an astonishing connection between completely prime primitive ideals of the enveloping algebra and unitary representations of the corresponding complex group. Conjecturally the simple socle of all such quotient algebras, viewed as modules for the group, are unitarizable and all unitary representations occur in this manner. For regular integral central characters the second assertion follows from work of Enright and Parthasarathy, whilst the first statement would be a consequence of a rather natural conjecture (4.6) for the Goldie rank polynomials.

We conclude by remarking that there are a very large number of questions which in view of the Kazhdan-Lusztig polynomials can be regarded as purely combinatorial. For the moment these seem to be of incredible difficulty and, in fact, except for Knuth's theorem and its generalization by Barbasch-Vogan, the only success in resolving such questions has been through their interpretation in the context of enveloping algebras and the use of deep theorems in non-commutative ring theory, algebraic geometry or algebraic topology. See Sec. 5 and ([25], Sect. 5) for some examples.

2. NOTATION AND CONVENTIONS

2.1. The base field is assumed to be the complex numbers $\mathbb{C}$ (though any algebraically closed field of characteristic zero would do). For any vector space V, let V^* denote its dual and $S(V)$ the symmetric algebra over V. If I is an ideal of $S(V)$ we denote by $\underline{V}(I)$ its variety of zeros in V^*. For any Lie algebra $\underline{a}$ we denote by $\mathfrak{U}(\underline{a})$ its enveloping algebra and $\mathfrak{Z}(\underline{a})$ the centre of $\mathfrak{U}(\underline{a})$. The subspaces $\mathfrak{U}^n(\underline{a}) = \mathbb{C}\{X^m : m \le n, X \in \underline{a}\}$, $\forall n \in \mathbb{N}$ form an increasing filtration of $\mathfrak{U}(\underline{a})$ and the graded algebra

$$\operatorname{gr} \mathfrak{U}(\underline{a}) := \bigoplus_{n\in\mathbb{N}} (\mathfrak{U}^n(\underline{a})/\mathfrak{U}^{n-1}(\underline{a})) : (\mathfrak{U}^{-1}(\underline{a}) = 0)$$

is by Poincaré-Birkhoff-Witt isomorphic to $S(\underline{a})$. Let M be a finitely generated $\mathfrak{U}(\underline{a})$ module, say with generating subspace M^0. Then

$$\mathrm{gr}\, M := \bigoplus_{n\in\mathbb{N}} (M^n/M^{n-1}) \; : \; M^n = \mathfrak{U}^n(\underline{a})M^0,$$

is a graded module for the graded ring $\mathrm{gr}\,\mathfrak{U}(\underline{a}) \cong S(\underline{a})$. After Bernstein $\underline{V}(\mathrm{Ann\ gr}\, M)$ is independent of the choice of generating subspace M^0 and is called the associated variety $\underline{V}(M)$ of M. It is a closed subvariety of $\underline{a}^*$, but need not be irreducible even if M is simple ([32], Sect. 9). We can conveniently define the Gelfand-Kirillov dimension $d(M)$ of M as the dimension of $\underline{V}(M)$. We call M, d-homogeneous (resp. d-critical) if $d(N) = d(M)$ (resp. $d(M/N) < d(M)$) for any non-zero submodule N of M.

2.2. Let A be a prime noetherian ring (for us always both left and right noetherian). Then by Goldie's theorem A embeds in a unique smallest ring Fract A in which the non-zero divisors of A are invertible. Furthermore, by the Artin-Wedderburn theorem Fract A is isomorphic to a matrix ring over a skew-field. We call the rank of this matrix ring the Goldie rank rk A of A and the skew-field the Goldie skew field of A. For example, if we call $A_m := \mathbb{C}[x_1, x_2, \ldots, x_m, \frac{\partial}{\partial x_1}, \ldots, \frac{\partial}{\partial x_m}]$ the Weyl algebra of index m over $\mathbb{C}$, then $\mathrm{rk}\, A_m = 1$, and so Fract A_m is a skew-field called the Weyl skew-field of index m.

2.3. Let $\underline{g}$ be a complex semisimple Lie algebra, $\underline{h}$ a Cartan subalgebra, $R \subset \underline{h}^*$ the set of non-zero roots, $B_0 \subset R$ a choice of simple roots, R^+ the corresponding positive system, $P(R)$ the lattice of weights, s_α the reflection corresponding to the root α and W the subgroup of Aut $\underline{h}^*$ generated by the s_α: $\alpha \in R$. It is called the Weyl group. Fix a Chevalley basis $\{X_\alpha, H_\beta\}_{\alpha\in R, \beta\in B}$ for $\underline{g}$ and set

$$\underline{n}^+ = \bigoplus_{\alpha\in R^+} \mathbb{C}X_\alpha, \quad \underline{n}^- = {}^t\underline{n}^+, \quad \underline{b} = \underline{n}^+ \oplus \underline{h}$$

where $u \mapsto {}^tu$ is the Chevalley antiautomorphism of $\mathfrak{U}(g)$ defined by setting ${}^tX_\alpha = X_{-\alpha}$, $\forall \alpha \in R$. Let $u \mapsto \check{u}$ be the principal antiautomorphism of $\mathfrak{U}(\underline{g})$ defined through $\check{X} = -X$, $\forall X \in \underline{g}$. Define

$j: \underline{g} \to \underline{g} \times \underline{g}$ through $j(X) = (X, {}^t\check{X})$ and set $\underline{k} = j(\underline{g})$.

2.4. For each $\lambda \in \underline{h}^*$ set $R_\lambda := \{\alpha \in R \mid (\alpha^\vee, \lambda) \in \mathbb{Z}\}$ (where $\alpha^\vee = 2\alpha/(\alpha,\alpha)$). This is again a root system whose Weyl group W_λ is generated by the $s_\alpha: \alpha \in R_\lambda$. Set $R_\lambda^+ = R_\lambda \cap R^+$ and let $B_\lambda \subset R_\lambda^+$ denote the corresponding simple roots. It is not always possible to find an element $w \in W$ such that $wB_\lambda \subset B_0$. This can cause some technical difficulties.

Call $\lambda \in \underline{h}^*$ dominant if $(\lambda,\alpha) \geq 0$, $\forall \alpha \in R_\lambda^+$ and regular if $(\lambda,\alpha) \neq 0$, $\forall \alpha \in R$. From now on we assume that λ is a fixed dominant and regular element of $\underline{h}^*$ and set $\Lambda = \lambda + P(R)$. Let Λ^+ (resp. Λ^{++}) denote the dominant (resp. dominant and regular) elements of Λ.

2.5. Let M, N be $\mathfrak{U}(\underline{g})$ modules. Then $\mathrm{Hom}_{\mathbb{C}}(M,N)$ is a $\mathfrak{U}(\underline{g}) \otimes \mathfrak{U}(\underline{g})$ module for the action: $((a \otimes b)\, x)m = {}^t\check{a}x\check{b}m$, $\forall a,b \in \mathfrak{U}(\underline{g})$, $x \in \mathrm{Hom}_{\mathbb{C}}(M,N)$, $m \in M$. It admits the important submodule (notation 2.3) $L(M,N) := \{x \in \mathrm{Hom}_{\mathbb{C}}(M,N) \mid \dim \mathfrak{U}(\underline{k})x < \infty\}$ (where we have canonically identified $\mathfrak{U}(\underline{g}) \otimes \mathfrak{U}(\underline{g})$ with $\mathfrak{U}(\underline{g} \times \underline{g})$). For M simple, Kostant suggested that the map $\mathfrak{U}(\underline{g}) \to L(M,M)$ defined by the action of $\mathfrak{U}(\underline{g})$ should always be surjective; but in fact this is seldom true. We call $L(M,M)$ the Kostant ring of M.

2.6. Let ρ denote the half sum of the positive roots. Let M be a $\mathfrak{U}(\underline{g})$ module. A highest weight vector $e_\mu \in M$ of weight $\mu-\rho \in \underline{h}^*$ is an element of M satisfying $Xe_\mu = 0$, $\forall X \in \underline{n}^+$, $He_\mu = (H,\mu-\rho)e_\mu$, $\forall H \in \underline{h}$. A highest weight module with highest weight $\mu-\rho$ is a module generated by a highest weight vector e_μ and which by the universality of the tensor product is an image of the Verma module $M(\mu) := \mathfrak{U}(\underline{g}) \otimes_{\mathfrak{U}(\underline{b})} \mathbb{C}e_\mu$. It is easily shown that $M(\mu)$ has a unique maximal submodule $\overline{M(\mu)}$ and that $L(\mu) := M(\mu)/\overline{M(\mu)}$, is just the unique simple highest weight module with highest weight $\mu-\rho$. We set $J(\mu) = \mathrm{Ann}\, L(\mu)$.

2.7. In studying highest weight modules it is natural to introduce the decomposition

$$\mathfrak{U}(\underline{g}) = \mathfrak{U}(\underline{h}) \oplus (\underline{n}^{-}\mathfrak{U}(\underline{g}) + \mathfrak{U}(\underline{g})\underline{n}^{+})$$

and let P denote the projection $\mathfrak{U}(\underline{g}) \to \mathfrak{U}(\underline{h})$ it defines. For example, for all $z \in Z(\underline{g})$ we see that $ze_\mu = P(z)e_\mu = (P(z),\mu-\rho)e_\mu$ where $P(z)$ is considered as a polynomial function on $\underline{h}^*$. In particular, $Z(\underline{g})$ acts on $M(\mu)$ by a scalar and more precisely a theorem of Harish-Chandra ([13], Chap. 7) asserts that this scalar depends exactly on the orbit $\hat{\mu}$ of μ under W. Again one easily verifies that $(ae_\mu, be_\mu) \mapsto (P({}^{t}ab),\mu-\rho)$ is a symmetric bilinear form (the contravariant form) on $M(\mu)$ satisfying $(am,n) = (m,{}^{t}an)$, $\forall m,n \in M(\mu)$, $\forall a \in \mathfrak{U}(\underline{g})$, with kernel $\overline{M(\mu)}$. As noted by Duflo, this gives that $J(\mu) := \text{Ann}\, L(\mu) = \{a \in \mathfrak{U}(\underline{g}) \mid (P({}^{t}bac),\mu-\rho) = 0,\ \forall b,c \in \mathfrak{U}(\underline{g})\}$ which has the important consequence that ${}^{t}J(\mu) = J(\mu)$.

Again given I a two-sided ideal of $\mathfrak{U}(\underline{g})$ we define its _characteristic variety_ $\underline{\text{Ch}}(I)$ through $\underline{\text{Ch}}(I) = \underline{V}(P(I)) + \rho$. One sees that $\underline{\text{Ch}}(I) = \{\mu \in \underline{h}^* \mid I \subset J(\mu)\}$ and so $\mu \in \underline{\text{Ch}}(J(\mu))$.

3. DUFLO'S THEOREM AND AN EQUIVALENCE OF CATEGORIES

3.1. Let $\underline{\mathcal{O}}$ denote the category of all $\mathfrak{U}(\underline{g})$ modules M satisfying, (i) $\dim Z(\underline{g})m < \infty$, $\forall m \in M$, (ii) $\dim \mathfrak{U}(\underline{b})m < \infty$, $\forall m \in M$, (iii) M is a direct sum of simple finite dimensional $\underline{h}$ weight subspaces. It is easy to see that each $M \in \text{Ob}\, \underline{\mathcal{O}}$ has a filtration by highest weight modules and further taking account of Harish-Chandra's theorem (2.7) it follows that each $M \in \text{Ob}\, \underline{\mathcal{O}}$ has finite length with composition factors amongst the $L(\mu) : \mu \in \underline{h}^*$. (See also [13], Chap. 7.)

For each $M \in \text{Ob}\, \underline{\mathcal{O}}$ give M^* a $\mathfrak{U}(g)$ module structure through $(a\xi,m) = (\xi,{}^{t}am)$, $\forall \xi \in M^*$, $m \in M$, $a \in \mathfrak{U}(\underline{g})$ (notation 2.3) and set

$\delta(M) = \{m \in M^* \mid \dim \mathfrak{U}(\underline{h})m < \infty\}$ which is a $\mathfrak{U}(\underline{g})$

submodule. One shows that $\delta(M) \in \text{Ob}\, \underline{\mathcal{O}}$ and that the functor

$\delta : M \mapsto \delta M$ on $\underline{O}$ is exact and contravariant. We note that a homomorphism $M \to \delta M$ is equivalent to a contravariant form on M and so from 2.7 there exists $\varphi : M(\mu) \to \delta M(\mu)$ such that $L(\mu) \cong \operatorname{Im} \varphi$.

3.2. It is technically advantageous to restrict to the full subcategory $\underline{O}_\Lambda$ of $\underline{O}$ of objects whose $\underline{h}$ weights lie in Λ. Let E denote the category of all finite dimensional $\mathfrak{U}(\underline{g})$ modules. We remark that

$$E \otimes M \in \mathrm{Ob}\, \underline{O}_\Lambda \qquad \forall E \in \mathrm{Ob}\, E, \quad \forall M \in \mathrm{Ob}\, \underline{O}_\Lambda .$$

We define the <u>Jantzen matrix</u> b_Λ (or simply, b) to be the matrix with entries $b(w,w') := [M(w\lambda) : L(w'\lambda)]$ (where $[M : L]$ denotes the number of times the simple factor L occurs in a composition series for M). Using the exact functor $M \mapsto E \otimes M : E \in \mathrm{Ob}\, E$, (translation functor) Jantzen ([18], Chap. 2) showed that b_Λ is independent of the choice of $\mu \in \Lambda^{++}$. This fails for $\mu \in \Lambda^{+}$; but the appropriate multiplicities can be rather simply deduced from the regular case. This phenomenon runs through the whole theory.

3.3. Let G denote the algebraic adjoint group of $\underline{g}$ and B (resp. H) the algebraic subgroup of G with Lie algebra $\underline{b}$ (resp. $\underline{h}$). Then W identifies with $N_G(H)/H$ and we have the <u>Bruhat decomposition</u>

$$G = \coprod_{w \in W} B n_w B$$

where $n_w \in N_G(H)$ is a representative of $w \in W$. In the sequel we shall often write w for n_w.

By continuity of multiplication the Zariski closure $\overline{Bn_wB}$ of Bn_wB is a union of B double cosets and we set $y \le w$ if $Bn_yB \subset \overline{Bn_wB}$. It is called the <u>Bruhat order</u> on W and can be described purely combinatorially. Using this combinatorial description and the translation functor, Bernstein, Gelfand and Gelfand ([13], Chap. 7) showed that $b(w,w') \neq 0 \Leftrightarrow w' \le w'$. (One has now a more intrinsic understanding of this result through a study of differential operators on the <u>flag variety</u> G/B.) It follows that b_Λ is an invertible matrix and we denote by a_Λ (or simply, a) its inverse and define

$$(*) \qquad a(w) := \sum_{w' \in W} a(w,w')w', \qquad \forall w \in W_\lambda .$$

The $a(w): w \in W_\lambda$ which form a basis for $\mathbb{Z}W_\lambda$ play a fundamental role in the study of primitive ideals. This stems from the following observation. Let $\underline{O}_\lambda^{\wedge}$ denote the full subcategory of $\underline{O}_\Lambda$ of objects with simple factors amongst the $L(w\lambda): w \in W_\lambda$. If we represent the element $[M(w\lambda)]$ of the Grothendieck group of $\underline{O}_\Lambda$ by w, then $[L(w\lambda)]$ is represented by $a(w)$. Furthermore this identification with $\mathbb{Z}W_\lambda$ gives the Grothendieck group a $W_\lambda - W_\lambda$ bimodule structure. Moreover, the right module structure can be implemented through an astute use of the translation functors (see [23] for example), an observation essentially due to Jantzen and developed by Vogan [39, 40] and the author. Miraculously the left module structure can also be implemented in a natural way, either through the Enright functor (see [28, 29] for example) or through what eventually comes to the same thing, the equivalence of categories theorem described below.

3.4. Let $\underline{H}$ denote the category of all $\mathfrak{U}(\underline{g} \times \underline{g})$ modules V satisfying,

(i) $\dim \mathfrak{Z}(\underline{g} \times \underline{g})v < \infty, \quad \forall v \in V$,

(ii) $\dim \mathfrak{U}(\underline{k})v < \infty, \quad \forall v \in V$ (notation 2.3),

(iii) $[V : E] < \infty$ for each finite dimensional simple $\underline{k}$ module E.

One calls $V \in \mathrm{Ob}\, \underline{H}$ a Harish-Chandra module (for the pair $(\underline{g} \times \underline{g}, \underline{k})$). It has been known for some time that the theory of Harish-Chandra modules follows closely that of highest weight modules. For example, the objects of $\underline{H}$ all have finite length and for each choice of central character the simple objects are indexed by W. This led eventually to the following result developed essentially independently in [17, 22] and in [5]. First observe that each object V in $\underline{H}$ can be considered as a $\mathfrak{U}(\underline{g}) - \mathfrak{U}(\underline{g})$ bimodule through $(a \otimes b)\cdot x = {}^t\check{a}x\check{b}$,

$\forall a,b \in \mathfrak{U}(\underline{g})$, $x \in V$ and let $\underline{H}_{\hat{\lambda}}$ denote the full subcategory of all objects in $\underline{H}$ for which the right action of $Z(\underline{g})$ is just that given by its action on $M(\lambda)$. One may fairly easily show that $T(N) := L(M(\lambda),N) \in \text{Ob}\ \underline{H}_{\hat{\lambda}}$ (notation 2.5) for all $N \in \text{Ob}\ \underline{O}_{\Lambda}$. Moreover, since λ is dominant, $M(\lambda)$ is projective on $\underline{O}_{\Lambda}$ and consequently the functor $T: N \mapsto T(N)$ is exact. Conversely given $V \in \text{Ob}\ \underline{H}_{\hat{\lambda}}$ one checks that $T'(V) := V \otimes_{\mathfrak{U}(\underline{g})} M(\lambda) \in \text{Ob}\ \underline{O}_{\Lambda}$ and that T' is a left adjoint to T.

THEOREM. <u>The unit</u> $\text{Id}_{\underline{H}_{\hat{\lambda}}} \to TT'$ <u>and counit</u> $T'T \to \text{Id}\ \underline{O}_{\Lambda}$ <u>maps are isomorphisms of functors. In particular,</u> T' <u>is exact and</u> T <u>is an equivalence of categories.</u>

We remark that the proof depends heavily on the fact that the map $\mathfrak{U}(\underline{g}) \mapsto L(M(\lambda),M(\lambda))$ is surjective. This in turn is (an easy) consequence of a classical theorem of Kostant.

3.5. The above result gives objects in the $\underline{O}_{\Lambda}$ category a $\mathfrak{U}(\underline{g})$ - $\mathfrak{U}(\underline{g})$ <u>bimodule</u> structure. It allows one to implement the left action of W_{λ} on the Grothendieck group of $\underline{O}_{\hat{\lambda}}$ (as discussed in 3.1) and implies, for example, the symmetry relation

$$(*) \qquad a(w,w') = a(w^{-1},w'^{-1}), \qquad \forall w,w' \in W_{\lambda}.$$

Again it shows that the simple objects in $\underline{H}_{\hat{\lambda}}$ are just the $V(-w\mu,-\lambda) := L(M(\lambda),L(w\mu)): w \in W_{\lambda},\ \mu \in \Lambda^{+}$. Finally it implies that the map $I \mapsto IM(\lambda)$ of ideals of $A_{\lambda} := \mathfrak{U}(\underline{g})/\text{Ann}\ M(\lambda)$ to submodules of $M(\lambda)$ is <u>bijective</u>. (This was conjectured by Dixmier.) In particular, it follows that $I = \text{Ann}(M(\lambda)/IM(\lambda))$ for any such ideal. Furthermore, if I is a prime ideal, it follows from Goldie's theorem that $\mathfrak{U}(\underline{g})/I$ is d-critical (see 2.1) and hence so is $M(\lambda)/IM(\lambda)$. Consequently, $M(\lambda)/IM(\lambda)$ admits a unique simple submodule which must take the form $L(\sigma\lambda): \sigma \in W_{\lambda}$, and furthermore $I = \text{Ann}\ L(\sigma\lambda)$. The latter being stable under the Chevalley antiautomorphism (2.7) implies eventually that σ is an involution. We have the

COROLLARY. The map $I \mapsto \mathrm{Soc}(M(\lambda)/IM(\lambda)) = L(\sigma\lambda)$ is a bijection of $\mathrm{Spec}\, A_\lambda$ onto a subset $\sum_\lambda^0$ of involutions of W_λ.

We call $\sum_\lambda^0$ the Duflo set. It coincides with the set of all involutions $\sum_\lambda$ of W_λ if and only if R_λ is a union of systems of type A_n.

3.6. For any $\mu \in \underline{h}^*$ a classical result of Kostant can be used to show ([13], Chap. 8) that $\mathrm{Ann}\, M(\mu)$ is generated by its intersection with $Z(\underline{g})$, which (as noted in 2.7) is a maximal ideal of $Z(\underline{g})$. Thus any primitive ideal of $U(\underline{g})$ contains some $\mathrm{Ann}\, M(\mu)$ and we can assume λ chosen such that $\mu \in \Lambda^+$. Let $\underline{X}_{\hat{\mu}}$ denote the set of all primitive ideals of $U(\underline{g})$ containing $\mathrm{Ann}\ M(\mu)$. Then map $I \mapsto I/\mathrm{Ann}\, M(\mu)$ identifies $\underline{X}_{\hat{\mu}}$ with $\mathrm{Spec}\, A_\mu$ described (for μ regular) by 3.5. This is Duflo's [14] description of $\mathrm{Prim}\, U(\underline{g})$ and we remark that after Borho-Jantzen ([8], Sec. 2) the $\underline{X}_{\hat{\mu}}$: $\mu \in \Lambda^{++}$ are isomorphic as ordered sets (ordered by inclusion). Apart from some degeneration (see 4.1) this last assertion extends to Λ^+.

3.7. The fundamental result stated in 3.4 can be regarded as a consequence of the following simple geometric fact. First note that the algebraic adjoint group of $\underline{g} \times \underline{g}$ coincides with $G \times G$ and we let K denote the algebraic subgroup of $G \times G$ with Lie algebra $\{(X,X) : X \in \underline{g}\}$ (we make a slight change of convention). The map $(x,y) \mapsto xy^{-1}$ of $G \times G \to G$ defines by passage to quotients a map of $G/B \times G/B \to B\backslash G/B$ which identifies the K orbits in the flag variety $G/B \times G/B$ of $G \times G$ with the B orbits in the flag variety G/B of G.

More generally let K be any algebraic subgroup of G such that $K\backslash G/B$ is finite. One may study the category $\underline{H}^K$ of $U(\underline{g})$ modules with an algebraic K action consistent with action of K on $U(\underline{g})$. The simple objects in this category have been classified by Beilinson and Bernstein [3] in a tour de force which represents a far reaching generalization of the Langslands classification (relevant to the case when G is a real Lie group and K a maximal compact subgroup). For simplicity let us just consider the subcategory $\underline{H}^K_{\hat{\rho}}$ of objects of $\underline{H}^K$ whose annihilators contain $\operatorname{Ann} M(\rho)$.

THEOREM. ([3], Sect. 3). _The simple objects in_ $H^K_{\hat{\rho}}$ _are in bijection with the set of pairs consisting of a_ K _orbit_ S _in_ G/B _and an irreducible character_ x _of the component group of the stabilizer of some_ $s \in S$.

In fact, the construction assigns to such pairs two objects in $H^K_{\hat{\rho}}$. The first, say $M_{S,x}$, generalizes the Verma module $M(w\rho)$ (assigned to the Bruhat cell Bn_wB/B, the stabilizer group of a point being connected). The second, say $L_{S,x}$, is constructed as a quotient of $M_{S,x}$ by generalizing δ.

One has then the general problem of computing $[M_{S,x} : L_{S',x'}]$ which has not yet been fully resolved; but which may be reduced to a special case of a problem in algebraic topology in the framework of the DGM intersection cohomology theory, where the simple objects $L_{S,x}$ correspond to the DGM sheaves.

4. THE ADDITIVITY PRINCIPLE, GOLDIE RANK AND WEYL GROUP REPRESENTATONS

4.1. Fix $w \in W_\lambda$. We define $\{J(w\mu)\}_{\mu\in\Lambda+}$ to be the t-_clan_ associated to w. Distinct elements in a given t-clan correspond to distinct central characters, i.e., the $Z(\underline{g}) \cap J(w\mu) : \mu \in \Lambda^+$ are pairwise distinct.

Now define a function p_w on Λ^+ through (notation 2.2)

$$p_w(\mu) := \operatorname{rk}(U(\underline{g})/J(w\mu)),$$

which is determined up to a scalar by the

THEOREM. ([25], 5.1). _For each_ $w \in W_\lambda$, _there exists a positive rational number_ c_w _such that_

(*) $$p_w = c_w a(w^{-1})\rho^m \qquad \text{(notation 3.3*)}$$

where m is the least integer ≥ 0 for which the right-hand side is non-zero. Furthermore $m = \text{card } R^+ - d(L(w\lambda))$ (notation 2.1).

Observe first of all that this result implies that p_w extends to a polynomial function (homogeneous of degree m) on $\underline{h}^*$. This in itself is a remarkable fact. Secondly note that (up to a scalar) p_w is completely determined by the Jantzen matrix, in fulfillment of the philosophy outlined in 1.2. We remark that the determination of the scale factor c_w is rather delicate. Obviously c_w must be sufficiently large so that $p_w(\mu)$ is an integer ≥ 0 for all $\mu \in \Lambda^+$. The natural conjecture is that it is the smallest rational number with this property, through this is difficult to check even in type A_n where these coefficients are known implicitly. Note that this property is immediately implied if $p_w(\mu) = 1$ for some $\mu \in \Lambda^+$, thus the importance of constructing completely prime (i.e., Goldie rank 1) primitive ideals. Part of the difficulty is that within a given t-clan it is relatively unusual to find such ideals. When $\lambda \in P(R)^{++}$, then

$$a(1) = \sum_{w \in W} w(\det w)$$

and substitution in (*) gives Weyl's dimension formula, namely

$$\dim L(\lambda) = \prod_{\alpha \in R^+} \frac{(\lambda,\alpha)}{(\rho,\alpha)},$$

where we have used the fact that $J(\rho)$ is the augmentation ideal of $\mathfrak{U}(\underline{g})$ and hence completely prime.

Though $p_w(\mu) > 0$, $\forall \mu \in \Lambda^{++}$ it can happen that $p_w(\mu) = 0$ if $\mu \in \Lambda^+$. It turns out that this corresponds to the degeneration of $J(w\mu)$ to $\mathfrak{U}(\underline{g})$ under translation to a wall as originally described by Borho and Jantzen ([8], 2. 12). Moreover if we set $\tau(w) = \{\alpha \in B_\lambda \mid w\alpha \in R^+\}$ then p_w is divisible by the polynomial

$$p_{R'} := \prod_{\alpha \in R' \cap R^+} \alpha$$

where R' is the root system generated by $\tau(w)$ (i.e. $R' = \mathbb{Z}\tau(w) \cap R$). Fix $\alpha \in B_\lambda$; if $(\mu,\beta) = 0$, $\beta \in B_\lambda$ implies $\beta = \alpha$, then $p_w(\mu) = 0$ $\Leftrightarrow \alpha \in \tau(w)$.

4.2. Let us just discuss briefly the step in the proof of 4.1 which implies the polynomial nature of the p_w. This is based on an _additivity principle_ for Goldie rank, namely

THEOREM. ([24], 5.11). _For each simple object_ $N \in \mathrm{Ob}\ \underline{\mathcal{O}}$ _and each_ $E \in \mathrm{Ob}\ \mathcal{E}$ _one has_

(*) $$\dim E.\ \mathrm{rk}\ L(N,N) = \sum \mathrm{rk}\ L(N_i,N_i)$$

where the sum runs over the simple factors N_i _of_ $E \otimes N$ _satisfying_ $d(N_i) = d(N)$.

Of course to use this result one must be able to relate $\mathrm{rk}(\mathfrak{U}(\underline{g})/\mathrm{Ann}\ N)$ to $\mathrm{rk}\ L(N,N)$. A key point here is that these integers are equal when $N \cong L(\sigma\lambda)$ for σ a Duflo involution (3.5). The proof of the above result relies partly on abstract ring theory—in particular Small's refinement of Goldie's theorem which shows that the left hand side equals $\sum y_i^{-1}\ \mathrm{rk}\ L(N_i,N_i)$ for some $y_i \in \mathbb{N}^+ \cup \{\infty\}$ ([24], Thm. 5.6) and partly on 3.4 which is used to show ([24], 3.2) that the endomorphism ring of $L(M(\lambda),L(w\lambda))$ considered as a right $\mathfrak{U}(\underline{g})$ module is just $L(L(w\lambda),L(w\lambda))$, a result which eventually implies that the y_i are all equal to 1. That (*) implies the p_w to be polynomial eventually results from Vogan's observation ([38], Sect. 4) that _any_ function p on Λ satisfying

$$\sum_{w \in W} p(\lambda + w\mu) = (\mathrm{card}\ W)p(\lambda), \qquad \forall \mu \in P(R)$$

is a polynomial (in fact a W-harmonic polynomial).

4.3. Theorem 4.1 has some important further consequences when one takes account of the $W_\lambda - W_\lambda$ module structure on the Grothendieck group of $\underline{O}_{\hat{\lambda}}$ implemented by the translation functors (3.2). For this it is convenient to set $p_w = p_J$ when $J = J(w\mu)$. Since by 3.6

$$J(w\mu) = J(y\mu) \iff J(w\mu') = J(y\mu'), \qquad \forall \mu,\mu' \in \Lambda^{++},$$

we may regard J as representing the whole t-clan associated to w.

Let τ be an irreducible representation of W. Because W is finite group and acts faithfully on $\underline{h}^*$, there exists a smallest integer $m \geq 0$ such that $[S_m(\underline{h}) : \tau] > 0$ (notation 2.1). Call τ <u>univalent</u> (with respect to $\underline{h}$) if $[S_m(\underline{h}) : \tau] = 1$. The significance of τ being univalent is that then $S_m(\underline{h})$ admits a unique simple submodule P_τ of type τ. We call P_τ the <u>univalent module</u> associated to τ. Not all Weyl group representations are univalent. However, we remark that if R' is a root subsystem then $p_{R'}$ (notation 4.1) generates a univalent submodule $P_{R'}$ of $S(\underline{h})$, though not all univalent representations are so obtained.

THEOREM. ([22], 5.5) <u>There exists a subset Ω_λ of the univalent representations of W_λ such that (notation 3.6)</u>

(i) $\underline{X}_{\hat{\lambda}} = \coprod_{\tau\in\Omega_\lambda} (\underline{X}_{\hat{\lambda}})_\tau$.

(ii) $\{p_J : J \in (\underline{X}_{\hat{\lambda}})_\tau\}$ <u>is a basis for the univalent module P_τ of W_λ (actually a $\mathbb{Z}$ basis)</u>.

(iii) WP_τ <u>is a univalent (hence irreducible) W module.</u>

We call $(\underline{X}_{\hat{\lambda}})_\tau$ an r-clan. Each r-clan is associated to a (distinct) irreducible representation of W. Univalence is a consequence of the fact that the left action of W_λ on the Grothendieck group of $\underline{O}_{\hat{\lambda}}$, as implemented through translation and use of 3.4, does not affect annihilators. Since Goldie rank is an invariant this eventually implies that $a(w^{-1})w'\rho^m$ is proportional to p_w for all $w' \in W_\lambda$. The fact (ii) that the Goldie rank polynomials are linearly independent for distinct annihilators results through a use of the characteristic variety

(2.7).

4.4. One has $\Omega_\lambda \subset \hat{W}_\lambda$ with equality iff R_λ has only type A_n factors. To determine Ω_λ in general it is enough to assume $\lambda \in P(R)^{++}$ (i.e., integral). This is because Ω_λ is implicitly determined by the Jantzen matrix and from the truth of the Kazhdan-Lusztig conjecture this depends only in the description of W_λ as a Coxeter group. Now suppose R' is a root subsystem of R with Weyl group W'. If $P \subset S(\underline{h})$ is a univalent module for W', then WP is a univalent module for W. This was pointed out by Lusztig and Spaltenstein [33]. The process $P \mapsto WP$ mirrors induction of primitive ideals and one may remark that WP is an irreducible component of $\mathrm{Ind}(P;W'\uparrow W)$. It is easy to show that Ω is stable under the above induction procedure where R' takes the form $R' = \mathbb{Z}B' \cap R$ for some subset $B' \subset B_0$. Noting that p_R (notation 4.1) generates the sign representation sg of W, one might also conjecture that Ω is stable under tensoring with sg. Remarkably this is very nearly but not quite true. In fact following a suggestion of Lusztig one defines an involution $*$ on $\hat{W}$ which is tensoring by sg except in three cases, one in type E_7 and two in type E_8, and one defines a representation to special if it obtains from the identity representation (of an appropriate Weyl subgroup of W) by a combination of $*$ and the above described induction.

THEOREM. ([1, 2]). Ω _is just the set of special representations of_ W.

Not surprisingly this result due to Barbasch and Vogan involves some case by case analysis. However one can understand that this results depends heavily on some symmetry property of the Jantzen matrix (5.3*) which can be thought of as implementing $*$.

4.5 It can be shown that the Goldie skew-field of $\mathfrak{U}(\underline{g})/J : J \in \mathrm{Prim}\,\mathfrak{U}(\underline{g})$ is constant on t-clans ([24], 4.8). One expects the Goldie skew-field (2.2) to be also constant on r-clans; but only a slightly weaker statement has been shown. Finally one expects the Goldie skew-field to always be a Weyl skew-field of index $\frac{1}{2} d(\mathfrak{U}(\underline{g})/J)$. For example this has been verified in type A_n ([24], 4.8).

4.6. For each $w \in W_\lambda$, set $S_\lambda(w) = \{\alpha \in R_\lambda^+ | w\alpha \in R_\lambda^-\}$ and $\ell_\lambda(w) = \text{card } S_\lambda(w)$ which coincides with the reduced length of w. Given $B' \subset B$, let $W_{B'}$ denote the subgroup of W_λ generated by the s_α: $\alpha \in B'$ and let $w_{B'}$ denote the unique longest element of $W_{B'}$. Set $w_\lambda = w_{B_\lambda}$.

LEMMA. ([30], 8.18). <u>For all</u> $w \in W_\lambda$ <u>one has</u> $\deg p_w \leq \ell_\lambda(w)$ <u>with equality iff</u> $w = w_{B'}$ <u>for some</u> $B' \subset B_\lambda$ <u>and then</u> p_w <u>is proportional to</u> $p_{R'}$ <u>where</u> $R' = \mathbb{Z}B' \cap R$.

Because p_w: $w \in W_\lambda$ is a W harmonic polynomial, it can be expressed as a sum formed from products of distinct positive roots (i.e. in R_λ^+). Since p_w must take positive values on Λ^{++}, it is natural to conjecture that the coefficients can always be chosen to be positive. If this holds we can obviously write $p_w = \sum c_K \alpha^K$: $c_K > 0$, and where α_K denotes a product of the elements of B_λ. Let supp p_w denote the set of $\alpha \in B_\lambda$ which occurs in the above expression. If $p_w(\mu) = 1$ for some $\mu \in \Lambda^{++}$ it follows from the positivity of the c_K that $p_w(\nu) = 0$ for all $\nu \in \Lambda^+$ satisfying (ν,α) for some $\alpha \in \text{Supp } p_w$. Consequently (4.1) $p_{R'}$ divides p_w where $R' = \mathbb{Z}(\text{Supp } p_w) \cap R$ and because $p_{R'}$ is the highest degree W-harmonic polynomial on $\mathbb{Q}(\text{Supp } p_w)$ it follows that p_w is proportional to $p_{R'}$.

The appearance of Goldie rank 1 for $\mu \in \Lambda^+$ is certainly much more subtle. We conjecture that for each $\tau \in \Omega_\lambda$ there exists $J \in (\underline{X}_{\hat{\lambda}})_\tau$ such that $p_J(\mu) = 1$ for some $\mu \in \Lambda^+$. When P_τ takes the form $P_{R'}$ for some $R' = \mathbb{Z}B' \cap R$ then for each $J \in (\underline{X}_{\hat{\lambda}})_\tau$ one can find $w \in W_\lambda$ such that

$$(*) \qquad w p_{R'} = \sum n_{J'} p_{J'} ,$$

with $n_{J'} \in \mathbb{N}$ and $n_J > 0$. If $p_J(\mu) = 1$ for some $\mu \in \Lambda^+$, we conjecture that it is possible to choose w in (*) such that $(wp_{R'})(\mu) = 1$ (so $n_J = 1$, and $p_{J'}(\mu) = 0$ for $n_{J'} \neq 0$, $J' \neq J$). For example, in type A_n this would imply that all the primitive ideals of Goldie rank 1 are induced, or in the language of ([8], Sect. 5) that the Dixmier map is surjective. This has been checked up to $n = 5$ by A. Klugman.

5. ORDERING OF PRIMITIVE IDEALS

5.1. If $x \in \mathbb{Q}W_\lambda$, we can write

$$x = \sum_{w \in W_\lambda} c_w a(w)$$

and we then let $[x]$ denote the set of $a(w)$ for which $c_w \neq 0$. For each subset $S \subset \mathbb{Q}W_\lambda$ we set

$$[S] = \bigcup_{x \in S} [x].$$

For each $w \in W_\lambda$ we set $D(w) := [W_\lambda a(w)]$. The set of all $w' \in W_\lambda$ such that $a(w') \in D(w)$ is called the <u>left cone</u> $\mathcal{D}(w)$ of W_λ generated by w. We similarly define $D'(w) := [a(w)W_\lambda]$ and let $\mathcal{D}'(w)$ denote the corresponding <u>right cone</u> generated by w. By 3.3(*) left and right cones are interchanged through the involution $w \mapsto w^{-1}$. Note that by 3.5(*) no ambiguity arises here.

Order the set of left (resp. right) cones of W_λ by inclusion. The <u>left cell</u> $\mathcal{C}(w)$ of W_λ generated by w is defined to be the complement in $\mathcal{D}(w)$ of the union of all the left cones strictly contained in $\mathcal{D}(w)$. A <u>right cell</u> $\mathcal{C}'(w)$ of W_λ is defined similarly with respect to $\mathcal{D}'(w)$. The left (or the right) cells of W_λ form a partition of W_λ. We set $C(w) = \{a(w) \mid w \in \mathcal{C}(w)\}$.

The above notations were introduced in [33] to try to understand the ordering of the primitive ideals. They also played a role in motivating the Kazhdan-Lusztig conjecture (see [41], Sect. 1, for example). In ([33], Sect. 5) it was shown that <u>Ch</u>$(J(w\lambda)) \supset \mathcal{D}(w)$. Vogan ([39])

established equality and thereby determined completely the ordering of the primitive ideals, a problem which had hitherto appeared intractable. We may express the result (which is rather easy to prove!) through the

THEOREM. <u>For all</u> $w, w' \in W_\lambda$,

(i) $J(w'\lambda) \supset J(w\lambda) \Leftrightarrow a(w') \in D(w)$.

(ii) $J(w'\lambda) = J(w\lambda) \Leftrightarrow a(w') \in C(w)$.

5.2. The above result shows that the Jantzen matrix completely determines the ordering of the primitive ideals. Knowledge of this matrix determines the cells $C(w)$ implicitly; but these have in fact been more explicitly determined. In type A_n this was obtained (cf. [20]) using the Robinson bijection and a combinatorial result of Knuth (see [21]). This has been extended to the remaining classical groups by Barbasch and Vogan [1, 2]. More generally we remark that $\mathbb{Q}D(w)$ is always a W_λ module for left multiplication. Consequently $\mathbb{Q}C(w)$ has a (quotient) W_λ module structure. It need not be irreducible; but it is easy to see from 4.1 that it contains the "Goldie rank representation" $\mathbb{Q}W_\lambda p(w)$ with multiplicity one. Barbasch and Vogan [1, 2] have determined exactly which representation occurs in each left cell; which is a little easier than determining the cells themselves.

5.3. The truth of the Kazhdan-Lusztig conjecture implies the remarkable relation

$$(*) \qquad a(w,w') = (\det ww')b(ww_\lambda, w'w_\lambda)$$

conjectured by Jantzen. It immediately implies that $D(w) \supset D(w') \Leftrightarrow D(ww_\lambda) \subset D(w'w_\lambda)$, $\forall w, w' \in W_\lambda$. Consequently (*) has three amusing consequences. First the map $J(w\lambda) \mapsto J(ww_\lambda\lambda)$ is an order antiautomorphism of $\underline{X}_{\hat{\lambda}}$. This symmetry was conjectured by Borho and Jantzen ([8], Sect. 2). Secondly (from (*)) we obtain that $\mathbb{Q}C(w) \otimes sg \cong \mathbb{Q}C(ww_\lambda)$ (this fact was pointed out to me by Barbasch—it can be read off from say the computation in [26], 4.7). It underlies the Barbasch-Vogan result discussed in 4.4; but notice the extraordinarily subtle fact that the

Goldie rank representation occurring in $QC(ww_\lambda)$ need not be sg tensored with the Goldie rank representation occurring in $QC(w)$.

Finally for any A_λ (notation 3.5) module N we may define its equalizer $E(N)$ to be the smallest ideal I such that $IN = N$ (recall that by 3.4, A_λ is Artinian for two-sided ideals). Equalizers are always idempotent ideals and every idempotent ideal is an equalizer of some A_λ module which can be chosen by 3.4 to belong to $\underline{\mathcal{O}}_\lambda$. We call an equalizer primal if it is the equalizer of a simple module. These can be characterized as the set of idempotent ideals of A_λ admitting a simple radical, or as the equalizer of some $L(w\lambda)$; and every idempotent ideal is a sum of such equalizers. One has [31] the

LEMMA. *For each* $w \in W_\lambda$,

$$(**) \qquad \underline{\mathrm{Ch}}(E(L(w\lambda))) = (W_\lambda \setminus \mathcal{D}(ww_\lambda))w_\lambda .$$

This indicates a duality between primal equalizers and primitive ideals.

5.4. The formula (**) of 5.3 can be regarded as a sum formula for a power of each prime ideal of A_λ in terms of primal equalizers (or dually as a formula for a primal equalizer as a power of an intersection of prime ideals of A_λ). This brings us naturally to the following

CONJECTURE. *Take* $\mu \in \Lambda^{++}$. *Every prime ideal of* A_μ *is idempotent (and hence itself a sum of primal equalizers).*

One knows ([19], Sect. 4) that the $J(w_\lambda s_\alpha \lambda) : \alpha \in B_\lambda$, the so-called minimal primitive ideals, have idempotent images in A_λ and in A_λ one even has $J(w_\lambda s_\alpha \lambda) = E(L(s_\alpha \lambda))$. Again for any $B' \cap B_\lambda$ a result ([16], 5.2) extending Duflo's sum formula ([14], Prop. 12) gives

$$\sum_{\alpha \in B'} J(w_\lambda s_\alpha \lambda) = J(w_\lambda w_{B'} \lambda)$$

and shows that the right-hand side (which is a primitive ideal) has an

idempotent image in A_λ. Apart from these examples, there is now sufficient knowledge about the structure of A_λ derived mainly from (3.4) and the truth of the Jantzen conjecture (6.3) to render the above conjecture very plausible. Finally we remark that the above conjecture may be regarded as an aspect of calculating generators (natural or otherwise) of a given primitive ideal. Given idempotence this is reduced via 5.3 to calculating generators for the $E(L(w\lambda))$. The latter having simple radicals are cyclic as bimodules
and the cyclic vector can be chosen to be a highest weight vector with respect to the diagonal action of $\underline{g}$ and belonging to a so-called "minimal $\underline{k}$-type" (which for complex groups is always unique).

5.5. The rather special nature of some of the above observations is underlined by the fact that they fail for non-regular μ. Thus in general for $\mu \in \Lambda^+ \setminus \Lambda^{++}$, the prime spectrum of A_μ (i.e., $\underline{X}_{\hat{\mu}}$) does not admit ([8], Sect. 2) an order reversing involution. Again A_μ can admit prime ideals ([19], Sect. 4) which are not idempotent.

6. THE JANTZEN FILTRATION AND KAZHDAN-LUSZTIG CONJECTURES

6.1. With respect to a Chevelley basis the structure constants for $\underline{g}$ take integer values and consequently the free $\mathbb{Z}$ module $\underline{g}_{\mathbb{Z}}$ generated over this basis admits a Lie algebra structure. Then for any ring A we may define $\underline{g}_A := \underline{g}_{\mathbb{Z}} \otimes_{\mathbb{Z}} A$. Analogous meanings are assigned to $\underline{h}_A$, $\underline{b}_A$, etc.

As in the case when A is a field we may define for each $\xi \in \underline{h}_A^* := \mathrm{Hom}_A(\underline{h}_A, A)$ the $\mathfrak{U}(\underline{b}_A)$ module A_λ to be A as an A module, with $H \in \underline{h}_A$ (resp. $X \in \underline{n}_A$) acting by multiplication by (ξ, H) (resp. by zero) and set

$$M(\xi) := \mathfrak{U}(\underline{g}_A) \otimes_{\mathfrak{U}(\underline{b}_A)} A_{\xi-\rho} .$$

Now consider the special case when $A = \mathbb{C}[t]$ and $\xi = \mu + t\delta$ for some $\mu, \delta \in \underline{h}^*$ with δ regular. Then $M(\xi) \otimes_{\mathbb{C}[t]} \mathbb{C}(t)$ is just a

Verma module for $\underline{g}$ defined over the field $\mathbb{C}(t)$. Because δ is regular, it is rather easy to check that the resulting module is irreducible and hence admits a non-degenerate contravariant form (2.7) with values in $\mathbb{C}(t)$. Restricted to $M(\xi)$ this form takes values in $\mathbb{C}[t]$. Consequently we may define for each $n \in \mathbb{N}$ a $U(\underline{g}_A)$ filtration of $M(\xi)$ through

$$M^n(\xi) = \{m \in M(\xi) \mid (M(\xi),m) \in (t^n)\}.$$

Now set $M^n(\mu) = M^n(\xi)_{t=0}$. This is a finite descending $U(\underline{g})$ filtration of $M(\mu)$ called a <u>Jantzen filtration</u> of $M(\mu)$. We set $M_n(\mu) = M^n(\mu)/M^{n+1}(\mu)$, $\forall n$. One has $M_0(\mu) \cong L(\mu)$.

Information on the $M_n(\mu)$ obtains by taking $A = S(\underline{h})$ and regarding $H \mapsto (\xi,H)$ as an element of $\underline{h}_A^*$. Then the contravariant form on $M(\xi)$ restricts to each $\underline{h}_A$ weight submodule $M(\xi)_{\xi-\nu-\rho}$: $\nu \in \mathbb{N}B$ on which the determinant D_ν is an element of $S(\underline{h})$.

THEOREM. ([34]; [18], Satz II). <u>For each</u> $\nu \in \mathbb{N}B$ <u>one has</u>

$$D_\nu(\xi) = \prod_{\alpha\in R^+} \prod_{r\in\mathbb{N}^+} ((\xi,\alpha^\vee) - r)^{P(\nu-r\alpha)}$$

<u>where</u> P <u>is the Kostant partition function defined through</u>

$$\sum_{\nu\in\mathbb{N}B} P(\nu)e^\nu = \prod_{\alpha\in R^+} \frac{1}{(1-e^\alpha)} .$$

This result is obtained by noting that $M(\xi)$ is reducible whenever $(\xi,\alpha^\vee) \in \mathbb{N}^+$ for some $\alpha \in R^+$. This gives enough information on the zeros of D_ν to conclude the above formula.

6.2. Taking $\xi = \mu + t\delta$ in 6.1, we may compute for each $\nu \in \mathbb{N}B$ the smallest integer k such that $D_\nu(\xi) \in (t^k)$ either by substitution in 6.1 or in terms of the Jantzen filtration. This eventually gives the Jantzen sum formula.

THEOREM. ([18], 5.3). <u>For each simple</u> $L \in \text{Ob}\ \mathcal{O}$, <u>one has</u>

$$(*) \qquad \sum_{n>0} [M^n(\mu) : L] = \sum_{\alpha\in R^+ | (\alpha^\vee,\mu)\in \mathbb{N}^+} [M(s_\alpha\mu) : L].$$

Observe that from (*) if $[M(\mu) : L] \neq 0$, then either $L \cong L(\mu)$ or $[M^n(\mu) : L] \neq 0$ for some $n > 0$. In the second case $[M(s_\alpha\mu) : L] \neq 0$, for some $\alpha \in R^+$ such that $(\alpha^\vee,\mu) \in \mathbb{N}^+$. This leads to a combinatorial criterium for determining when $[M(\mu) : L] \neq 0$ and is in fact just the Bernstein-Gelfand-Gelfand result based on the Bruhat order discussed in 3.1. The above proof due to Jantzen is particularly elegant and short.

6.3. Now take $\mu = w\lambda$ with $w \in W_\lambda$ and choose $\alpha \in B_\lambda$ such that $ws_\alpha < w$ (Bruhat order). It is relatively easy to show that dim $\mathrm{Hom}_{\mathfrak{U}(\mathfrak{g})}(M(w\lambda), M(ws_\alpha\lambda)) = 1$ and Jantzen ([18], 5.18) made the seemingly innocuous conjecture (based on sum formulae extending 6.2) that with respect to the embedding $M(w\lambda) \hookrightarrow M(ws_\alpha\lambda)$ one must have

$$(*) \qquad M^n(w\lambda) = M(w\lambda) \cap M^{n+1}(ws_\alpha\lambda), \qquad \forall n \in \mathbb{N}.$$

Motivated by the Kazhdan-Lusztig conjecture for the Jantzen matrix it was shown ([17], 4.8) that

THEOREM. _Assume_ (*) _holds for all_ $\alpha \in B_\lambda$. _Take_ $w \in W_\lambda$, $\alpha \in B_\lambda$ _such that_ $ws_\alpha < w$. _Then for all_ $n \in \mathbb{N}$ _one has_

(i) $M_n(w\lambda)$ _is semi-simple_.

(ii) $[M_{n+1}(ws_\alpha\lambda) : L(y\lambda)] = [M_n(w\lambda) : L(y\lambda)]$ _if_ $y > ys_\alpha$.

(iii) $[\mathfrak{U}_\alpha M_n(w\lambda) : L(y\lambda)] = [M_n(ws_\alpha\lambda) : L(y\lambda)] + [M_{n+1}(w\lambda) : L(y\lambda)]$ _if_ $y < ys_\alpha$, _where_ $\mathfrak{U}_\alpha$ on _$\mathcal{O}$ is exact and_ $\mathfrak{U}_\alpha^2 = 0$, $\mathfrak{U}_\alpha L(y\lambda) \neq 0 \Leftrightarrow y > ys_\alpha$.

Given that $M(w_\lambda\lambda) \cong L(w_\lambda\lambda)$ the above formulae determine $[\mathfrak{U}_\alpha L : L']$ and $[M_n(w\lambda) : L']$ inductively (where L, L' are simple objects in $\mathcal{O}$). Consequently the Jantzen matrix $[M(w\lambda) : L(w'\lambda)]$ can be calculated and this agreed with the formulae conjectured by Kazhdan and Lusztig ([32], Sect. 1). At least from a computational point of view the above theorem probably represents the most transparent formulation of their conjecture. Note that the multiplicities in each filtration step of $M(w\lambda)$ are determined and we call this information the JKL data.

6.4. Bernstein informed me that a proof of the Jantzen conjecture 6.3(*) (and that all Jantzen filtrations are equivalent) derives from Gabber's purity theorem ([12]). The proof is based partly on the theory of $\mathcal{D}$ modules over G/B and partly on a reduction modulo p, in which roughly speaking the different steps in the Jantzen filtration are separated as distinct eigenspaces of the Frobenius. Eventually one might hope for a simpler and purely algebraic proof. To understand the difficulty we remark that for $n = 0$ the question is equivalent to showing the following. Take $\mu = w\lambda$ and suppose $\alpha \in B_0$ satisfies $s_\alpha w < w$. Consider $M(\mu)$ as a module for the $sl(2)$ subalgebra of $\underline{g}$ generated by X_α, $X_{-\alpha}$ and show with respect to the contravariant form that $M(\mu)$ is an orthogonal direct sum of indecomposable modules. Although $M(\mu)$ admits essentially only two types of indecomposable summands, this question involves a rather delicate choice of basis. This question has also a "t analogue" equivalent to the Jantzen conjecture.

6.5. Vogan gave a somewhat different interpretation of the Kazhdan-Lusztig polynomials using the extension groups $\text{Ext}^j(M(w\lambda), L(w'\lambda))$ defined in the $\underline{\mathcal{O}}$ category. Unfortunately he was only able to show ([40], 3.5) that their conjecture was equivalent to the semi-simplicity hypothesis

$$(*) \qquad \mathfrak{U}_\alpha L \text{ is \underline{semisimple}}, \qquad \forall \alpha \in B_\lambda$$

for each <u>simple object</u> $L \in \text{Ob}\,\underline{\mathcal{O}}$.

We remark that (*) follows from the Jantzen conjecture through the analysis of 6.3.

Beilinson-Bernstein and independently Brylinski-Kashiwara showed how to use the theory of $\mathcal{D}$ modules over G/B to reduce the Kazhdan-Lusztig conjecture to a question on DGM sheaves solved essentially by deep results of Deligne. A similar and slightly more refined machinery is necessary for the proof the Jantzen conjecture discussed in 6.4. Finally we remark that is an obvious conjecture ([17], Sect. 5) for the ranks of the extension groups $\mathrm{Ext}^j(M(w\lambda),M(w'\lambda))$; but this has so far remained unproven.

6.6. One of the technical advantages of 6.2 over the Kazhdan-Lusztig conjecture is that it gives a way of computing the Duflo set $\sum_\lambda^0$. Indeed for each $J \in \underline{X}_{\hat{\lambda}}$ set $C_J = \{w \in W_\lambda | J = J(w\lambda)\}$ and define $k_J = \min_{w\in C_J} \min_{\ell\in\mathbb{N}} \{\ell | [M_\ell(\lambda) : L(w\lambda)] > 0\}$. Then using 6.2, one easily shows ([26], 4.9) that $L(\sigma\lambda)$ is the unique simple submodule of $M_{k_J}(\lambda)$ with annihilator J, where σ is the unique element of $C_J \cap \sum_\lambda^0$. In virtue of 5.1 this shows that the JKL data and hence Kazhdan-Lusztig polynomials implicitly determine the Duflo set $\sum_\lambda^0$.

7. THE KOSTANT RING

7.1. Let N be a simple $\mathfrak{U}(\underline{g})$ module. We have already seen (4.2) the importance of the Kostant ring $L(N,N)$ (notation 2.5) in the study of Goldie rank when N is a highest weight module. Here we discuss some general results for arbitrary N, together with some very refined results when $N \in \mathrm{Ob}\ \underline{\mathcal{O}}$. First of all we remark that $L(N,N) \in \mathrm{Ob}\ \underline{H}$ (notation 3.4). Set $J = \mathrm{Ann}\ N$. Then the action of $\mathfrak{U}(\underline{g})$ on N gives a ring embedding $\mathfrak{U}(\underline{g})/J \hookrightarrow L(N,N)$ which is not always surjective even if $\underline{g} \cong sl(2)$ ([27]).

Our previous remark shows that $L(N,N)$ is finitely generated as a $\mathfrak{U}(\underline{g})$ - $\mathfrak{U}(\underline{g})$ bimodule, hence as say a left $\mathfrak{U}(\underline{g})$ module. It follows that $L(N,N)$ is a (primitive) noetherian ring and hence admits a ring of fractions Fract $L(N,N)$. Set $A = \mathfrak{U}(\underline{g})/\mathrm{Ann}\ N$ and let S denote the set of non-zero divisors of A. Then (by Goldie's theorem) S is an Ore set in A. One can further show that S in an Ore set in $L(N,N)$ and that Fract $L(N,N) = S^{-1}L(N,N)$. Consequently we have an embedding $S^{-1}A \hookrightarrow$ Fract $L(N,N)$, which gives Fract $L(N,N)$ a $S^{-1}A - S^{-1}A$ bimodule structure and in particular a $\mathfrak{U}(\underline{g})$ - $\mathfrak{U}(\underline{g})$ bimodule structure.

PROPOSITION. ([31]). $L(N,N)$ _is the unique largest_ $\mathfrak{U}(\underline{g})$ - $\mathfrak{U}(\underline{g})$ _submodule of_ Fract $L(N,N)$ _which is an object of_ $\underline{H}$.

In principle this result greatly simplifies the classification of the possible Harish-Chandra modules which are Kostant rings of a simple $\mathfrak{U}(\underline{g})$ module. For example, if the embedding $S^{-1}A \hookrightarrow \text{Fract}\, L(N,N)$ is an isomorphism it follows that $L(N,N) \cong L(L(\sigma\mu),L(\sigma\mu))$ where $\sigma \in \sum_\lambda^0$, $\mu \in \Lambda^+$ are chosen such that $J = \text{Ann}\, L(\sigma\mu)$. Conjecturally in this case one even has that the map $\mathfrak{U}(\underline{g}) \to L(N,N)$ defined by the action of $\mathfrak{U}(\underline{g})$ is surjective.

7.2. Since $L(N,N) \in \text{Ob}\, \underline{H}$ it admits a unique smallest non-zero ideal I. We can always assume that $\mu \in \Lambda^+$ is chosen such that $I \supset \text{Ann}\, M(\mu)$. Set $M = I \otimes_{\mathfrak{U}(\underline{g})} M(\mu)$. One has $M \in \text{Ob}\underline{O}_{\hat{\mu}}$. Consider $L(N,N)$ as a left $\mathfrak{U}(\underline{g})$ module. One has the

PROPOSITION. ([31]). $L(M,M) = \text{End}_{\mathfrak{U}(\underline{g})} L(N,N)$.

Conjecturally $I = \text{Soc}\, L(N,N)$. (This is true for highest weight modules ([26], 4.13) and follows from the truth of the Vogan semisimplicity hypothesis (6.5).) In this case M is a direct sum of simple objects in $\underline{O}_{\hat{\mu}}$ and so $L(M,M)$ which is completely determined by I can be considered to be known. Obviously the above result shows that $L(N,N)$ completely determines $L(M,M)$ as a ring. What one is aiming for is to prove the converse and thus to show that as a ring $L(N,N)$ is completely determined by the $\mathfrak{U}(\underline{g})$ - $\mathfrak{U}(\underline{g})$ module structure of its socle.

7.3. More precise results have been obtained when N is a highest weight module. For example, we have (notation 5.1) the

LEMMA. ([16], 3.8). $L(L(w\lambda),L(y\lambda)) \neq 0 \Leftrightarrow C(w^{-1}) = C(y^{-1})$.

7.4. To describe $L(L(w\lambda),L(y\lambda))$ it is convenient to make two technical hypotheses on $\sum_\lambda^0$ which may be regarded as purely combinatorial conjectures involving the Kazhdan-Lusztig polynomials.

Take $\sigma \in \Sigma_\lambda^0$ and recall that $\mathbb{Q}C(\sigma)$ has the structure of a left $\mathbb{Q}W_\lambda$ module.

(*) $$\mathbb{Q}C(\sigma) = \mathbb{Q}W_\lambda a(\sigma).$$

Secondly given $w,y \in W_\lambda$ one may consider the product $a(w)a(y)$ in $\mathbb{Q}W_\lambda$. Since $\mathbb{Q}W$ has basis $a(z) : z \in W_\lambda$ we can write

$$a(w)a(y) \doteq \sum_{z \in W_\lambda} c_z a(z)$$

where the symbol $\doteq$ indicates that only the terms lying in $C'(w) \cap C(y)$ have been retained.

(**) $$a(\sigma)^2 = n_\sigma a(\sigma) \quad \underline{\text{for some}} \quad n_\sigma \in \mathbb{N}^+.$$

THEOREM. Assume $w,y \in W_\lambda$ satisfy $C(x^{-1}) = C(y^{-1})$. Then

(i) $L(L(w\lambda),L(y\lambda)) = \bigoplus_i L(L(\sigma\lambda),L(x_i\lambda))$ where σ is the unique element of $\Sigma_\lambda^0 \cap C(w)$.

(ii) If (*), (**) above hold, the $x_i \in W_\lambda$ are determined through the relation

$$a(w^{-1})a(y) \doteq n_\sigma(\sum a(x_i)).$$

7.5. The above result shows that to a large extent $L(L(w\lambda),L(y\lambda))$ has been completely determined. We remark that through the non-degenerate convariant form on $L(y\lambda)$ one may regard $L(L(w\lambda),L(y\lambda))$ as the largest Harish-Chandra module of $(L(y\lambda) \otimes L(w\lambda))^*$. This leads us to the following general question. Let $\underline{k}$ be a maximal compact subalgebra of $\underline{g}$. (More generally the Lie algebra of subgroup K satisfying the conditions of 3.7.) Given any simple object $L \in \mathrm{Ob}\,\mathcal{O}$ compute the largest Harish-Chandra module (which respect to the pair $(\underline{g},\underline{k})$) of L^*. It should be possible to answer this question in terms of the geometric set-up described in 3.7.

7.6. One of the interests in studying the Kostant ring $L(N,N)$ of a simple module N is that it is a more natural object than $U(\underline{g})/\mathrm{Ann}\,N$.

However, from 7.5 and the remarks in 7.2 we are now coming closer to an understanding to the relationship between these two objects. This brings us to the question of determining the Jordan-Hölder composition factors of $\mathfrak{U}(\underline{g})/\text{Ann } N$ (as a $\mathfrak{U}(\underline{g})$ - $\mathfrak{U}(\underline{g})$ bimodule) or if one prefers of $L(N,N)$. We can always choose $\mu \in \Lambda^+$ such that $\text{Ann } N \supset \text{Ann } M(\mu)$ and then by translation principles it is enough to take μ regular. Duflo ([14], Prop. 6) determined the composition factors of $\text{Ann } N/\text{Ann } M(\mu)$; but this turns out to be rather meagre information unless of course one can determine the multiplicities. When $\text{Ann } N = J(w_\lambda w_{B'}, \lambda)$ for some $B' \subset B_\lambda$ a complete solution is given in ([26], 4.8). Unfortunately the result is misleadingly simple ([29], Sect. 5).

To attack the above problem, we formulate a conjecture. For this we define for each $J \in \underline{X}_\lambda^\wedge$ the element

$$e_J := \sum_{w \in W_\lambda} [M(\lambda)/JM(\lambda) : L(w\lambda)]a(w)$$

of $\mathbb{Z}W_\lambda$. Knowledge of the $e_J : J \in \underline{X}_\lambda^\wedge$ obviously determines the ordering of the primitive ideals. Yet this problem was also solved by 5.1. This motivates the following procedure. Let us first define an involution $*$ on $\mathbb{Z}W_\lambda$ by extending the map $w \mapsto w^{-1}$ linearly. It is the adjoint for the inner product $(y,z) \mapsto \text{tr } y^{-1}z$ on $\mathbb{Q}W_\lambda$. By 3.5(*) it follows that $a(w)^* = a(w^{-1})$. Through the existence of a non-degenerate form (2.7) on any simple object in $\underline{\mathcal{O}}$ we obtain $e_J = e_J^*$ (see also 3.5). Now recall that $\mathbb{Q}D(w)$ is a left ideal of $\mathbb{Q}W_\lambda$ (5.1). Since $\mathbb{Q}W_\lambda$ is a semi-simple Artinian ring we can write $\mathbb{Q}D(w) = \mathbb{Q}W_\lambda e(w)$ for some unique self-adjoint projection $e(w)$. Nothing could be more natural than the

CONJECTURE. <u>For each</u> $w \in W_\lambda$, $e_{J(w\lambda)}$ <u>is a multiple of</u> $e(w)$

We remark that ([29], 5.4) this holds if $w = w_\lambda w_{B'} : B' \subset B_\lambda$; but this is rather special. Again from 5.1 we do have that $e_J \in \mathbb{Q}D(w)$ and so $e_J = e_J e(w) = e(w)e_J$ where the last relation obtains on taking adjoints.

Since $e_J \doteq a(\sigma)$ where σ is the unique element of $C(w) \cap \Sigma_\lambda^0$ it follows that the above conjecture implies (*), (**) of 7.3 and furthermore that $e_J = n_\sigma e(\sigma)$. Since $e(\sigma)$ (and n_σ) are completely determined by the Jantzen matrix, one may consider that a positive answer to this conjecture determines completely the composition factors in $\mathfrak{U}(\underline{g})/J : J \in \underline{X}_{\hat{\lambda}}$. Yet this conjecture fails and even in (**) we must allow $n_\sigma \in Z(\mathbb{Q}W_\lambda)$.

8. ASSOCIATED VARIETIES AND SINGULAR SUPPORTS

In 8.1-8.4 $\underline{g}$ may denote an arbitrary finite dimensional Lie algebra.

8.1. Let M be a finitely generated $\mathfrak{U}(\underline{g})$ module. Its associated variety $\underline{V}(M)$ (notation 2.1) is a closed conical (i.e., $x \in \underline{V}(M) \Rightarrow \alpha x \in \underline{V}(M)$, $\alpha \in \mathbb{C}$) subvariety in $\underline{g}^*$. Given J a two-sided ideal of $\mathfrak{U}(\underline{g})$ we set $\underline{V}(J) := \underline{V}(\mathfrak{U}(\underline{g})/J)$ where $\mathfrak{U}(\underline{G})/J$ is considered as a left or as a right $\mathfrak{U}(\underline{g})$ module. We may then define $\underline{VA}(M) := \underline{V}(\text{Ann } M)$.

The interest of studying associated varieties comes from the possibility of comparing the representation theory of $\underline{g}$ to the algebraic geometry of $\underline{g}^*$. In this last respect we recall the algebraic adjoint group G acts on $\underline{g}$ and by transposition on $\underline{g}^*$. For a two-sided ideal J of $\mathfrak{U}(\underline{g})$ it is easy to show that $\underline{V}(J)$ is G stable and hence a union of G-orbits. In particular for any finitely generated module M we have that $\underline{VA}(M) \supset \overline{G\underline{V}(M)}$. A basic question is to show that equality holds.

8.2. Since $S(\underline{g})$ is commutative and identifies with $\text{gr}\,\mathfrak{U}(\underline{g})$, it admits a Poisson bracket $\{\ ,\ \}$ structure. This is defined on homogeneous elements and extended by linearity. We may write any homogeneous element in the form $\text{gr}_m(a) : m \in \mathbb{N}$, $a \in \mathfrak{U}^m(\underline{g})$, and then we have

$$\{\text{gr}_m(a), \text{gr}_n(b)\} := \text{gr}_{m+n-1}[a,b].$$

The Poisson bracket may be regarded as the first approximation to the commutator bracket $[a,b]$ and first arose in the description of the classical limit of quantum mechanics. We call a subvariety V of $\underline{g}^*$ <u>involutive</u> if its ideal of definition $I(V) := \{f \in S(\underline{g}) \mid f(V) = 0\}$ is stable under the Poisson bracket. Note that each component of an involutive variety is again involutive. One has the

THEOREM. *Let* M *be a finitely generated* $\mathfrak{U}(\underline{g})$ *module; then* $V(M)$ *is involutive.*

This is deep fact with a long history. Suffice to say that in its most general form a proof has been given by Gabber [15]. It is nontrivial because although $\operatorname{gr} J$ for any left ideal J of $\mathfrak{U}(\underline{g})$ is obviously stable under the Poisson bracket, this property is not conserved by taking radicals. One often quoted application is to show that $2d(M) \geq d(\mathfrak{U}(\underline{g})/\operatorname{Ann} M)$ ($\underline{g}$ algebraic); but this has a more elementary proof. A more impressive consequence is that for any $\lambda \in P(R)^{++}$ one has

$$\dim(M(\lambda)/\sum_{\alpha \in B_0} M(s_\alpha \lambda)) < \infty.$$

This result had previously only a rather roundabout (though subtle) proof using the Weyl group.

8.3. A further general (and deep) result concerning varieties is the following theorem due to Gabber [46]. Its proof derives from a similar result of Kashiwara for $\mathcal{D}$ modules. Recall 2.1.

THEOREM. *Let* M *be a finitely generated* d-*homogeneous* $\mathfrak{U}(\underline{g})$ *module. Then* $\underline{V}(M)$ *is equidimensional* (i.e., *each component of* $\underline{V}(M)$ *has dimension* $d(M)$).

A simple $\mathfrak{U}(\underline{g})$ module M is trivially d-homogeneous. Moreover for M simple $\mathfrak{U}(\underline{g})/\operatorname{Ann} M$ is d-homogeneous as a left $\mathfrak{U}(\underline{g})$ module. In particular the associated variety of a primitive ideal is always equidimensional.

8.4. There are two fundamental and interrelated problems concerning associated varieties. The first is to determine the associated variety of any primitive ideal and of any simple module (or at least one belonging to a good class such as the highest weight modules). Secondly to show that every closed irreducible conical G stable (resp. involutive) subvariety of $\underline{g}^*$ is the associated variety of some primitive ideal (resp. simple $\mathfrak{U}(\underline{g})$ module). We remark that for $\underline{g}$ solvable it is possible to construct an irreducible representation from any $f \in \underline{g}^*$ and thus obtain a bijection between G orbits in $\underline{g}^*$ and primitive ideals in $\mathfrak{U}(\underline{g})$; whilst the passage to the associated variety is very far from being injective and can at most map onto very few orbits (i.e., the conical ones), so we are taking a different viewpoint from that of the solvable case. Experience has shown that the calculation of the associated variety is, in the semisimple case, the more relevant viewpoint.

8.5. Here and in the remainder of the section we assume $\underline{g}$ semisimple. Then $\underline{g}$ identifies with $\underline{g}^*$ through the Killing form and we call $X \in \underline{g}$ nilpotent if $\operatorname{ad}_{\underline{g}} X$ is a nilpotent endomorphism. The set N of all nilpotent elements is a closed conical G-stable subvariety of $\underline{g}^*$. More precisely if $Y(\underline{g})$ denotes the algebra of G invariant polynomial functions on $\underline{g}^*$ and Y_+ the augmentation ideal of $Y(\underline{g})$ then ([13], 8.1.3) one has $\underline{V}(Y_+ S(\underline{g})) = N$. Now if $J \in \operatorname{Prim} \mathfrak{U}(\underline{g})$, then $J \cap Z(\underline{g}) \in \operatorname{Max} Z(\underline{g})$ and so $\operatorname{gr} J \supset Y_+ S(\underline{g})$. Consequently $\underline{V}(J) \subset N$. On the other hand N is a finite union of orbits and so a fortiori is $\underline{V}(J)$. This gives the following observation of Borho and Kraft ([9], 7.2).

LEMMA. Take $J \in \operatorname{Prim} \mathfrak{U}(\underline{g})$. Then $\underline{V}(J)$ is irreducible if and only if it is closure of a nilpotent orbit (i.e., G orbit of nilpotent elements).

8.6. Giving modules in the $\underline{\mathcal{O}}_\Lambda$ category a bimodule structure (see 3.5) allows one to compare properties of left and right annihilators. This leads to the following result (notation 4.3).

LEMMA. $\underline{V}(J)$ is independent of the choice of $J \in (\underline{X}_{\hat{\mu}})_\tau$ and the

<u>choice of</u> $\mu \in \Lambda^+$.

In more picturesque words the associated variety of a primitive ideal J depends only on the t-clan and the r-clan to which it belongs. Now by 4.3(iii) each r-clan $(\underline{X}_{\hat{\lambda}})_\tau$ is associated to a unique simple W module WP_τ, and if $\underline{V}(J)$ is irreducible it determines by 8.5 a unique nilpotent orbit in $\underline{g}^*$. On the other hand, Springer established and studied in a number of papers [36] a correspondence between nilpotent orbits and Weyl group representations and a natural conjecture is that this correspondence assigns to WP_τ exactly the dense nilpotent orbit in $\underline{V}(J)$. This was established in the integral case (i.e., $\lambda \in P(R)^{++}$) by Borho and Brylinski ([7], Sect. 6) using the result for induced ideals and some case by case considerations of Barbasch and Vogan ([1],[2]). However, this computation gives little insight into this remarkable phenomenon. Two much better viewpoints are discussed below through at present they unfortunately involve some (undoubtedly correct!) conjectures.

8.7. We have seen that closures of nilpotent orbits are the appropiate candidates for varieties of primitive ideals. Now let L be a simple highest weight module. One easily sees that $\underline{V}(L) \subset \underline{n}^+$ and is B stable and from our previous remarks it is also closed, equidimensional and involutive. Any irreducible subvariety of $\underline{n} := \underline{n}^+$ with these properties takes (see [30], 7.4 for example) the form $\underline{V}(w) := \overline{B(\underline{n} \cap w(\underline{n}))}$ for some $w \in W$, where $w(\underline{n})$ denotes the image of $\underline{n}$ under w (which is well-defined as $\underline{n}$ is H stable). The $\underline{V}(w) : w \in W$ need not all be distinct.

Now let $I(w)$ denote the ideal of definition of $\underline{V}(w)$ (defined in $S(\underline{n}^-)$). The quotient algebra $S(\underline{n}^-)/I(w)$ is $\underline{h}$ stable and so each $\nu \in P(R)^{++}$ defines in an obvious fashion an H stable gradation of this algebra. If we let $r_w(\nu)$ denote the leading coefficient of the associated Hilbert-Samuel polynomial, then it is quite easy to show that r_w extends to rational function on $\underline{h}^*$ with denominator the product of the positive roots and numerator a homogeneous polynomial q_w of degree card R^+ - dim $\underline{V}(w)$.

Because G/B is a complete variety ([37], p. 68), $G\underline{V}(w)$ is a closed irreducible subvariety of N; hence the closure of a uniquely determined nilpotent orbit $\underline{St}(w)$. After Steinberg [37] the map $w \mapsto \underline{St}(w)$ is surjective and after Spaltenstein [35] the irreducible components of $\underline{St}(w) \cap \underline{n}$ are amongst the $\underline{V}(w') : w' \in W$. It is quite easy to show ([30], Sect. 3) that the corresponding $q_{w'}$, span a representation of W which we can naturally associate to the orbit $\underline{St}(w)$. A basic question is to show that this representation is just that defined by Springer. This is a question in pure algebraic geometry. A positive answer would imply that these representations are irreducible and pairwise non-isomorphic for distinct orbits and in particular the q_w would be linearly independent for distinct $\underline{V}(w)$. Although this can be checked in most cases as yet no satisfactory general proof has yet to be given; but R. Hotta has recently reported a proof [43].

8.8. Now even when L is a simple highest weight module $\underline{V}(L)$ need not be irreducible. It is therefore convenient to assume that we have included in the definition of $\underline{V}(L)$ the multiplicity of each component (see [30], Sect. 5). Now since $\underline{V}(L)$ is also an H stable subvariety of $\underline{n}$, we may define a polynomial p_L on $\underline{h}^*$ relative to $\underline{V}(L)$ by the procedure described in 8.7. Because $\underline{V}(L)$ is equidimensional p_L is a linear combination of the q_w occurring with coefficient equal to the multiplicity of $\underline{V}(w)$ in $\underline{V}(L)$. On the other hand, if $L = L(y\lambda)$ for some $y \in W_\lambda$ it is not difficult ([30], 5.1) to show that p_L is proportional to $p_{y^{-1}}$ (notation 4.1), that is to the Goldie rank polynomial associated to Ann $L(y^{-1}\lambda)$. These observations explain rather nicely the relationship between representations of the Weyl group coming from the Goldie rank polynomials and those coming from Springer's correspondence. Counting multiplicities in $\underline{V}(\cdot)$ one has ([30], 5.2)

LEMMA. $\mathbb{Z}\underline{V}(L(w\lambda)) = \mathbb{Z}\underline{V}(L(y\lambda)) \Leftrightarrow J(w^{-1}\lambda) = J(y^{-1}\lambda), \ \forall w,y \in W_\lambda$.

8.9. A basic technical advance in the study of $\mathfrak{U}(\underline{g})$ modules has been made possible through the theory of $\mathcal{D}_X$ modules. Here X denotes the flag variety G/B and $\mathcal{D}_X$ the sheaf of differential operators on X. Through the left action of G on G/B any $Y \in \underline{g}$ may be considered as a vector field on G/B and so as a global section $\varphi(Y) \in \Gamma(X,\mathcal{D}_X)$. Furthermore φ lifts to a ring homomorphism $\mathfrak{U}(\underline{g}) \to \Gamma(X,\mathcal{D}_X)$ which has kernel equal to Ann $M(\rho)$ (and is surjective). Consequently $\mathcal{D}_X$ admits say a right $\mathfrak{U}(\underline{g})$ module structure and a left $\mathcal{D}_X$ module structure. Using translation principles (again!) Bernstein and Beilinson [3] proved that the functor $M \mapsto \mathcal{D}_X \otimes_{\mathfrak{U}(\underline{g})} M$ from the category of finitely generated left $A_\rho := \mathfrak{U}(\underline{g})/\text{Ann } M(\rho)$ modules to the category of coherent $\mathcal{D}_X$ modules is an equivalence of categories. This result is important for many reasons. First it lies behind the proof of 3.7. Secondly it relates the intersection cohomology groups on X to extension groups in $\mathcal{O}_{\hat{\rho}}$, which eventually yields a proof of the Kazhdan-Lusztig conjectures ([3]; [10]). Thirdly it quite simply gives a new invariant for any finitely generated A_ρ module M; namely its singular support $\underline{S}(M)$. This is defined in a manner analogous to $\underline{V}(M)$. Namely one filters $\mathcal{D}_X$ by the order of the differential operators and one defines $\underline{S}(M)$ to be the zero set (in the cotangent space $T^*(X)$) of $\text{Ann}_{\text{gr }\mathcal{D}_X}(\text{gr } \mathcal{M})$, where $\mathcal{M} = \mathcal{D}_X \otimes_{\mathfrak{U}(\underline{g})} M$. As for $\underline{V}(M)$ it is convenient to include multiplicites of irreducible components.

A notable property of $\underline{S}(M)$ is that if $M \neq 0$ then $\dim \underline{S}(M) \geq \frac{1}{2} \dim T^*(X) = \text{card } R^+$; even though M itself may be very small—say finite dimensional. By the usual arguments this gives $K \dim M \leq \dim \underline{S}(M) - \text{card } R^+$ and in particular $K \dim A_\rho \leq \text{card } R^+$. The opposite inequality (which is a little more elementary) was obtained by Levasseur [42]. A slight technical improvement then gives

THEOREM. For any $\mu \in \underline{h}^*$ one has $K \dim(\mathfrak{U}(\underline{g})/\text{Ann } M(\mu)) = \text{card } R^+$.

Conjecturally for any $J \in \operatorname{Spec} \mathfrak{U}(\underline{g})$

$$K \dim(\mathfrak{U}(\underline{g})/J) = \frac{1}{2} d(\mathfrak{U}(\underline{g})/J) + d(\mathfrak{Z}(\mathfrak{U}(\underline{g})/J)).$$

Surprisingly although it is easy to check that $K \dim \mathfrak{U}(\underline{g}) \geq \dim \underline{b}$, it is not known if equality holds.

8.10. Given $x \in X = G/B$, let $\underline{g}_x$ denote the Lie subalgebra of $\underline{g}$ corresponding to the stabilizer subgroup G_x of G. The left action of G on X determines a surjective map of $\underline{g}$ to the tangent space $T_x(X)$ of X at x with kernel $\underline{g}_x$. Taking duals gives an injective map $\pi_x: T_x^*(X) \to \underline{g}^*$ with image $\underline{g}_x^\perp$. The map $\pi: (x,v) \mapsto \pi_x(v)$ of $T^*(X)$ into $\underline{g}^*$ is called the <u>moment map</u> of the homogeneous space X. Identifying $T_x^*(X)$ with $\underline{g}_x^\perp$ through π_x it becomes projection onto the second factor in $T^*(X)$. Noting that $\underline{g}_{gx}^\perp = g\underline{g}_x^\perp$ and that $\underline{g}_x^\perp = \underline{b}^\perp = \underline{n}$ when $x = \bar{1} \in G/B$, we can deduce an isomorphism $T^*(X) \xrightarrow{\sim} G \times_B \underline{n}$ of algebraic vector bundles under which π identifies with the map $(g,X) \mapsto gX$ of $G \times_B \underline{n}$ onto $G\underline{n} = N$ (notation 8.5). This identifies the moment map of the flag variety with Springer's desingularization (as pointed out to me by Borho) of N. A further important property of π is that it relates associated varieties with singular supports through the following result of Borho and Brylinski: [7].

LEMMA. <u>(Notation</u> 8.9). <u>For each</u> $M \in \operatorname{Ob} \underline{\mathcal{O}}_{\hat{\rho}}$ <u>one has</u> $\underline{V}(M) = \pi(\underline{S}(M))$.

8.11. Let us write $L_w = L(w\rho)$, $M_w = M(w\rho)$, $\mathfrak{U}_w = \mathfrak{U}(\underline{g})/J(w\rho)$ (considered as a left $\mathfrak{U}(\underline{g})$ module), $\forall w \in W$. Under the conventions of 8.9 we have the further result of Borho and Brylinski [7].

THEOREM. <u>For each</u> $w \in W$ <u>one has</u>

$$\underline{S}(\mathcal{U}_w) = G\underline{S}(L_w) = G \times_B \underline{V}(L_{w^{-1}}).$$

Recalling (8.8) it follows (counting multiplicities in $\underline{S}(\cdot)$) that $\mathbb{Z}\underline{S}(\mathcal{U}_w) = \mathbb{Z}\underline{S}(\mathcal{U}_y)$ if and only if $J(w\rho) = J(y\rho)$, $\forall w,y \in W$. (Actually it is conjectured in the integral case that $\underline{V}(L_{w^{-1}})$ and hence $\underline{S}(\mathcal{U}_w)$ is always irreducible and so multiplicities can probably be ignored.) We see that the singular support of $\mathcal{U}_w$ which is more refined than the associated variety of $J(w\rho)$ separates primitive ideals (at least in the integral case).

8.12. Let $\mathcal{U}$ denote the variety of all unipotent elements in G and $\mathcal{B}$ the variety of all Borel subgroups of G (which identifies with G/B). The <u>Steinberg variety</u> S of G is defined through

$$S := \{(u,B_1,B_2) \in \mathcal{U} \times \mathcal{B} \times \mathcal{B} \mid u \in B_1 \cap B_2\}.$$

It is an equidimensional algebraic variety of dimension $2r$: $r = \text{card } R^+$ whose irreducible components are the closures of the $S_w := \{(u,gwB,gB) \in S,\ g \in G\}$, $w \in W$. Kazhdan and Lusztig [41] have given a topological construction of a left and right action of W on $H_{4r}(S,\mathbb{Q})$ (which has basis $[\overline{S}_w]$: $w \in W$) and shown that then as a $W - W$ bimodule it is isomorphic to $\mathbb{Q}W$. We define the matrix with entries $B(y,w)$ through

$$y[\overline{S}_1] = \sum_{w\in W} B(y,w)[\overline{S}_w].$$

Similar to the Jantzen matrix, this satisfies $B(y,y) = 1$ and $B(y,w) = 0$ unless $y \geq w$. Consequently this matrix is invertible and we denote the entries of the inverse matrix by $A(y,w)$: $y,w \in W$. The analogy with the Goldie rank polynomials leads us to the following

CONJECTURE. (<u>Notation</u> 8.7). <u>For each</u> $y \in W$ <u>one has</u> (<u>up to a scalar</u>)

$$q_y = \sum_{w\in W} A(y,w)w\rho^m$$

<u>where</u> m <u>is the least non-negative integer such that right hand side is non-zero</u> (<u>and so equal to</u> $\text{card } R^+ - \dim \underline{V}(y)$).

One can show ([30], 9.8) that this would imply that q_y generates the representation associated by Springer to the nilpotent orbit

$\underline{St}(y)$ (notation 8.7). One may also conclude that the module QWq_y is always univalent.

8.13. For any non-singular subvariety Y of X one defines the conormal bundle $T^*_Y(X)$ through

$$T^*_Y(X) = \{(y,v) \mid y \in Y, \quad v \in T_y(Y)^\perp\}.$$

One always has $\dim T^*_Y(X) = \dim X$. For example, if we set $X_w = NwB/B$, then $T_{wB/B}(X)$ identifies with $\underline{g}/w(\underline{b})$ and $T_{wB/B}(X_w)$ with the subspace $\underline{n}/n \cap w(\underline{n})$, and hence $T_{wB/B}(X_w)^\perp$ with the subspace $\underline{n} \cap w(\underline{n})$ of $\underline{g}$. Then $T_w := T^*_{X_w}(X)$ becomes $\{(bwB, b(\underline{n} \cap w(\underline{n}))) \mid b \in B\}$. From this we obtain that $\pi(\overline{T}_w) = \underline{V}(w)$ (notation 8.7). Taking $\underline{S}(\cdot)$ to include multiplicities we suggest (notation 8.13) the

CONJECTURE. <u>For all</u> $y \in W$, <u>one has</u>

$$(*) \qquad \underline{S}(M_y) = \sum_{w \in W} B(y,w)\overline{T}_w.$$

Let w_0 denote the longest element of W, then after Brylinski-Kashiwara ([10]).

$$\underline{S}(M_{w_0}) = \bigcup_{w \in W} \overline{T}_w \qquad \text{(no multiplicities counted)},$$

which is consistent with our conjecture. Again from (*) it is immediate that

$$(**) \qquad \underline{S}(L_z) = \sum_{y \in W} \sum_{w \in W} a(z,y)B(y,w)\overline{T}_w.$$

In type A_n, Kazhdan and Lusztig have suggested that $B(y,w)$ coincides with the Jantzen matrix and so we obtain simply that $\underline{S}(L_z) = \overline{T}_z$, as conjectured by Borho and Brylinski, which for example gives $\underline{V}(L_z) = \underline{V}(z)$ a result which has been reported by them. In general one has $\underline{V}(L_z) \subset \underline{V}(z)$ ([30], Sect. 8) and this inclusion may be strict (for reasons of dimension). Correspondingly outside type A_n the matrix defined by $B(y,w)$ certainly differs from the Jantzen matrix. Finally by 8.10 we obtain from (**) that

$$\underline{V}(L_z) = \sum_{y\in W} \sum_{w\in W} a(z,y)B(y,w)\underline{V}(w)$$

and so from the remarks in 8.8 we obtain (up to scalars) that

$$p_{z^{-1}} = \sum_{y\in W} \sum_{w\in W} a(z,y)B(y,w)q_w$$

and we remark that this is consistent with 4.1 and conjecture 8.12.

9. HOMOLOGICAL DIMENSION OF PRIMITIVE QUOTIENTS

9.1. Fix $\mu \in \Lambda^+$ and set $A_\mu = \mathfrak{U}(\underline{g})/\mathrm{Ann}\, M(\mu)$. Either from the study of $\mathcal{D}_X$ modules or from our classification of primitive ideals (and hence of Spec A_μ) one can now hope to say rather a lot about the module theory of A_μ. For example consider the category of (finitely generated) right A_μ modules. How may we construct projective modules in this category? Obviously A_μ is projective as a right A_μ module. Yet A_μ has a left $\mathfrak{U}(\underline{g})$ module structure so we may tensor by finite dimensional $\mathfrak{U}(\underline{g})$ modules on the left and taking direct summands gives further projective right A_μ modules. Recognizing that we may identify A_μ with $L(M(\mu),M(\mu))$ (cf. [ ], Sect. 3 for a complete discussion of this question) and carrying the tensor product operation onto the right-hand factor of $M(\mu)$ leads to the following result [31]. (In this if M is a right module over a ring A, then M^* denotes $\mathrm{Hom}_A(M,A)$ and we define $MM^* \subset \mathrm{End}_A M$ through $m\xi\colon m' \mapsto m\xi(m')$, $\forall m,m' \in M$, $\forall \xi \in M^*$.)

LEMMA. Let P be a direct summand of $E \otimes M(\mu)$ for some finite dimensional $\mathfrak{U}(\underline{g})$ module E, and set $Q := L(M(\mu),P)$. Then

(i) Q is projective as a right A_μ module.

(ii) $Q^* \cong L(P,M(\mu))$.

(iii) $QQ^* = \mathrm{End}_{A_\mu} Q \cong L(P,P)$.

<u>Remark</u>. If μ is regular then any projective object P in $\underline{\mathcal{O}}_\Lambda$ so obtains from $M(\mu)$. This is false in general and for a P not so obtained it can even happen that Q has infinite homological dimension [31].

9.2. As pointed out to me by Bernstein, the equivalence of categories theorem alluded to in 8.9 implies that A_ρ has global dimension $2\dim X = \operatorname{card} R$. By an appropriate version of the translation principle this also holds for $A_\mu : \mu \in P(R)^{++}$. Yet as noted above A_μ for μ non-regular generally has infinite global dimension.

9.3. An advantage of the above presentation of projective A_μ modules is that it enables one to compute the trace ideal $Q^*Q = \{\xi(q) : \xi \in Q^*, q \in Q\}$ of such a module. Recalling that $\lambda \in \underline{h}^*$ is assumed dominant and regular we have [31]:

PROPOSITION. For each $w \in W$, set $Q_w := L(M(\lambda),P(w\lambda))$, where $P(w\lambda)$ denotes the projective cover of $L(w\lambda)$. Then (notation 5.3)

$$Q_w^*Q_w = E(L(w^{-1}\lambda)).$$

An amusing corollary of this result is that $L(P(w_\lambda\lambda),P(w_\lambda\lambda))$ is always a simple ring.

9.4. Let K_0 denote the Grothendieck group of projective right A_ρ modules. If $P(\mu)$ denotes the projective cover of $L(\mu) : \mu \in P(R)$, then by 9.1, $Q(\mu) := L(M(\rho),P(\mu))$ is a projective right A_ρ module. Let K_0' denote the subgroup of K_0 generated by the $[Q(\mu)]$. From the relations obtained from tensoring by finite dimensional modules one easily shows that $\operatorname{card} K_0' \leq \operatorname{card} W$. Bernstein has informed me that he can prove equality (through comparison with the Grothendieck group of the category of coherent sheaves on G/B). Conjecturally $K_0' = K_0$.

9.5. The notion "clan" arises from ring theory. For example, for any $\mu \in \Lambda^{++}$ and any $w \in W_\lambda$ there exists a simple $A_\mu - A_\lambda$ bimodule V_w such that $\ell(V_w) = J(w\mu)$, $r(V_w) = J(w\lambda)$ (where ℓ, r denote left, respectively right, annihilator). This follows from the fact that $L(L(w\lambda),L(w\mu)) \neq 0$ which in turn results from translation principles. This extends in an appropriate way to Λ^+ and shows that elements lying in the same t-clan are linked ("module link") in the above sense. A similar assertion follows for an r-clan from 7.3. In the case when $\lambda = \mu$ one may also define an "ideal link" between elements in the same r-clan as follows. The ideals $J, K \in \operatorname{Spec} A_\lambda$ are ideal linked if $J = \ell(J \cap K/K)$, $K = r(J \cap K/K)$. (Since ${}^tJ = J$ this "relation" is symmetric; but conjecturally it is never reflective). Writing $J = \overline{J(w\lambda)}$, $K = \overline{J(y\lambda)}$ (where bar denotes the projection $\mathfrak{U}(\underline{g}) \to A_\lambda$) it follows easily that J, K are ideal linked if $\operatorname{Ext}^1(L(w\lambda), L(y\lambda)) \neq 0$ (again because $\delta L(w\lambda) \cong L(w\lambda)$ the order of factors here is unimportant). As remarked by Borho, it follows from the Vogan semisimplicity hypothesis (6.5(*)) that the relation generated by ideal links defines exactly the r-clans, and this further defines the module links in $\operatorname{Spec} A_\lambda$. (An example of Stafford shows that the relation generated by ideal links can be strictly weaker than that generated by module links in general.)

The significance of ideal links is that if $J, K \in \operatorname{Spec} A$ (where A is a noetherian ring) are ideal linked then A does not admit localization at the prime ideal J, that is $S(J) := \{a \in A \mid \bar{a}$ non-zero divisor in $A/J\}$ is not an Ore set in A. If this were so $S^{-1}J$ would be the unique maximal ideal of the local ring $S^{-1}A$ and this eventually contradicts the existence of an ideal link. See [44].

9.6. Recently Levasseur ([42], Thm. 1) has shown that $A_\mu : \mu \in \Lambda^+$ has injective dimension equal to card $R - d(L(\mu))$. This was a surprise as Roos had earlier conjectured it to be always equal to $\frac{1}{2}$ card R.

9.7. Through the analysis of ([11], Sect. 5) and 9.1, 9.2 one may show that the Conze embedding of A_μ (notation 9.1) in the Weyl algebra $A_n : n = \text{card } R^+$ is a flat extension for μ regular. This was obtained through discussions with Smith, who with Hodges describes how to adjust this embedding to get faithfulness. This provides a ring theoretic alternative to the Bernstein-Beilinson equivalence of categories theorem [3].

9.8. We have recently [45] established the Borho-Kraft conjecture in general (see 1.2, 8.6). The proof uses Hotta's result [43] which allows one to compare the representations of the Weyl group defined by Goldie rank polynomials to those defined by the Springer correspondence.

REFERENCES

[1] D. Barbasch and D. Vogan, Primitive ideals and orbital integrals in complex classical groups, Math. Ann., 259 (1982), 153-199.

[2] ______, Primitive ideals and orbital integrals in exceptional groups, preprint, MIT-Rutgers (1981).

[3] A. Beilinson and J. Bernstein, Localisation de g modules, Comptes Rendus, 292 (1981), 15-18.

[4] J. N. Bernstein, Modules over a ring of differential operators, Funct. Anal. Prilož., 5 (1970), 89-101.

[5] J. N. Bernstein and S. I. Gelfand, Tensor products of finite and infinite dimensional representations of semisimple Lie algebras, Compos. Math., 41 (1980), 245-285.

[6] W. Borho and J.-L. Brylinski, Differential operators on homogeneous spaces I: Irreducibility of the associated variety for annihilators of induced modules, Invent. Math., 6 (1982), 437-476.

[7] ______, to appear.

[8] W. Borho and J. C. Jantzen, Über primitive Ideale in der Einhüllenden einer halbeinfacher Lie-Algebra, Invent. Math., 39 (1977), 1-53.

[9] W. Borho and H. Kraft, Über die Gelfand-Kirillov Dimension, Math. Ann., 220 (1976), 1-24.

[10] J.-L. Brylinski and M. Kashiwara, Kazhdan-Lusztig Conjecture and Holonomic Systems, Invent. Math., 64 (1981), 387-410.

[11] N. Conze-Berline and M. Duflo, Sur les représentations induites des groupes semi-simples complexes, Compos. Math., 34 (1977), 307-336.

[12] P. Deligne, Preprint, I.H.E.S. (1982).

[13] J. Dixmier, Algèbres enveloppantes, Cahiers Scientifiques, XXXVII, Gauthier-Villars, Paris, 1974.

[14] M. Duflo, Sur la classification des idéaux primitifs dans l'algèbre enveloppante d'une algèbre de Lie semi-simple, Ann. of Math., 105 (1977), 107-130.

[15] O. Gabber, The integrability of the characteristic variety, Amer. J. Math., 103 (1981), 445-468.

[16] O. Gabber and A. Joseph, On the Bernstein-Gelfand-Gelfand resolution and the Duflo sum formula, Compos. Math., 43 (1981), 107-131.

[17] ______, Towards the Kazhdan-Lusztig conjecture, Ann. Éc. Norm. Sup., 14 (1981), 261-302.

[18] J.-C. Jantzen, Moduln mit einem höchsten Gewicht, LN 750, Springer-Verlag, Berlin/Heidelberg/New York, 1980.

[19] A. Joseph, On the annihilators of simple subquotients of the principal series, Ann. Éc. Norm. Sup., 10 (1977), 419-439.

[20] A. Joseph, Sur la classification des idéaux primitifs dans l'algèbre enveloppante de sl(n+1,C), Comptes Rendus, A 287 (1978), 302-306.

[21] _____, Towards the Jantzen conjecture, Compos. Math., 40 (1980), 35-67.

[22] _____, Dixmier's problem for Verma and Principal series submodules, J. London Math. Soc., 20 (1979), 193-204.

[23] _____, W-module structure in the primitive spectrum of the enveloping algebra of a semisimple Lie algebra, in LN 728, Springer-Verlag, Berlin/Heidelberg/New York, 1979, pp. 116-135.

[24] _____, Goldie rank in the enveloping algebra of a semisimple Lie algebra, I, J. Algebra, 65 (1980), 269-283.

[25] _____, Goldie rank in the enveloping algebra of a semisimple Lie algebra, II, J. Algebra, 65 (1980), 284-306.

[26] _____, Goldie rank in the enveloping algebra of a semisimple Lie algebra, III, J. Algebra, 73 (1981), 295-326.

[27] _____, Kostant's problem and Goldie rank, in LN 880, Springer-Verlag, Berlin/Heidelberg/New York, 1981, pp. 249-266.

[28] _____, The Enright functor in the Bernstein-Gelfand-Gelfand $\mathcal{O}$ category, Invent. Math., 67 (1982), 423-445.

[29] _____, Completion functors in the $\mathcal{O}$ category, to appear in "Non-commutative Harmonic Analysis Proceedings," Marseille-Luminy, 1982.

[30] _____, On the associated variety of a highest weight module, J. Algebra, to appear.

[31] A. Joseph and T. Stafford, to appear.

[32] D. A. Kazhdan and G. Lusztig, Representations of Coxeter groups and Hecke algebras, Invent. Math., 53 (1979), 165-184.

[33] G. Lusztig and N. Spaltenstein, Induced unipotent classes, J. Lond. Math. Soc., 19 (1979), 41-52.

[34] N. N. Shapovalov, On a bilinear form on the universal enveloping algebra of a complex semisimple Lie algebra, Funct. Anal. Prilož. 6 (1972), 307-311.

[35] N. Spaltenstein, Classes unipotentes et sous-groupes de Borel, LN 946, Springer-Verlag, Berlin/Heidelberg/New York, 1982.

[36] T. A. Springer, A construction of representations of Weyl groups, Invent. Math., 44 (1978), 279-293.

[37] R. Steinberg, Conjugacy classes in algebraic groups, LN 366, Springer-Verlag, Berlin/Heidelberg/New York, 1974.

[38] D. A. Vogan, Gelfand-Kirillov dimension for Harish-Chandra modules, Invent. Math., 48 (1978), 75-98.

[39] _____, Ordering of the primitive spectrum of a semi-simple Lie algebra, Math. Ann., 248 (1980), 195-203.

[40] D. A. Vogan, Irreducible characters of semisimple Lie groups II, The Kazhdan-Lusztig conjectures, Duke Math. J., 46 (1979), 805-859.

[41] D. A. Kazhdan and G. Lusztig, A topological approach to Springer's representations, Adv. Math., 38 (1980), 222-228.

[42] T. Levasseur, Dimension injective des quotients primitif minimaux de l'algèbre enveloppante d'une algèbre de Lie semi-simple, Comptes Rendus, 292 (1981), 385-387.

[43] R. Hotta, On Joseph's construction of Weyl group representations, preprint, Tokyo 1982.

[44] K. A. Brown, Ore sets in enveloping algebras, Preprint, Glasgow (1982).

[45] A. Joseph, The associated variety of a primitive ideal, Preprint (1983).

[46] T. Levasseur, Preprint (1983).

UNITARY REPRESENTATIONS AND BASIC CASES

A. W. Knapp*

In attempting to classify the irreducible unitary representations of linear semisimple Lie groups, one knows that it is enough to decide which of certain standard representations in Hilbert space admit new inner products with respect to which they are unitary. In this context B. Speh and the author [3] introduced a notion of basic case and gave a conjecture that would if true reduce the classification problem to a study of finitely many basic cases in each group. The paper [3] did not, however, tell how to calculate what the basic cases are. The present paper will address this question, giving some theorems that usually make it a simple matter to identify the basic cases.

The paper is organized as follows: In §1 we review the setting of the classification problem and restate the existence-uniqueness theorem for basic cases. In §2 we give two reduction theorems for calculating basic cases and show how to apply them. The proof of the second reduction theorem is in §3.

The development of the theory of basic cases has been influenced extensively by conversations with David Vogan. Vogan's paper [5] may be viewed as a related but different attempt to isolate the phenomena that lead to unitary representations.

1. Definition of basic cases

Let G be a connected linear semisimple group with maximal compact subgroup K. We assume as in [3] that $\operatorname{rank} G = \operatorname{rank} K$.

* Supported by NSF Grant MCS 80-01854 and by a Guggenheim Fellowship.

Let $P = MAN$ be a parabolic subgroup of G with $\operatorname{rank} M = \operatorname{rank}(K \cap M)$, let σ be a discrete series or limit of discrete series representation of M, and let e^{ν} be a homomorphism of A into $\mathbb{C}^{\times}$. We denote by $U(P,\sigma,\nu)$ the unitarily induced representation

$$U(P,\sigma,\nu) = \operatorname{ind}_P^G(\sigma \otimes e^{\nu} \otimes 1) .$$

This representation may be regarded as acting in a closed subspace of L^2 functions on K with values in the space on which σ operates. When $\operatorname{Re} \nu$ is in the closed positive Weyl chamber relative to N and when a certain computable finite group (known as an "R group") is trivial, this representation has a unique irreducible quotient called the Langlands quotient and denoted $J(P,\sigma,\nu)$. We shall assume these conditions on ν are satisfied; they are always satisfied when $\operatorname{Re} \nu$ is in the open positive Weyl chamber.

The representations $J(P,\sigma,\nu)$ act in quotient Hilbert spaces, and the classification question for the unitary dual comes down to deciding which of the J's admit new inner products that make them unitary. In fact, by an observation of Vogan's recited in [2], it is enough to handle ν real-valued. For ν real-valued (and $\operatorname{rank} G = \operatorname{rank} K$), the representation $U(P,\sigma,\nu)$ (and hence also $J(P,\sigma,\nu)$) always admits a nonzero invariant Hermitian form, and the question is whether the known operator that relates this form to the L^2 inner product is semidefinite.

We think of σ as fixed and ν as varying, and we look for those real ν in the closed positive Weyl chamber for which $J(P,\sigma,\nu)$ can be made unitary. Then it appears from examples that there are only finitely many distinct pictures of unitary points for a given G and that most of these pictures are associated to subgroups of G. The idea behind "basic cases" is to pick out finitely many σ's whose pictures ought to include all the pictures that are new for G. Then we want to associate to a general (G,σ) a pair (L,σ^L) with $L \subseteq G$ and σ^L basic such that the pictures of unitary points for σ and σ^L ought to match exactly.

To define basic cases σ, we restrict attention to a class of σ's for which some minimal K-type of $U(P,\sigma,\nu)$ depends coherently upon σ, and then the basic case is the σ of smallest parameter in the class.

In more detail, let $\mathfrak{b} \subseteq \mathfrak{k}$ be a compact Cartan subalgebra of $\mathfrak{g}$, let Δ be the roots of $(\mathfrak{g}^{\mathbb{C}},\mathfrak{b}^{\mathbb{C}})$, and let Δ_K be the subset of compact roots. We may assume that the Lie algebra $\mathfrak{a}$ of A is built by Cayley transform from strongly orthogonal noncompact roots $\{\alpha_1, \ldots, \alpha_\ell\}$. We decompose $\mathfrak{b} = \mathfrak{b}_- \oplus \mathfrak{b}_r$, where $\mathfrak{b}_-$ is the common kernel of the α_j, and we let

$$\Delta_r = \Delta \cap \Sigma\, \mathbb{R}\alpha_j .$$

From $\mathfrak{b}_r$ and Δ_r we can construct a split semisimple subalgebra $\mathfrak{g}_r$ of $\mathfrak{g}$. Let G_r be the corresponding analytic subgroup, and choose $K_r = K \cap G_r$ as maximal compact subgroup. Since each $\pm\alpha_j$ is in Δ_r, we have $\mathfrak{a} \subseteq \mathfrak{g}_r$. In fact, $\mathfrak{a}$ can be taken as the Iwasawa $\mathfrak{a}$ of $\mathfrak{g}_r$. The M of a corresponding minimal parabolic subgroup of G_r is then $M_r = Z_{K_r}(\mathfrak{a})$; M_r is a direct sum of two-element groups, and it is a subgroup of the center Z_M of M.

The roots of M can be naturally identified with the subset Δ_- of Δ orthogonal to all α_j. If σ is a discrete series or limit of discrete series representation of M, then we know that σ is induced from some $\sigma^{\#}$ on

$$M^{\#} = M_0 Z_M = M_0 M_r .$$

We let χ denote the scalar value of σ on the subgroup M_r of Z_M, and we let (λ_0, C) be a Harish-Chandra parameter of $\sigma^{\#}$. Here λ_0

is dominant for the Weyl chamber C of $i\mathfrak{b}_-$, and we let $(\Delta_-)^+$ be the corresponding positive system in Δ_-.

The paper [1] shows how to obtain a positive system Δ^+ such that

(i) λ_0 is Δ^+ dominant

(ii) $\Delta^+ \supseteq (\Delta_-)^+$

(iii) Δ_r is generated by the Δ^+ simple roots that it contains

(iv) some other properties hold.

The theorems of [1] then identify the (highest weights of the) minimal K-types of $U(P,\sigma,\nu)$ as all Δ_K^+ dominant expressions of the form

$$\Lambda = \lambda - E(2\rho_K) + 2\rho_{K_r} + \mu . \tag{1.1}$$

Here ρ refers to a half-sum of positive roots, λ is the Blattner parameter of $\sigma^\#$ given by

$$\lambda = \lambda_0 - \rho_{-,c} + \rho_{-,n} , \tag{1.2}$$

and E is the orthogonal projection on $\sum \mathbb{R}\alpha_j$. The linear functional μ is any minimal (= fine) K_r-type for the principal series representations of G_r with M_r parameter the translate of χ given by

$$\omega = \chi \cdot \exp(E(2\rho_K) - 2\rho_{K_r})|_{M_r} .$$

We say σ has $(\{\alpha_j\},\Delta^+,\chi,\mu)$ as a <u>format</u> if Λ in (1.1) is a minimal K-type of $U(P,\sigma,\nu)$, i.e., if Λ is Δ_K^+ dominant. We consider simultaneously all σ's with a common format and pick out a smallest one. Theorem 3.1 of [3], reproduced below, gives the sense in which there exists a unique smallest one.

<u>Theorem 1.1</u>. Suppose $G^{\mathbb{C}}$ is simply connected. Among all infinitesimal characters λ_0 of discrete series or limits of discrete series of M with a particular format for G, there exists a unique one $\lambda_{0,b}$ such that any other λ_0 for that format has $\lambda_0 - \lambda_{0,b}$ dominant for Δ^+ and G-integral.

We call $\lambda_{0,b}$ or its associated σ_b the <u>basic case</u> for the format. When $G^{\mathbb{C}}$ is not simply connected, we pass to the appropriate cover of G in order to use Theorem 1.1 to define "basic case"; back in the original G, the parameter $\lambda_{0,b}$ continues to make sense, but σ_b may no longer be single-valued. In any event there are only finitely many basic cases for each G.

Some detailed examples appear in [3], all attached to minimal parabolic subgroups. For the double cover of $SO(2n,1)$, σ_b is the trivial representation or the spin representation. For $SU(n,1)$ and $SU(N,2)$ the basic cases are finitely many one-dimensional representations of M close to the trivial representation. For $Sp(n,1)$ with $n \geq 2$, M is $SU(2)\times Sp(n-1)$. The basic cases $(k\times \text{fundamental})\otimes 1$, with $0 \leq k \leq 2n-1$, were listed in [3]; there is one other basic case—given by $\sigma_b = 1\otimes\sigma_0$, where σ_0 is the fundamental representation attached to the long simple root of $Sp(n-1)$.[1]

Returning to the general (G,σ), we recall how [3] associates to (G,σ) a basic case (L,σ^L) for a certain subgroup L of G. Let λ_0 be the infinitesimal character of σ, let $(\{\alpha_j\},\Delta^+,\chi,\mu)$ be a compatible format, and let $\lambda_{0,b}$ be the basic case for this format. Let $\mathfrak{q} = \mathfrak{l}^{\mathbb{C}}\oplus\mathfrak{u}$ be the parabolic subalgebra of $\mathfrak{g}^{\mathbb{C}}$ defined by the Δ^+ dominant form $\lambda_0 - \lambda_{0,b}$:

$\mathfrak{q}$ is built from $\mathfrak{b}^{\mathbb{C}}$ and all $\beta\in\Delta$ with $\langle\lambda_0-\lambda_{0,b},\beta\rangle \geq 0$,

$\mathfrak{l}^{\mathbb{C}}$ is built from $\mathfrak{b}^{\mathbb{C}}$ and all $\beta\in\Delta$ with $\langle\lambda_0-\lambda_{0,b},\beta\rangle = 0$,

$\mathfrak{u}$ is built from all $\beta\in\Delta$ with $\langle\lambda_0-\lambda_{0,b},\beta\rangle > 0$.

---

[1] This additional basic case for $Sp(n,1)$ was inadvertently omitted from the list in [3]. For it the induced representation has two minimal K-types, and $J(P,\sigma_b,t\rho_A)$ is not infinitesimally unitary for any $t > 0$.

Set $\mathfrak{l} = \mathfrak{l}^{\mathbb{C}} \cap \mathfrak{g}$, and let L be the corresponding analytic subgroup of G. The root system of $(\mathfrak{l}^{\mathbb{C}}, \mathfrak{b}^{\mathbb{C}})$, namely

$$\Delta^L = \{\beta \in \Delta \mid \langle \lambda_0 - \lambda_{0,\mathfrak{b}}, \beta \rangle = 0\},$$

contains all $\pm\alpha_j$, and thus $\mathfrak{l}^{\mathbb{C}}$ contains $\mathfrak{a}^{\mathbb{C}}$. Then it follows that

$$P^L = (M \cap L)A(N \cap L)$$

is a parabolic subgroup of L. We define σ^L by

$$\lambda_0^L = \lambda_0 - \rho(\mathfrak{u}) \tag{1.3a}$$

$$\chi^L = \chi \cdot [\exp E(2\rho(\mathfrak{u} \cap \mathfrak{l}^{\mathbb{C}}))]|_{M_r}. \tag{1.3b}$$

The propositions in §4 of [3] establish the following.

Theorem 1.2. The definitions (1.3) consistently define σ^L, and $(\{\alpha_j\}, \Delta^+ \cap \Delta^L, \chi^L, \mu)$ is a compatible format for σ^L. Moreover, σ^L is the basic case for this format.

Remark. The group L is reductive, not necessarily semisimple, and the statement that σ^L is a basic case is more precisely a statement about the restriction of σ^L within the derived group of L.

Conjecture 5.1 of [3] expects that $J(P,\sigma,\nu)$ is infinitesimally unitary for G if and only if $J(P^L,\sigma^L,\nu)$ is infinitesimally unitary for L.

2. Reduction theorems

Even in ostensibly easy examples, it is a bit subtle to determine the basic cases without a guess as to what they are.[2] In this section we give two reduction theorems to make this determination easier. We apply the theorems to give formulas for the basic cases attached to maximal and minimal parabolic subgroups.

Throughout this section we work with a fixed format $(\{\alpha_j\},\Delta^+,\chi,\mu)$. Following [1], we say that

$$\text{a root in } \Delta \text{ is } \begin{cases} \underline{\text{real}} & \text{if in } \Sigma\,\mathbb{R}\alpha_j \\ \underline{\text{imaginary}} & \text{if orthogonal to } \Sigma\,\mathbb{R}\alpha_j \\ \underline{\text{complex}} & \text{otherwise.} \end{cases}$$

The first theorem is a kind of localization theorem for the calculation of basic cases. Fix a complex or imaginary Δ^+ simple root β, and let

$$\begin{aligned} \Delta^H &= \text{root system generated by } \beta \text{ and } \Delta_r \\ \mathfrak{b}^H &= \mathbb{R}iH_\beta + \mathfrak{b}_r \\ \mathfrak{h}^{\mathbb{C}} &= (\mathfrak{b}^H)^{\mathbb{C}} + \sum_{\gamma\in\Delta^H} \mathbb{C}X_\gamma \\ \mathfrak{h} &= \mathfrak{g} \cap \mathfrak{h}^{\mathbb{C}} \\ H &= \text{(semisimple) connected subgroup of } G \text{ corresponding} \\ &\quad\ \text{to } \mathfrak{h}\,. \end{aligned} \tag{2.1}$$

We use a superscript H to denote the usual subgroups, subalgebras, etc., associated with H. Note that $\Delta_r^H = \Delta_r$ and thus $G_r^H = G_r$. Let

---

[2] Cf. Footnotes 1 and 3 elsewhere in this paper.

$$(\Delta^H)^+ = \Delta^+ \cap \Delta^H$$

$$\mathfrak{b} = \sum_{\substack{\gamma \in \Delta^+ \\ \gamma \notin \Delta^H}} \mathbb{C} X_\gamma \, . \qquad (2.2)$$

Theorem 2.1. Fix a format $(\{\alpha_j\},\Delta^+,\chi,\mu)$ and a complex or imaginary Δ^+ simple root β, and make the corresponding definitions (2.1) and (2.2). Let $\lambda_0 = \lambda_{0,b}$ be the basic case for the format $(\{\alpha_j\},\Delta^+,\chi,\mu)$ for G, and define

$$\lambda_0^H = \lambda_0 - \rho(\mathfrak{b}) \qquad \text{(restricted to } \mathfrak{b}^H)$$

$$\chi^H = \chi \cdot [\exp E(2\rho(\mathfrak{b} \cap \mathfrak{k}^{\mathbb{C}}))]|_{M_r} \, .$$

Then $(\{\alpha_j\},(\Delta^H)^+,\chi^H,\mu)$ is a format for H, and the basic case for this format is exactly λ_0^H.

Proof. This is proved in the same way as Propositions 4.1 and 4.2 of [3] but with H in place of L.

Corollary 2.2. Let $\lambda_{0,b}$ be the basic case for the format $(\{\alpha_j\},\Delta^+,\chi,\mu)$ of G, and let β be an imaginary Δ^+ simple root. Then

$$\frac{2\langle \lambda_{0,b}, \beta\rangle}{|\beta|^2} = \begin{cases} 1 & \text{if } \beta \text{ is in } \Delta_K^+ \\ 0 & \text{if not}. \end{cases}$$

Proof. We apply Theorem 2.1. Since $\pm\beta$ are orthogonal to the other members of Δ^H, we may think of Δ^H as being just $\{\pm\beta\}$. Then H is locally $SU(2)$ or $SL(2,\mathbb{R})$, and the corollary results from direct calculation.

Corollary 2.3. Let $\lambda_{0,b}$ be the basic case for a format $(\{\alpha\},\Delta^+,\chi,\mu)$ of G that corresponds to a maximal parabolic subgroup other than in a real form of $G_2^{\mathbb{C}}$, and let β be a Δ^+ simple root. Then

$$\frac{2\langle\lambda_{0,b},\beta\rangle}{|\beta|^2} = \begin{cases} 1 & \text{if } \beta \text{ is compact imaginary} \\ 0 & \text{if } \beta \text{ is real or noncompact imaginary} \\ \text{correction}(\beta) & \text{if } \beta \text{ is complex}. \end{cases}$$

Here correction(β) is always 0, $\frac{1}{2}$, or 1, depending on the form of β. With ε denoting a member of $(ib_-)'$, the formula for correction(β) is

$$\text{correction}(\beta) = \begin{cases} \frac{1}{2} & \text{if } \beta=\varepsilon-\frac{1}{2}\alpha,\ |\beta|=|\alpha|, \text{ and } \mu=0 \\ \frac{1}{2}(1-\operatorname{sgn}\langle\mu,\gamma\rangle) & \text{if } \beta=\varepsilon-\frac{1}{2}\alpha,\ |\beta|=|\alpha|,\ \mu=\pm\frac{1}{2}\alpha, \text{ and a sign } \pm \text{ is fixed so that } \gamma=\varepsilon\pm\frac{1}{2}\alpha \text{ is compact} \\ \frac{1}{2}\pm\left|\frac{2\langle\mu,\beta\rangle}{|\beta|^2}\right| & \text{if } \beta=\varepsilon-\alpha,\ |\beta|^2=2|\alpha|^2, \text{ and the sign } \pm \text{ is fixed to be } + \text{ if } \beta \text{ is compact for } G \text{ and } - \text{ if } \beta \text{ is noncompact for } G \\ \max\{0,-\frac{2\langle\mu,\gamma\rangle}{|\gamma|^2}\} & \text{if } \beta=\varepsilon-\frac{1}{2}\alpha,\ |\alpha|^2=2|\beta|^2, \text{ and a sign } \pm \text{ is fixed so that } \gamma=\varepsilon\pm\frac{1}{2}\alpha \text{ is compact}. \end{cases}$$

Proof. For β real the formula is trivial, and for β imaginary the formula comes from Corollary 2.2. For β complex we apply Theorem 2.1 and are led to a group H of rank 2, where we make an explicit computation. The group H is locally $SU(2,1)$ for the first two forms of β, $SO(4,1)$ for β compact of the third form, and $Sp(2,\mathbb{R})$ in the remaining cases.

If we try to use Theorem 2.1 to handle the general case, we find that H can still be fairly complicated. Specifically the subgroup G_r of H is split with rank G_r = rank K_r, and its simple components are of type A_1, B_n, C_n, D_{2n}, E_7, E_8, F_4, or G_2. However, not all of G_r is needed to handle the projection $E(\beta)$ of the simple root β, and we seek a second reduction theorem that allows us to discard the unnecessary part of G_r.

The success of such a reduction depends upon the nature of β. The full list of possibilities is enumerated in the following lemma.

Lemma 2.4. Apart from indexing and signs, the following expressions $\beta = \epsilon + \Sigma c_j \alpha_j$ (with ϵ in $(ib_-)'$) are the only possibilities for a complex root in Δ other than in a factor of type $G_2^{\mathbb{R}}$. Each such possibility has

$$c_j = \frac{\langle \beta , \alpha_j \rangle}{|\alpha_j|^2} \quad \text{and} \quad \Sigma \frac{4\langle \beta , \alpha_j \rangle^2}{|\beta|^2 |\alpha_j|^2} = n < 4 .$$

(1) $n = 1$. β and α_1 of equal length; $\beta = \epsilon + \frac{1}{2}\alpha_1$.

(2) $n = 2$.

a) $\beta, \alpha_1, \alpha_2$ of same length; $\beta = \epsilon + \frac{1}{2}\alpha_1 + \frac{1}{2}\alpha_2$.

b) β long relative to α_1; $\beta = \epsilon + \alpha_1$.

c) β short relative to α_1; $\beta = \epsilon + \frac{1}{2}\alpha_1$.

(3) $n = 3$.

a) $\beta, \alpha_1, \alpha_2, \alpha_3$ of same length; $\beta = \epsilon + \frac{1}{2}\alpha_1 + \frac{1}{2}\alpha_2 + \frac{1}{2}\alpha_3$.

b) β and α_2 long, α_1 short; $\beta = \epsilon + \alpha_1 + \frac{1}{2}\alpha_2$.

c) β and α_2 short, α_1 long; $\beta = \epsilon + \frac{1}{2}\alpha_1 + \frac{1}{2}\alpha_2$.

Proof. We apply Parseval's equality to the expansion of β in terms of the orthogonal elements $\epsilon, \alpha_1, \ldots,$ and we are led to the list of possibilities in the statement of the lemma.

Taking advantage of the reduction in Theorem 2.1, let us now suppose that G and Δ^+ are such that there are no imaginary simple roots and there is exactly one complex simple root, which we call β. Fix a subset S of the indices $1, \ldots, \ell$ such that

$$\{\alpha_j \mid j \in S\} \supseteq \{\alpha_j \mid \langle \alpha_j, \beta \rangle \neq 0\},$$

and let

$$\begin{aligned}
\Delta^H &= \Delta \cap [\mathbb{R}\beta + \sum_{j \in S} \mathbb{R}\alpha_j] \\
\mathfrak{b}^H &= \mathbb{R}iH_\beta + \sum_{j \in S} \mathbb{R}iH_{\alpha_j} \\
\mathfrak{h}^{\mathbb{C}} &= (\mathfrak{b}^H)^{\mathbb{C}} + \sum_{\gamma \in \Delta^H} \mathbb{C} X_\gamma \\
\mathfrak{h} &= \mathfrak{g} \cap \mathfrak{h}^{\mathbb{C}} \\
H &= \text{(semisimple) connected subgroup of } G \text{ corresponding to } \mathfrak{h}.
\end{aligned} \tag{2.3}$$

Again we use a superscript H to denote the usual subgroups, subalgebras, etc., associated with H. In terms of the given format for G, we define

$$\begin{aligned}
(\Delta^H)^+ &= \Delta^+ \cap \Delta^H \\
\mu^H &= \mu|_{\mathfrak{b}^H} \\
\omega_0^H &= \omega|_{M_r^H \cap \exp \mathfrak{b}_-} \\
\chi_0^H &= \omega_0^H \cdot \exp(E^H(2\rho_K^H) - 2\rho_{K_r^H})|_{M_r^H \cap \exp \mathfrak{b}_-}.
\end{aligned} \tag{2.4}$$

Theorem 2.5. Fix a format $(\{\alpha_j\},\Delta^+,\chi,\mu)$ for which the only nonreal Δ^+ simple root is the complex root β, and make the corresponding definitions (2.3) and (2.4). Assume further that

(i) if ϵ is in Δ_- and $\epsilon \pm \alpha_{j_0}$ are roots, then j_0 is in S

(ii) in the notation of Lemma 2.4, β is of type (1), (2a), or (3a), or else β is compact and is of type (2b).

Then there exists an extension ω^H of ω_0^H to a character of M_r^H such that if μ is a fine K_r-type for ω then μ^H is a fine K_r^H-type for ω^H. For any such extension ω^H and corresponding χ^H, if $\lambda_0 = \lambda_{0,b}$ is the basic case for the format $(\{\alpha_j\},\Delta^+,\chi,\mu)$, then $(\{\alpha_j\}_{j\in S}, (\Delta^H)^+,\chi^H,\mu^H)$ is a format for H and the basic case $\lambda_{0,b}^H$ for this format is exactly $\lambda_0^H = \lambda_0|_{b^H}$.

Remarks. We may ignore real forms of $G_2^{\mathbb{C}}$, since otherwise $\Delta^H = \Delta$. Then we see that Assumptions (i) and (ii) are satisfied if all roots in Δ have the same length. In particular the theorem reduces calculations of basic cases for Δ of type E_6, E_7, or E_8 to a classical root system Δ^H of rank at most four. Moreover, the corollary below shows that the theorem handles formats associated to minimal parabolic subgroups. With a little additional work, one can weaken Assumption (ii) in the general case, but we shall not do so here.

Theorem 2.5 will be proved in §3. The proof uses the following lemma, which we need also when we apply Theorem 2.5 to obtain Corollary 2.7.

Lemma 2.6. In the notation of Theorem 2.5, if the component of β orthogonal to all α_j is ϵ, then every member of Δ^+ is of the form $\gamma = n\epsilon + \sum c_j\alpha_j$ with $n = 0, 1$, or 2.

Proof. Given γ in Δ^+, expand γ in terms of the simple roots and then regroup to see that $\gamma = n\epsilon + \Sigma c_j \alpha_j$ with n an integer ≥ 0. Then

$$2|\beta|^2 \geq |\gamma|^2 \geq |n\epsilon|^2 = n^2|\epsilon|^2 = n^2 \times (\tfrac{1}{4} \text{ or } \tfrac{1}{2} \text{ or } \tfrac{3}{4})|\beta|^2 \qquad (2.5)$$

by Lemma 2.4. So $2 \geq \frac{1}{4}n^2$ and $n \leq 2$.

Corollary 2.7.[3] Let $\lambda_{0,b}$ be the basic case for a format $(\{\alpha_j\}, \Delta^+, \chi, \mu)$ of G that corresponds to a minimal parabolic subgroup, and let β be a Δ^+ simple root. Then

$$\frac{2\langle \lambda_{0,b}, \beta\rangle}{|\beta|^2} = \begin{cases} 1 & \text{if } \beta \text{ imaginary (and compact)} \\ 0 & \text{if } \beta \text{ real} \\ \text{correction}(\beta) & \text{if } \beta \text{ complex}. \end{cases}$$

Here correction(β) is always 0, $\frac{1}{2}$, or 1, depending on the form of β. With ϵ denoting a member of $(i\mathfrak{b}_-)'$, the formula for correction(β) is

$$\text{correction}(\beta) = \begin{cases} \frac{1}{2} & \text{if } \beta = \epsilon - \frac{1}{2}\alpha_j,\ |\beta| = |\alpha_j|,\ \mu \perp \alpha_j \\ \frac{1}{2}(1 - \operatorname{sgn}\langle \mu, \gamma\rangle) & \text{if } \beta = \epsilon - \frac{1}{2}\alpha_j,\ |\beta| = |\alpha_j|,\ \mu \not\perp \alpha_j, \\ & \text{and a sign} \pm \text{ is fixed so that} \\ & \gamma = \epsilon \pm \frac{1}{2}\alpha_j \text{ is compact} \\ \frac{1}{2} + \left|\frac{2\langle \mu, \beta\rangle}{|\beta|^2}\right| & \text{if } \beta = \epsilon - \alpha_j \text{ with } |\beta|^2 = 2|\alpha_j|^2 \\ c + \left|\frac{2\langle \mu, \gamma\rangle}{|\gamma|^2}\right| & \text{if } \beta = \epsilon - \frac{1}{2}\alpha_i - \frac{1}{2}\alpha_j,\ |\beta| = |\alpha_i| \\ & = |\alpha_j|,\ c = \frac{1}{2} \text{ or } 0 \text{ according as} \\ & \frac{1}{2}(\alpha_i + \alpha_j) \text{ is or is not a root, and} \\ & \text{a sign} \pm \text{ is fixed so that} \\ & \gamma = \epsilon - \frac{1}{2}\alpha_i \pm \frac{1}{2}\alpha_j \text{ is compact}. \end{cases}$$

---

[3] A result of this sort was announced in [3]. However, the formula for correction(β) in [3] contains some misprints and an omission.

Proof. For β real the formula is trivial, and for β imaginary the formula comes from Corollary 2.2. For β complex we apply Lemma 2.4. Our assumption about a minimal parabolic subgroup means that $\Delta_{-,n} = \emptyset$. If β were of type (2c) or (3a) or (3b), then 2ε would be in $\Delta_{-,n}$; if β were of type (2b) with β noncompact, then ε would be in $\Delta_{-,n}$. So β is of none of these types.

We apply Theorem 2.1 and prepare to apply Theorem 2.5. Normally we let $S = \{j \mid \langle\beta,\alpha_j\rangle \neq 0\}$. But if there is an index j_0 such that $\varepsilon' + \alpha_{j_0}$ is in Δ with ε' orthogonal to all α_j, then we adjoin j_0 to S. (Such an index j_0 is necessarily unique.) Then Assumption (i) is certainly satisfied.

In view of what we have already proved, Assumption (ii) will be satisfied if we show that β cannot be of type (3c). If, on the contrary, β is of type (3c), then the component Δ_β of Δ to which β belongs is of type B_n, C_n, or F_4. For the case of B_n, we note that β and α_2 are short and nonorthogonal, contradiction. For the case of C_n or F_4, α_1 shows the existence of long noncompact roots, and it follows that the group corresponding to Δ_β is split over $\mathbb{R}$. Since the format is assumed attached to a minimal parabolic subgroup, Δ_β can contain no complex roots, in contradiction to the assumed form of β.

Thus we can apply Theorem 2.5. We consider the possibilities for S. First suppose Δ^+ contains some member $\gamma = \varepsilon' - \alpha_{j_0}$ with $\varepsilon' \neq 0$ orthogonal to all α_j, and let ε be the projection of β orthogonal to all α_j. Lemma 2.6 shows that $\varepsilon' = 2\varepsilon$ or $\varepsilon' = \varepsilon$. If $\varepsilon' = 2\varepsilon$, examination of (2.5) shows that $|\varepsilon|^2 = \frac{1}{4}|\beta|^2$, i.e., β is of type (3), in contradiction with what we already know about β.

So γ must be of the form $\varepsilon - \alpha_{j_0}$. For ε to be a root, β must be of type (2a) or (2b). If β is of type (2a), say

$\beta = \varepsilon - \frac{1}{2}\alpha_i - \frac{1}{2}\alpha_j$, then $\frac{1}{2}\alpha_i + \frac{1}{2}\alpha_j$ is in Δ. If β is of type (2b), then β must be γ. In any event, the only way S can have more than one element is if β is of type (2a).

When Δ^+ does not contain a member $\varepsilon' - \alpha_{j_0}$, then the only way S can have more than one element is if β is of type (2a), since β is not of type (3).

Thus if β is not of type (2a), we can apply Corollary 2.3 to H to handle matters. If β is of type (2a), say with $\beta = \varepsilon - \frac{1}{2}\alpha_i - \frac{1}{2}\alpha_j$, then we make an explicit calculation. If $\frac{1}{2}(\alpha_i + \alpha_j)$ is a root, the calculation is in a group locally isomorphic to $SO(5,2)$ or $SO(6,3)$; if $\frac{1}{2}(\alpha_i + \alpha_j)$ is not a root, the calculation is in a group locally isomorphic to $SU(2,2)$. There are no other possibilities, and the corollary follows.

3. Proof of Theorem 2.5

Throughout this section the notation and assumptions in Theorem 2.5 will be in force. We write $\beta = \varepsilon - \Sigma b_i\alpha_i$ with ε orthogonal to all α_j.

Lemma 3.1. $\Delta_-^H = \Delta_-$ and $\Delta_{-,c}^H = \Delta_{-,c}$.

Proof. Let γ be in Δ_- and write $\gamma = c\beta + \Sigma c_i\alpha_i$. Here $c \neq 0$ to make γ orthogonal to all α_j. Taking the inner product with α_j, we have

$$0 = \langle\gamma,\alpha_j\rangle = c\langle\beta,\alpha_j\rangle + c_j|\alpha_j|^2.$$

Thus $c_j \neq 0$ implies $\langle\beta,\alpha_j\rangle \neq 0$, which implies j is in S. Thus γ is in Δ^H, and we obtain $\Delta_-^H = \Delta_-$. The equality $\Delta_{-,c}^H = \Delta_{-,c}$ then follows from Assumption (i).

Lemma 3.2. $\rho - \rho^H - \rho_r + \rho_r^H$ is analytically integral on $\mathfrak{b}^H$.

Remark. Assumption (ii) is used only in this lemma and in Lemma 3.3.

Proof. The expression in question is half the sum of the members of Δ^+ that are in neither Δ^H nor Δ_r, and it is enough to show that the expression is actually a sum of members of Δ^+. The roots contributing to the expression are all complex, and we write them out as in Lemma 2.4. Then we consider together roots differing only in the signs attached to each α. If there are at least two α's, then half the sum of the members of a class is an integer multiple of the sum of the root with all plus signs and the root with all minus signs. So the only problem is with roots of types (1), (2b), and (2c). We show the only ones that contribute are of type (2b).

For type (1) let $\gamma = \varepsilon' \pm \frac{1}{2}\alpha_k$. Consideration of lengths shows that $\beta = \varepsilon' - \frac{1}{2}\alpha_j$, and then $j = k$ to force $2\langle\gamma,\beta\rangle/|\beta|^2$ to be an integer. Thus γ is already in Δ^H. For type (2c) let $\gamma = \varepsilon' \pm \frac{1}{2}\alpha_k$. Then $2\varepsilon'$ is a root, and Lemma 2.6 implies $\varepsilon' = \varepsilon$. Assumption (ii) forces β to be of type (2a), (2b), or (3a), and (2b) is ruled out since 2ε is a root. If $\beta = \varepsilon - \frac{1}{2}\alpha_i - \frac{1}{2}\alpha_j$ is of type (2a), then $\frac{1}{2}(\alpha_i + \alpha_j) \pm \frac{1}{2}\alpha_k$ is a pair of orthogonal roots whose difference is a root. Then the sum $\alpha_i + \alpha_j$ must be a root, in contradiction to strong orthogonality. Finally if β is of type (3a), then $\langle\beta,\alpha_k\rangle \neq 0$ in order to avoid $2\langle\beta,\gamma\rangle/|\beta|^2 = 1/2$. In short, no root γ of type (1) or (2c) makes a contribution.

For type (2b) let $\gamma = \varepsilon' \pm \alpha_k$. Half the sum of these two roots is ε', which is a root. This completes the proof.

Lemma 3.3. Every member of Δ_K^+ is the sum of members of $(\Delta_K^H)^+$ and Δ_r^+.

Proof. Let γ in Δ_K^+ be given. In view of Lemma 2.6, we can write $\gamma = c\varepsilon + \Sigma c_i\alpha_i$ with $c = 1$ or 2. Lemma 3.1 shows we may assume $\Sigma c_i\alpha_i \neq 0$.

First suppose $c = 2$. Then $|\gamma|^2 > 4|\varepsilon|^2$, and (2.5) shows that $|\varepsilon|^2 = \frac{1}{4}|\beta|^2$ and $2|\beta|^2 = |\gamma|^2$. That is, β is of type (3) in Lemma 2.4, hence of type (3a) by Assumption (ii). Also γ is of type (2); being long, it must be of type (2a) or (2b). If γ is of type (2a), then the α's in it are all long, while the α's in β are short; thus we obtain $2\langle\gamma,\beta\rangle/|\gamma|^2 = 1/2$, contradiction. So γ is of type (2b), and then Assumption (i) and Lemma 3.1 show that γ is already in $(\Delta_K^H)^+$.

Thus $c = 1$. Suppose now that $|\beta|^2 = 2|\gamma|^2$. Then β is of type (3), and γ is of type (2). Hence β is of type (3a) by Assumption (ii), and γ (being short) must be of type (2a) or (2c). If γ is of type (2a), then the α's in it are all short, while the α's in β are all long; thus we obtain $2\langle\gamma,\beta\rangle/|\beta|^2 = 1/2$, contradiction. So γ is of type (2c): $\gamma = \varepsilon \pm \frac{1}{2}\alpha_i$. The requirement $2\langle\gamma,\beta\rangle/|\beta|^2 \neq 1/2$ forces α_i to occur in β, and we conclude that γ is already in $(\Delta_K^H)^+$.

With $c = 1$, suppose next that $|\gamma|^2 = 2|\beta|^2$. Then β is of type (2) and γ is of type (3). Since β is short and γ is long, β is of type (2a) or (2c), and γ is of type (3a) or (3b). Assumption (ii) says β is not of type (2c). So β is of type (2a), say $\beta = \varepsilon - \frac{1}{2}\alpha_i - \frac{1}{2}\alpha_j$. Then $2\langle\beta,\gamma\rangle/|\gamma|^2 \neq 1/2$ implies γ is of type (3b) of the form $\gamma = \varepsilon \pm \alpha_i \pm \frac{1}{2}\alpha_k$. It follows that $\frac{1}{2}(\alpha_i + \alpha_j) \pm \frac{1}{2}\alpha_k$ are orthogonal roots whose difference is a root; hence the sum $\alpha_i + \alpha_j$ is a root, in contradiction with strong orthogonality.

Consequently we may assume that $c = 1$ and $|\beta| = |\gamma|$. Then β and γ are both of type (n) for $n = 1, 2,$ or 3. If $n = 1$, write

$\beta = \epsilon - \frac{1}{2}\alpha_i$. Then $2\langle\beta,\gamma\rangle/|\gamma|^2 \neq 3/2$ implies $\gamma = \epsilon \pm \frac{1}{2}\alpha_i$; hence γ is in $(\Delta_K^H)^+$.

Suppose $n=2$. If β is of type (2b) (and is thus compact, by assumption), say $\beta = \epsilon - \alpha_i$, then $\gamma = \epsilon \pm \alpha_i$ or else $\langle\gamma,\alpha_i\rangle = 0$, by consideration of the lengths of the α's in γ. In the first case γ is in $(\Delta_K^H)^+$, and in the second case $\langle\gamma,\beta\rangle > 0$ and β simple imply that $\gamma = \beta + (\gamma - \beta)$ is the required decomposition of γ.

If β is of type (2a), write $\beta = \epsilon - \frac{1}{2}\alpha_i - \frac{1}{2}\alpha_j$. If γ is of type (2b), Assumption (i) shows γ is in $(\Delta_K^H)^+$. If γ is of type (2c), say $\gamma = \epsilon \pm \frac{1}{2}\alpha_k$, then $\frac{1}{2}(\alpha_i+\alpha_j) \pm \frac{1}{2}\alpha_k$ are orthogonal roots, and we are led to conclude that $\alpha_i + \alpha_j$ is a root, contradiction. So we may assume γ is of type (2a). We may assume that γ is not $\epsilon \pm \frac{1}{2}\alpha_i \pm \frac{1}{2}\alpha_j$, and then the fact that $2\langle\gamma,\beta\rangle/|\beta|^2$ is an integer means that $\gamma = \epsilon \pm \frac{1}{2}\alpha_k \pm \frac{1}{2}\alpha_m$ with $\{i,j\} \cap \{k,m\}$ empty. Let us assume that i precedes j; we choose a sign $\pm$ so that $\tilde{\beta} = \epsilon - \frac{1}{2}\alpha_i \pm \frac{1}{2}\alpha_j$ is compact. Then $\gamma = \tilde{\beta} + (\gamma - \tilde{\beta})$ is the required decomposition; $\gamma - \tilde{\beta}$ is positive because $\gamma - \beta$ is positive and because i precedes j.

Finally suppose $n=3$. By Assumption (ii) β is of type (3a), say $\beta = \epsilon - \frac{1}{2}\alpha_i - \frac{1}{2}\alpha_j - \frac{1}{2}\alpha_k$. Then 2ϵ is noncompact, and so γ is of type (3a) or (3b). If γ is of type (3b), then $2\langle\beta,\gamma\rangle/|\gamma|^2 \neq 1/2$ implies γ is of the form $\epsilon \pm \frac{1}{2}\alpha_i \pm \alpha_m$. Then $\alpha_m - \frac{1}{2}(\alpha_j + \alpha_k)$ is a root, necessarily negative for β to be simple. Hence j or k precedes m. Let us say that j precedes k, for definiteness. Then the roots $\frac{1}{2}\alpha_j \pm \frac{1}{2}\alpha_k \pm \alpha_m$ are all positive. Choose the coefficient of α_m to match that in γ, choose the coefficient of $\frac{1}{2}\alpha_k$ to make the whole root compact, and call the result δ. Then $\gamma = \delta + (\gamma - \delta)$ is the required decomposition of γ.

For γ of type (3a), we argue in the same way. Since

$2\langle\beta,\gamma\rangle/|\gamma|^2 \neq 1/2$, we may assume $\gamma = \varepsilon \pm \frac{1}{2}\alpha_i \pm \frac{1}{2}\alpha_{j'} \pm \frac{1}{2}\alpha_{k'}$. Say j precedes k and j' precedes k'. Since β is simple, j precedes j'. Then the roots

$$\tfrac{1}{2}\alpha_j \pm \tfrac{1}{2}\alpha_k \pm \tfrac{1}{2}\alpha_{j'} \pm \tfrac{1}{2}\alpha_{k'}$$

are all positive. We choose the signs for $\alpha_{j'}$ and $\alpha_{k'}$ to match those in γ and the sign for α_k to make the whole root compact. If we call the result δ, then $\gamma = \delta + (\gamma - \delta)$ is the required decomposition of γ. This proves the lemma.

<u>Proof of Theorem 2.5</u>. We may assume $G^{\mathbb{C}}$ is simply connected. It is clear that λ_0^H is dominant for $(\Delta_-^H)^+$. From (1.2) and Lemma 3.1 it follows that $\lambda^H = \lambda$ on $\mathfrak{b}^H$ and hence that λ^H is analytically integral on $\mathfrak{b}_-^H$ and is dominant for $(\Delta_{-,c}^H)^+$. Then it follows that λ_0^H is the infinitesimal character of a discrete series or limit of discrete series of $(M \cap H)_0$.

Next we show that $\exp \lambda^H$ and any extension χ^H of χ_0^H agree on $(\exp \mathfrak{b}_-^H) \cap (\exp \mathfrak{b}_r^H)$, so that we obtain a well defined representation of $(M \cap H)^\#$, then of $M \cap H$. By Lemma 3.2, the restriction to $\mathfrak{b}^H$ of the linear functional

$$2\rho_K - 2\rho_K^H - (\rho - \rho^H - \rho_r + \rho_r^H) \tag{3.1}$$

is analytically integral on $\mathfrak{b}^H$. By Lemma 3 of [1], we have

$$2\rho_K - 2\rho_{-,c} - \rho + \rho_- + \rho_r - E(2\rho_K) = 0 \tag{3.2}$$

and a similar identity on H. Restricting (3.2) to $\mathfrak{b}^H$, subtracting the corresponding identity on H, and taking into account Lemma 3.1, we obtain

$$2\rho_K - 2\rho_K^H - (\rho - \rho^H - \rho_r + \rho_r^H) = E(2\rho_K) - E^H(2\rho_K^H) \tag{3.3}$$

on $\mathfrak{b}^H$. The integrality of (3.1) means that the right side of (3.3) is analytically integral on $\mathfrak{b}^H$. Therefore

$$\exp(E^H(2\rho_K^H) - 2\rho^H_{K_r}) = \exp(E(2\rho_K) - 2\rho_{K_r}) \qquad (3.4)$$

on $\exp \mathfrak{b}^H_-$. The required consistency

$$\chi_0^H = \exp \lambda^H \qquad \text{on} \quad (\exp \mathfrak{b}^H_-) \cap (\exp \mathfrak{b}^H_r)$$

follows immediately by combining (3.4) with the identities

$$\begin{aligned}
\omega_0^H &= \omega && \text{on } M_r^H \cap \exp \mathfrak{b}_- \\
\omega_0^H &= \chi_0^H \cdot \exp(E^H(2\rho_K^H) - 2\rho^H_{K_r}) && \text{on } M_r^H \cap \exp \mathfrak{b}^H_- \\
\omega &= \chi \cdot \exp(E(2\rho_K) - 2\rho_{K_r}) && \text{on } M_r \\
\chi &= \exp \lambda && \text{on } (\exp \mathfrak{b}_-) \cap (\exp \mathfrak{b}_r) \\
\lambda^H &= \lambda && \text{on } \mathfrak{b}^H .
\end{aligned}$$

It follows from the definition of "fine" in [4] that every irreducible constituent of $\tau_\mu|_{K_r^H}$ is fine. Since τ_{μ^H} is such a constituent, μ^H is fine. Fix ω^H as some constituent of $\tau_{\mu^H}|_{M_r^H}$. Since $M_r^H \cap \exp \mathfrak{b}^H_-$ centralizes G_r, τ_μ is scalar on $M_r^H \cap \exp \mathfrak{b}^H_-$, and thus ω^H has to agree with ω on $M_r^H \cap \exp \mathfrak{b}^H_-$. Thus ω^H is an extension of ω_0^H, and μ^H is a fine K_r^H-type for ω^H.

Now we check that Λ^H is $(\Delta_K^H)^+$ dominant. The formula (1.1) for Λ is

$$\Lambda = \lambda - E(2\rho_K) + 2\rho_{K_r} + \mu ,$$

and there is a similar formula for Λ^H. Restricting the two formulas to $\mathfrak{b}^H$ and subtracting, we obtain

$$\Lambda^H - \Lambda = [E(2\rho_K) - 2\rho_{K_r}] + [E^H(2\rho_K^H) - 2\rho^H_{K_r}] . \qquad (3.5)$$

Both bracketed terms on the right are orthogonal to Δ_K^H, and the $(\Delta_K^H)^+$ dominance of Λ^H therefore follows from the Δ_K^+ dominance of Λ.

This establishes that $(\{\alpha_j\}_{j\in S}, (\Delta^H)^+, \chi^H, \mu^H)$ is a format for H and that λ_0^H is compatible with it. We now want to see that λ_0^H is the basic case for this format. Assume the contrary. Then there exists a Δ_H^+ dominant H-algebraically integral form ξ^H on $\mathfrak{b}^H$ not identically 0 such that $\lambda_0^H - \xi^H$ corresponds to a nonzero representation (corresponding to a cover of H) compatible with our format for H. One of the conditions on ξ^H is that $\langle \xi^H, \alpha_j \rangle = 0$ for j in S. We extend ξ^H to ξ on $\mathfrak{b}$ by requiring $\langle \xi, \alpha_j \rangle = 0$ for all j (and also $\langle \xi, \beta \rangle = \langle \xi^H, \beta \rangle$). Then ξ is a multiple $n\Lambda_\beta$ of the fundamental form Λ_β for G corresponding to β. Since $G^{\mathbb{C}}$ is simply connected, ξ is analytically integral on $\mathfrak{b}$. From the fact that $\lambda_0^H - n\Lambda_\beta|_{\mathfrak{b}^H}$ corresponds to a nonzero representation compatible with the format of H, we shall prove that $\lambda_0' = \lambda_0 - n\Lambda_\beta$ corresponds to a nonzero representation compatible with the format of G, in contradiction with the fact that λ_0 is a basic case. This contradiction will prove that λ_0^H is a basic case and will complete the proof of the theorem.

The integrality conditions are no problem. We need to see that λ_0' is dominant for $(\Delta_-)^+$, that its λ' is dominant for $\Delta_{-,c}^+$, and that its Λ' is dominant for Δ_K^+. The required dominance for λ_0' and λ' follows from Lemma 3.1 and the corresponding properties in H of $\lambda_0^H - n\Lambda_\beta|_{\mathfrak{b}^H}$.

For Λ', formula (3.5) shows that $\langle \Lambda, \gamma \rangle \geq 0$ for γ in $(\Delta_K^H)^+$, and each γ in $\Delta_{K_r}^+$ satisfies

$$\langle \Lambda', \gamma \rangle = \langle \Lambda, \gamma \rangle - n\langle \Lambda_\beta, \gamma \rangle = \langle \Lambda, \gamma \rangle \geq 0 .$$

Thus Λ' is Δ_K^+ dominant by Lemma 3.3. This completes the proof of the theorem.

References

[1] A. W. Knapp, Minimal K-type formula, "Non Commutative Harmonic Analysis and Lie Groups," Springer-Verlag Lecture Notes in Math. 1021, (1983), 107-118.

[2] A. W. Knapp and B. Speh, Status of classification of irreducible unitary representations, "Harmonic Analysis Proceedings, Minneapolis 1981," Springer-Verlag Lecture Notes in Math. 908 (1982), 1-38.

[3] A. W. Knapp and B. Speh, The role of basic cases in classification: Theorems about unitary representations applicable to SU(N,2), "Non Commutative Harmonic Analysis and Lie Groups," Springer-Verlag Lecture Notes in Math. 1021, (1983), 119-160.

[4] D. A. Vogan, Fine K-types and the principal series, mimeographed notes, Massachusetts Institute of Technology, 1977.

[5] D. A. Vogan, Understanding the unitary dual, these Proceedings.

Department of Mathematics
Cornell University
Ithaca, New York 14853, U.S.A.

LEFT CELLS IN WEYL GROUPS

G. Lusztig*
Department of Mathematics, M.I.T.
Cambridge, Massachussetts 02139

In [KL_1], Kazhdan and myself have defined a partition of an arbitrary Coxeter group into subsets called left cells. These subsets enter in an essential way in the classification of primitive ideals in the enveloping algebra of a semisimple Lie algebra. In this paper we shall generalize the definition of [KL_1] to include the case where the simple reflections are given different weights. We shall give an application of this to Schubert varieties. We shall also give some examples concerning a Weyl group of type B_n.

1. Let (W,S) be a Coxeter group and let $\varphi: W \to \Gamma$ be a map of W into an abelian group Γ such that $\varphi(s_1 s_2 \dots s_p) = \varphi(s_1)\varphi(s_2)\dots \varphi(s_p)$ for any reduced expression $s_1 s_2 \dots s_p$ in W. We shall set $\varphi(w) = q_w^{1/2}$, $(w \in W)$. Let H_φ be the Hecke algebra of W with respect to φ; this is an algebra over the group ring $\mathbb{Z}[\Gamma]$. As a $\mathbb{Z}[\Gamma]$-module, it is free with basis T_w, $(w \in W)$. The multiplication is defined by

$$(T_s + 1)(T_s - q_s) = 0, \qquad (s \in S)$$

$$T_{s_1 s_2 \dots s_p} = T_{s_1} T_{s_2} \dots T_{s_p}, \quad \text{if } s_1 s_2 \dots s_p \text{ is a reduced expression in } W.$$

The unit element is T_e. It will be convenient to introduce a new basis $\tilde{T}_w = q_w^{-1/2} T_w$, $(w \in W)$. We then have

$$(\tilde{T}_s + q_s^{-\frac{1}{2}})(\tilde{T}_s - q_s^{\frac{1}{2}}) = 0, \qquad (s \in S)$$

---

*Supported in part by the National Science Foundation.

$$\tilde{T}_{s_1 s_2 \ldots s_p} = \tilde{T}_{s_1} \tilde{T}_{s_2} \ldots \tilde{T}_{s_p}, \qquad \text{if } s_1 s_2 \ldots s_p \text{ is a reduced expression in } W.$$

Let $a \to \bar{a}$ be the involution of the ring $\mathbb{Z}[\Gamma]$ which takes γ to γ^{-1} for any $\gamma \in \Gamma$. We extend it to an involution $h \to \bar{h}$ of the ring H_φ by the formula

$$\overline{\sum_w a_w \tilde{T}_w} = \sum_w \bar{a}_w \tilde{T}^{-1}_{w^{-1}}, \qquad (a_w \in \mathbb{Z}[\Gamma]).$$

(Note that $\tilde{T}_s^{-1} = \tilde{T}_s + (q_s^{-\frac{1}{2}} - q_s^{\frac{1}{2}})$, $s \in S$, hence $\tilde{T}_w$ is ivertible for all $w \in W$.) Let us define elements $R^*_{x,y} \in \mathbb{Z}[\Gamma]$, $(x,y \in W)$, by

$$\tilde{T}^{-1}_{y^{-1}} = \sum_x \bar{R}^*_{x,y} \tilde{T}_x.$$

It is easy to see that $R^*_{x,y} = 0$ unless $x \leq y$ in the standard partial order of W. Using the fact that $h \to \bar{h}$ is an involution, we see that

(1.1) $$\sum_{x \leq y \leq z} \bar{R}^*_{x,y} R^*_{y,z} = \delta_{x,z}$$

for all $x \leq z$ in W. Note also that $q_x^{-1/2} q_y^{1/2} R^*_{x,y} \in \mathbb{Z}[\Gamma^2]$. For example, $R^*_{x,x} = 1$ for all $x \in W$. Let ℓ be the length function on W.

(1.2) If $x < y$, $\ell(y) = \ell(x) + 1$, then x is obtained by dropping $s \in S$ in a reduced expression of y, and we have $R^*_{x,y} = q_s^{1/2} - q_s^{-1/2}$.

(1.3) If $x < y$, $\ell(y) = \ell(x) + 2$, then x is obtained by dropping $s \in S$ and $t \in T$ in a reduced expression of y, and we have $R^*_{x,y} = (q_s^{1/2} - q_s^{-1/2})(q_t^{1/2} - q_t^{-1/2})$.

We now assume given a total order on Γ compatible with the group structure on Γ. Let Γ_+ be the set of elements of Γ which are strictly positive for this total order and let $\Gamma_- = (\Gamma_+)^{-1}$. We shall assume that $q_s^{1/2} \in \Gamma_+$ for all $s \in S$. We have

2. Proposition. *Given* $w \in W$, *there is a unique element* $C'_w \in H_\varphi$ *such that*

$$\bar{C}'_w = C'_w$$

$$C'_w = \sum_{y \le w} P^*_{y,w} \tilde{T}_y$$

where $P^*_{w,w} = 1$ *and, for any* $y < w$, $P^*_{y,w}$ *is a* $\mathbb{Z}$-*linear combination of elements in* Γ_-. *Moreover,* $q_y^{-1/2} q_w^{1/2} P^*_{y,w} \in \mathbb{Z}[\Gamma^2]$.

(When f is constant on S, this is the same as (1.1.c) of $[KL_1]$.)

We must show that the system of equations

(2.1) $$P^*_{w,w} = 1$$

(2.2) $$\bar{P}^*_{x,w} - P^*_{x,w} = \sum_{x<y\le w} R^*_{x,y} P^*_{y,w} \quad , (\forall x < w),$$

with unknowns $P^*_{x,w}$,has a unique solution such that $P^*_{x,w}$ is a $\mathbb{Z}$-linear combination of elements in Γ_-, for $x < w$. This is shown by induction on $\ell(w) - \ell(x)$. The uniqueness is clear. To show existence, we shall use a suggestion of O. Gabber, which simplifies somewhat the original proof in $[KL_1]$. We fix $x < w$ and assume that for all y, $x < y \le w$, the $P^*_{y,w}$ have been already constructed and have the required property. It is then enough to show that $\overline{\sum_{x<y\le w} R^*_{x,y} P^*_{y,w}} = - \sum_{x<y\le w} R^*_{x,y} P^*_{y,w}$. But we have

$$\overline{\sum_{x<y\le w} R^*_{x,y} P^*_{y,w}} = \sum_{x<y\le z\le w} \overline{R^*_{x,y}} R^*_{y,z} P^*_{z,w}$$

$$= \sum_{x<z\le w} \Big(\sum_{x\le y\le z} \bar{R}^*_{x,y} R^*_{y,z} P^*_{z,w} - R^*_{x,z} P^*_{z,w} \Big)$$

and, using (1.1), this equals $- \sum_{x<z\le w} R^*_{x,z} P^*_{z,w}$, as required. The last assertion follows from (2.2). This completes the proof of the proposition.

3. Now let $s \in S$, $w \in W$ be such that $w < sw$. For each y such that $sy < y < w$, we define an element

$$M^s_{y,w} \in \mathbf{Z}[\Gamma]$$

by the inductive condition

(3.1) $\sum_{\substack{y \le z < w \\ sz < z}} P^*_{y,z} M^s_{z,w} - q_s^{1/2} P^*_{y,w}$ is a combination of elements in Γ_-

and by the symmetry condition

$$\bar{M}^s_{y,w} = M^s_{y,w}. \tag{3.2}$$

The condition (3.1) determines uniquely the coefficient of γ in $M^s_{y,w}$ for all $\gamma \in \Gamma - \Gamma_-$; the condition (3.2) determines the remaining coefficients. We have $q_s^{-1/2} q_y^{-1/2} q_w^{1/2} M^s_{y,w} \in \mathbf{Z}[\Gamma^2]$.

4. <u>Proposition</u>. <u>Let</u> $s \in S$ <u>and let</u> $w \in W$. <u>Then</u>:

(4.1) $$(\tilde{T}_s + q_s^{-1/2}) C'_w = C'_{sw} + \sum_{\substack{z < w \\ sz < z}} M^s_{z,w} C'_z, \qquad \text{if } w < sw$$

(4.2) $$(\tilde{T}_s - q_s^{1/2}) C'_w = 0, \qquad \text{if } w > sw$$

(compare with (2.3.a), (2.3.c) in [KL_1]).

<u>Proof</u>. If $w = e$, then (4.1) is clearly true. Now assume that $w \neq e$ and that the proposition is already proved for all $w' < w$. Using (4.2), we see that

(4.3) $$P^*_{u,z} = q_s^{-1/2} P^*_{su,z} \qquad \text{if } u < su \le z, \quad sz < z < w.$$

<u>Case 1</u>: $w < sw$. Consider the left hand side minus the right hand side of (4.1). The coefficient of $\tilde{T}_y$ in that expression is

$$f_y = q_s^{\varepsilon/2} P^*_{y,w} + P^*_{sy,w} - P^*_{y,sw} - \sum_{\substack{y \le z < w \\ sz < z}} P^*_{y,z} M^s_{z,w}$$

where $\varepsilon = \begin{cases} 1 & \text{if } sy < y \\ -1 & \text{if } sy > y \end{cases}$ and $P^*_{x,x'}$ is defined to be zero whenever $x \not\leq x'$. If $sy < y$, then (3.1) shows that f_y is a $\mathbb{Z}$-linear combination of elements in Γ_-. If $sy > y$, then applying (4.3) we see that

$$f_y = q_s^{-1/2} P^*_{y,w} + P^*_{sy,w} - P^*_{y,sw} - q_s^{-1/2} \sum_{\substack{sy \le z < w \\ sz < z}} P^*_{sy,z} M^s_{z,w}.$$

It follows that

$$f_y = q_s^{-1/2} f_{sy} + q_s^{-1/2} P^*_{sy,sw} - P^*_{y,sw}$$

hence, again, f_y is a $\mathbb{Z}$-linear combination of elements in Γ_-. But $(\tilde{T}_s + q_s^{-1/2}) = C'_s$, C'_w, $M^s_{z,w}$ are each fixed by the involution $h \to \bar{h}$. Hence $\overline{\sum_y f_y \tilde{T}_y} = \sum_y f_y \tilde{T}_y$. Assume that some f_{y_0} is non-zero. We can take y_0 to have maximal possible length subject to the property $f_{y_0} \neq 0$. Then the coefficient of $\tilde{T}_{y_0}$ in $\overline{\sum_y f_y \tilde{T}_y}$ is equal to $\overline{f_{y_0}}$. Thus $f_{y_0} = \bar{f}_{y_0}$. This contradicts the fact that f_{y_0} is a non-zero combination of elements in Γ_-. Thus we have $f_y = 0$ for all $y \in W$ and (4.1) is proved for w.

<u>Case 2</u>: $w > sw$. Applying (4.1) to sw, we see that

$$C'_w = (\tilde{T}_s + q_s^{-1/2}) C'_{sw} - \sum_{\substack{z < sw \\ sz < z}} M^s_{z,sw} C'_z.$$

Clearly, $(\tilde{T}_s - q_s^{1/2})(\tilde{T}_s + q_s^{-1/2}) = 0$ and, by the induction hypothesis, $(\tilde{T}_s - q_s^{1/2}) C'_z = 0$ for all z, $(z < sw,\ sz < z)$. Hence $(\tilde{T}_s - q_s^{1/2}) C'_w = 0$, as required.

5. <u>Proposition</u>. <u>Let</u> $y < w$ <u>be such that</u> $\ell(w) = \ell(y) + 1$. <u>Then</u> y <u>is obtained by dropping a simple reflection</u> s <u>in a reduced expression of</u> w.

(a) <u>We have</u> $P^*_{y,w} = q_s^{-1/2}$

(b) <u>Let</u> t <u>be a simple reflection such that</u> $ty < y < w < tw$. <u>Then</u>

$$M^t_{y,w} = \begin{cases} 0, & \text{if } q_t^{1/2} < q_s^{1/2} \\ 1, & \text{if } q_t^{1/2} = q_s^{1/2} \\ q_s^{1/2}q_t^{-1/2} + q_s^{-1/2}q_t^{1/2}, & \text{if } q_t^{1/2} > q_s^{1/2} \end{cases}$$

<u>Proof</u>. From (1.2) and (2.2) we see that

$$\bar{P}^*_{y,w} - P^*_{y,w} = R^*_{y,w} = q_s^{1/2} - q_s^{-1/2}$$

and (a) follows. If t is as in (b), then by (3.1) and (a), $M^t_{y,w} - q_t^{1/2}q_s^{-1/2}$ must be a $\mathbb{Z}$-linear combination of elements in Γ_-. From this and from (3.2), the desired formula for $M^t_{y,w}$ follows.

6. Let $j: H_\varphi \to H_\varphi$ be the ring involution defined by $j(\sum_w a_w\tilde{T}_w) = \sum_w \bar{a}_w\varepsilon_w\tilde{T}_w$, where $\varepsilon_w = (-1)^{\ell(w)}$. It commutes with the involution $h \to \bar{h}$. Let $C_w = \varepsilon_w j(C'_w)$. Then

$$\bar{C}_w = C_w \quad \text{and} \quad C_w = \sum_{y \le w} \varepsilon_y\varepsilon_w \bar{P}^*_{y,w}\tilde{T}_y. \qquad \text{(Compare [KL}_1\text{, 1.1].)}$$

Applying j to (4.1) and (4.2) we get:

$$(6.1) \qquad (\tilde{T}_s - q_s^{1/2})C_w = C_{sw} - \sum_{\substack{z<w \\ sz<z}} \varepsilon_z\varepsilon_w M^s_{z,w}C_z, \qquad \text{if } w < sw$$

$$(6.2) \qquad (\tilde{T}_s + q_s^{-1/2})C_w = 0, \qquad \text{if } w > sw.$$

Let $j': H_\varphi \to H_\varphi$ be the anti-automorphism of the ring H_φ defined by $j'(\tilde{T}_w) = \tilde{T}_{w^{-1}}$ and $j'(a) = a$ for $a \in \mathbb{Z}[\Gamma]$. It is easy to see that $j'(C_w) = C_{w^{-1}}$. Therefore, from (6.1) and (6.2) we can deduce

$$(6.3)\quad C_w(\tilde{T}_s - q_s^{1/2}) = C_{ws} - \sum_{\substack{z<w\\ zs<z}} \varepsilon_z\varepsilon_w M^s_{z^{-1},w^{-1}} C_z, \qquad \text{if } w < ws$$

$$(6.4)\quad C_w(\tilde{T}_s + q_s^{-1/2}) = 0, \qquad \text{if } w > ws.$$

Let $\underset{L,\varphi}{\le}$ be the preorder relation on W generated by the relation "$x' \underset{L,\varphi}{\le} x$ if there exists $s \in S$ such that $C_{x'}$ appears with non-zero coefficient in $\tilde{T}_s \cdot C_x$ (expressed in the C_w-basis)." We call it the left preorder. The equivalence relation associated to $\underset{L,\varphi}{\le}$ is denoted $\underset{L,\varphi}{\sim}$ and the corresponding equivalence classes in W are called the left cells of W (with respect to φ). Given $x,y \in W$, we say that $x \underset{LR,\varphi}{\le} y$ if there exists a sequence $x = x_0, x_1, \dots, x_n = y$ of elements in W such that for $i = 0,1,\dots,n-1$, we have either $x_i \underset{L,\varphi}{\le} x_{i+1}$ or $x_i^{-1} \underset{L,\varphi}{\le} x_{i+1}^{-1}$. The equivalence relation on W corresponding to the preorder $\underset{LR,\varphi}{\le}$ is denoted $\underset{LR,\varphi}{\sim}$ and the correspondint equivalence classes on W are called the two sided cells of W. (These notions were introduced in [KL_1] in the case where φ is constant on S.)

For any $x \in W$, we denote I_x^L (resp. $\hat{I}_x^L$) the $\mathbb{Z}[\Gamma]$-submodule of H_φ spanned by the elements C_y, $y \underset{L,\varphi}{\le} x$, (resp. by the elements C_y, $y \underset{L,\varphi}{\le} x$, $y \underset{L,\varphi}{\not\sim} x$). We define similarly I_x^{LR} and $\hat{I}_x^{LR}$, by replacing $\underset{L,\varphi}{\le}$, $\underset{L,\varphi}{\sim}$ by $\underset{LR,\varphi}{\le}$, $\underset{LR,\varphi}{\sim}$ in the previous definition. It is clear from (6.1)-(6.4) that I_x^L, $\hat{I}_x^L$ are left ideals of H_φ and that I_x^{LR}, $\hat{I}_x^{LR}$ are two-sided ideals of H_φ. Hence $I_x^L/\hat{I}_x^L$ is a left H_φ-module with a natural basis given by the images of C_y for y in the left cell of x; $I_x^{LR}/\hat{I}_x^{LR}$ is a two sided H_φ-module with a natural basis given by the images of C_y, for y in the two sided cell of x. With respect to this basis, the action of $\tilde{T}_s$ $(s \in S)$ is given by a matrix which is completely determined by the elements $M^s_{y,w}$.

From now on, we assume that Γ is the infinite cyclic group with generator $q^{1/2}$ with the order relation $q^{i/2} \le q^{j/2} \Leftrightarrow i \le j$. We then have $q_w^{1/2} = q^{m(w)/2}$ where $m\colon W \to \{1,2,3,\dots\}$. In this case, we have

$P^*_{y,w} \in \mathbb{Z}[q^{-1/2}]$ and $q^{\frac{m(w)-m(y)}{2}} P^*_{y,w} \in \mathbb{Z}[q]$ for any $y \leq w$. Moreover $P^*_{y,w}$ has no constant term if $y < w$. From (3.1) we see that, when defined, $M^s_{y,w}\, q^{\frac{m(s)-1}{2}} \in \mathbb{Z}[q^{1/2}]$. In particular, $M^s_{y,w}$ is a constant whenever $m(s) = 1$. (This is the case considered in [KL_1].)

Consider, for example, the case where (W,S) is a Weyl group of type B_2 with $S = \{s_1,s_2\}$, $(s_1s_2)^4 = 1$, and let $m(s_1) = 1$, $m(s_2) = c \geq 2$. We have:

$$P^*_{s_2,s_2s_1s_2} = q^{-\frac{c+1}{2}} - q^{-\frac{c-1}{2}}, \qquad P^*_{e,s_2s_1s_2} = q^{-\frac{2c+1}{2}} - q^{-\frac{2c-1}{2}}$$

$$P^*_{s_1,s_1s_2s_1} = q^{-\frac{c+1}{2}} + q^{-\frac{c-1}{2}}, \qquad P^*_{e,s_1s_2s_1} = q^{-\frac{c+2}{2}} + q^{-\frac{c}{2}}$$

and $P^*_{y,w} = q^{\frac{m(y)}{2} - \frac{m(w)}{2}}$ for all other pairs $y \leq w$. (In particular, $P^*_{y,w}$ may have negative coefficients.) We have

$$M^{s_2}_{s_2s_1,s_1s_2s_1} = M^{s_2}_{s_2,s_1s_2} = q^{\frac{c-1}{2}} + q^{\frac{-c+1}{2}}, \qquad M^{s_1}_{s_1s_2,s_2s_1s_2} = M^{s_1}_{s_1,s_2s_1} = 0.$$

The left cells are:

$$\{e\},\ \{s_1\},\ \{s_2,s_1s_2\},\ \{s_2s_1s_2\},\ \{s_2s_1,s_1s_2s_1\},\ \{s_1s_2s_1s_2\}.$$

The corresponding H_φ-modules $I^L_x/\hat{I}^L_x$ (with scalars extended to an algebraic closure of $\mathbb{Q}(q^{1/2})$) are all irreducible. (This is in contrast with the situation when $m(s_1) = m(s_2) = 1$ in which case there are only four left cells.) The two-sided cells are $\{e\}$, $\{s_1\}$, $\{s_2s_1s_2\}$, $\{s_2,s_1s_2,s_2s_1,s_1s_2s_1\}$, $\{s_1s_2s_1s_2\}$.

7. If we specialize $q^{1/2}$ to 1, and take coefficients in $\mathbb{Q}$, the H_φ-modules $I^L_x/\hat{I}^L_x$ become left W-modules; they give a direct sum decomposition of the left regular representation of W; they are said to be the W-modules carried by the left cells (with respecto to φ). Simi-

larly, $I_x^{LR}/\hat{I}_x^{LR}$ become two sided W-modules; they give a direct sum decomposition of the two sided regular representation of W. Hence the two sided cells give rise to an equivalence relation on the set of irreducible representations of W: two representations are equivalent if they can be connected by a chain such that any two consecutive ones appear in the same $I_x^{LR}/\hat{I}_x^{LR}$. The equivalence relation on the representations is known in the case where $\varphi(s) = q$ $(\forall s \in S)$, by the work of Barbasch-Vogan [BV]. It coincides with the equivalence relation described in [L_1]. It is likely that, in general, the equivalence relation should still be that in [L_1], except that instead of the a-function used there one should use an a-function which depends on φ: for any irreducible W-module E, we define $a_\varphi(E)$ to be the order at 0 of the rational function in q giving the formal degree of the Hecke algebra H_φ corresponding to E. In particular, it should be true that the a_φ-function should be constant on each equivalence class.

8. Let G be a simple adjoint group defined over $\mathbb{C}$ and let $\alpha: G \to G$ be an outer automorphism which leaves stable a Borel subgroup $B \subset G$ and a maximal torus $T \subset B$. We assume that the corresponding map α: $W \to W$ (W = Weyl group of G) is non-trivial. Let W_1 be the fixed point set of α on W. It is well known that W_1 is a Coxeter group with a set of generators S_1 corresponding to the orbits of α on S (= the simple reflections of W); to an orbit $\mathcal{O}$, there corresponds the longest element in the subgroup generated by $\mathcal{O}$. Let $\varphi: W_1 \to \{q^{\mathbb{Z}}\}$ be the function defined by $\varphi(w_1) = q^{\ell(w_1)}$ where $\ell(w_1)$ is the length of w_1 with respect to (W,S). Let y, w be two elements of W_1 such that $y \le w$. We shall give an interpretation of $P^*_{y,w}$ (defined with respect to φ) in terms of Schubert varieties, analogous to [KL_2]. Let $\bar{B}_w \subset G/B$ be the Schubert variety corresponding to w and let B_y be the Bruhat cell corresponding to y. Then α acts naturally on $\bar{B}_w$ and on its subvariety B_y. Let $\mathcal{H}^{2i}_{B_y}(\bar{B}_w)$ be the stalk of the 2i-th

intersection cohomology sheaf of $\bar{B}_w$ at a point of B_y which is fixed by α. Then α acts naturally on $H^{2i}_{B_y}(\bar{B}_w)$ and we have

$$(8.1) \qquad \sum_{i\geq 0} \mathrm{Tr}(\alpha, H^{2i}_{B_y}(\bar{B}_w))q^i = q^{\frac{\ell(w)-\ell(y)}{2}} P^*_{y,w}.$$

(Note that ℓ is here the length function on W, not on W_1.) The proof is similar to that of [KL_2]. The formula (8.1) explains why the coefficients of $P^*_{y,w}$ may be negative.

9. The multiplication in the Hecke algebra can be interpreted in terms of a multiplication of complexes in a derived category of constructible sheaves over the flag manifold. (See [LV], [S].) This interpretation together with (8.1) allows us to deduce the following.

Let $y,w \in W_1$ and let us write $C_y \cdot C_w = \sum_{z_1 \in W_1} N_{y,w,z_1} C_{z_1}$, $N_{y,w,z_1} \in \mathbb{Z}[q^{1/2},q^{-1/2}]$ (an identity in the Hecke algebra of W_1, with respect to φ). We can also consider y, w as elements in W and attach to them elements $\tilde{C}_y$, $\tilde{C}_w$ in the Hecke algebra of W, with respect to $\varphi(w) = q^{\ell(w)}$. ($\tilde{C}_w$ is just C_w with respect to W.) We then have $\tilde{C}_y \cdot \tilde{C}_w = \sum_{z\in W} \tilde{N}_{y,w,z}\tilde{C}_z$, $\tilde{N}_{y,w,z} \in \mathbb{Z}[q^{1/2},q^{-1/2}]$. The coefficients of $\varepsilon_y\varepsilon_w\varepsilon_z\tilde{N}_{y,w,z}$, ($\varepsilon_w = (-1)^{\ell(w)}$), are ≥ 0 and can be interpreted as dimensions of certain vector spaces on which α acts whenever $z \in W_1$. Moreover the trace of α on that vector space is the corresponding coefficient of $N_{y,w,z}$. It follows that

(9.1) If $z \in W_1$ and $N_{y,w,z} \neq 0$ then $\tilde{N}_{y,w,z} \neq 0$.

From the definition of the left preorder $\underset{L,\varphi}{\leq}$ it now follows easily that

(9.2) If $y,w \in W_1$ satisfy $y \underset{L,\varphi}{\leq} w$ with respect to the left preorder of W_1,φ then they satisfy the similar inequality with respect to the left preorder of W,φ. Hence any left cell of W_1 (with respect to φ) is contained in a left cell of W (with respect to φ).

10. Assume, for example that (W,S) is a Weyl group of type A_n $(n \geq 3)$ and that α is the unique automorphism of order 2 of (W,S). Then (W_1,S_1) is a Weyl group of type $B_{n/2}$ if n is even and $B_{(n+1)/2}$ if n is odd. The restriction of $\varphi(w) = q^{\ell(w)}$ to S_1 has values $q^2,q^2,\ldots,q^2,q^3$ if n is even and $q^2,q^2,\ldots,q^2,q$ if n is odd. It is known that each left cell of W contains a unique involution and it carries an irreducible representation of W. It is clear that $\alpha\colon W \to W$ permutes among themselves the left cells of W, and it maps each two sided cell of W into itself (since α is an inner automorphism of W). Let $n(W_1)$ be the number of left cells of W_1 (with respect to φ) and let $n(W)$ be the number of α-stable left cells of W. We have

$$n(W_1) \geq n(W). \tag{10.1}$$

Indeed, each α-stable left cell of W contains some element of W_1 (for example, it contains a unique involution which is necessarily fixed by α), hence by (9.2) it contains a left cell of W_1 (with respect to φ). Let $n'(W_1)$ be the sum of the dimensions of the irreducible representations of W_1. Since the representations carried by the left cells of W_1 (with respect to φ) give a direct sum decomposition of the left regular representation of W_1, we see that

$$n'(W_1) \geq n(W_1) \tag{10.2}$$

with equality if and only if each left of cell of W_1,φ carries an irreducible representation of W_1.

Now let E be an irreducible representation of W. According to [KL_1], E admits a natural basis (e_i) in 1-1 correspondence with the set of left cells contained in the two-sided cell Ω of W corresponding to E. This basis has the following property: the permutation defined by α on the set of left cells in Ω corresponds to the permutation of the basis (e_i) defined by the action of $\pm w_0$ on E, where w_0 is the longest element in W. Thus, the number of α-stable left cells in Ω is equal to $|Tr(w_0,E)|$, hence

$$n(W) = \sum_E |Tr(w_0,E)|, \tag{10.3}$$

with the sum over all irreducible representations E of W.

Next, we note that there exists an imbedding $W_1^\vee \subset W^\vee$ of the set of irreducible representations of W_1 into the analogous set for W with the following property: if $E_1 \in W_1^\vee$ corresponds to $E \in W^\vee$, then $Tr(w_0,E) = \pm\dim E_1$, and if $E \in W^\vee$ is not in the image of our imbedding, then $Tr(w_0,E) = 0$. This can be seen by direct computation, using Murnaghan's rule, or one can argue as follows. We can regard W as the Weyl group of a unitary group over a finite field F_{p^r} and W_1 as the relative Weyl group. The unipotent representations of the unitary group are parametrized by the elements of $W^\vee$ and the unipotent representation corresponding to $E \in W^\vee$ has degree given by a polynomial in p^r which for $r \to 0$ becomes $\pm Tr(w_0,E)$. This polynomial is divisible by $(p^r - 1)$ unless the unipotent representation is in the principal series. The unipotent representations in the principal series are parametrized by the irreducible representations of W_1 and the representation corresponding to $E_1 \in W_1^\vee$ has degree given by a polynomial in p^r which for $r \to 0$ becomes $\dim E_1$. Hence our assertion follows. We seen then from (10.3) that

$$n(W) = \sum \dim(E_1),$$

with the sum over all irreducible representations E_1 of $W_1^\vee$, hence

$$n(W) = n'(W_1).$$

Comparing with (10.1) and (10.2) it follows that

$$n(W_1) = n(W) = n'(W_1)$$

hence we have proved the following.

11. Theorem. Each left cell of W_1 (with respect to φ) carries an irreducible representation of W_1. It contains a unique involution and is the intersection of W_1 with a left cell of W (= symmetric group $\mathfrak{S}_{n+1}$).

12. In the case where (W,S) is a Weyl group and $\varphi(s) = q$ for all $s \in S$, one can use the character formulas [L_2] for the unipotent representations of a semisimple group over a finite field, to get some new information on the structure of left cells in W. For example, when (W,S) is of type B_n or D_n one can prove [L_2] that

(12.1) Any left cell in W carries a representation of W which is multiplicity free and has a number of irreducible components equal to a power of 2. Moreover the set of irreducible components can be organized in a natural way as a vector space over the field with 2 elements.

REFERENCES

[BV]. D. Barbasch, D. Vogan: Primitive ideals and orbital integrals in complex classical groups, Math. Ann. 259 (1982), 153-199.

[KL_1]. D. Kazhdan, G. Lusztig: Representations of Coxeter groups and Hecke algebras, Invent. Math. 53 (1979), 165-184.

[KL_2]. ______: Schubert varieties and Poincaré duality, Proc. Symp. Pure Math. vol. 36 (1980), 185-203, Amer. Math. Soc.

[L_1]. G. Lusztig: A class of irreducible representations of a Weyl group, II, Proc. Kon. Nederl. Akad. Series A. vol. 85(2), 1982, 219-226.

[L_2]. ______: Characters of reductive groups over a finite field, to appear.

[LV]. G. Lusztig, D. Vogan: Singularities of closures of K-orbits on flag manifolds, Invent. Math. 71 (1983), 365-379.

[S]. T. A. Springer: Applications of intersection cohomology, Séminaire Bourbaki, Fév. 1982, Paris.

The Selberg Trace Formula IV: Inner Product Formulae (Final Considerations)

by

M. Scott Osborne*

and

Garth Warner*

University of Washington
Seattle, Washington 98195

Contents

*Research of both authors supported in part by the National Science Foundation.

§1. Introduction

This is the fourth in a projected series of papers in which we plan to come to grips with the Selberg trace formula, the ultimate objective being a reasonably explicit expression. Our primary purpose here will be to complete the investigation commenced in [2-(c)]. As there, then, the basic problem is the study of the (L^2) inner product of two truncated Eisenstein series. Of course, thanks to well-known results of Langlands (cf. [2-(b)]), the main case of interest is that of Eisenstein series associated with residual forms. It turns out, however, that the question is deep and multifaceted; moreover, it can be approached and looked at in a variety of ways. Because of this, we have tried to steer a course on middle ground, paying attention to generalities whenever it seems best to do so, but, at the same time, keeping the particulars always in view. Naturally, these goals need not be compatible with one another, thus some compromising on our part has proved to be necessary.

Supposing that the pair (G,Γ) is as usual, the theory spelled out below depends heavily on the truncation procedure and its consequences; see [2-(b)] for the details. Combinatorial ideas are central to this process. Certain refinements and supplements to what has been said along these lines in §2 of [2-(b)] are the theme of the present §2. Additional properties of the truncation operator are discussed in §5. §§3-4 are also preliminary in character; they will be used elsewhere as well. The important notion of "exponent" is the subject of §6, the results of which are at the basis for all that follows. §7 is devoted to an examination of the effect of taking residues on the inner product formula established in §7 of [2-(c)]. The upshot, Proposition 7.3, while less explicit than one might like, carries, nevertheless, a great deal of information. It is applied systematically in §8 in conjunction with the concept of "control". §9 is tangential, albeit instructive. Our main conclusions are arrived at in §10. The most important result is Theorem 10.5, providing, as it does, a very specific grip on the inner product formula for truncated residual Eisenstein series. The significance of such a theorem for the Selberg trace formula hardly requires elaboration; it will be employed by us in another place when we come to the contribution from the continuous spectrum.

Besides the theory of truncation, the paper at hand also relies heavily on certain facts from [2-(c)]. These will be mentioned as the need arises. Furthermore, as a general reference and suggested overall introduction to the subject, we shall use our monograph:

The Theory of Eisenstein Systems, Academic Press, N.Y., 1981.

Throughout the sequel, the title of this work has been abbreviated to TES.

To terminate, a few remarks on the relationship between this work and that of Arthur ([1-(a)], [1-(b)]) seem to be appropriate. Of course, Arthur operates adelically while we operate at infinity. Some aspects of the theory, though, are common to both settings. The definition of Γ-function (modulo a sign change) is taken from Arthur [1-(a)]; on the other hand, our proofs of its properties are, at several key points, very different (and much simpler) than his. Also, the idea of differentiating in the truncation parameter can be found in Arthur [1-(b)]; again, the methods utilized in its exploitation are quite distinct. The treatment of residues infra is due to us, as is the explicit inner product formula (Theorem 8.3) and its consequences (the theorems of control). Finally, Arthur's papers contain no results comparable to ours on the imaginary axis, i.e., explication via the Ю-function and the Я-function.

§2. Γ-Functions

The purpose of this § is to study the properties of a certain class of combinatorial functions. These functions arise naturally in the theory of the truncation operator, hence are best discussed in the setting of [2-(b)], to which we refer for all unexplained notation and additional information (see in particular §2 of that paper). It should be noted that some of our results have been obtained by Arthur [1-(a)], at least within the context of parabolic subgroups alone; the approach below though, is quite different and, of course, the situation is more general.

Let us recall that the investigation centers on the following data:

(1) A finite dimensional inner product space

$$(V,(?,?))$$

of dimension ℓ, say;

(2) A basis $\{\lambda_1,\ldots,\lambda_\ell\}$ of V subject to the condition

$$(\lambda_i,\lambda_j) \leq 0 \qquad (i \neq j).$$

[The corresponding dual basis $\{\lambda^1,\ldots,\lambda^\ell\}$ then has the property that

$$(\lambda^i,\lambda^j) \geq 0 \qquad (i \neq j).]$$

Given any subset F of $L = \{1,\ldots,\ell\}$, introduce, as usual, the subspaces $V(F)$ and V_F, so that $V = V(F) \oplus V_F$. Bear in mind too that associated with F are sets

$$\begin{cases} \{\lambda_1^F,\ldots,\lambda_\ell^F\} \\ \{\lambda_F^1,\ldots,\lambda_F^\ell\}, \end{cases}$$

the first being a basis of V, the second being its dual.

Keeping always to the notation of [2-(b)], let F_1,F_2 be subsets of L with $F_1 \subset F_2$. As there, we agree to write

$$\chi_{F_1,F_2}$$

for the characteristic function of the set

$$\{H\in V : (\lambda_i^{F_1},H) > 0 \quad (i\in F_2 - F_1)\}$$

and, dually,

$$\chi^{F_1,F_2}$$

for the characteristic function of the set

$$\{H\in V : (\lambda_{F_2}^i,H) > 0 \ (i\in F_2 - F_1)\}.$$

In the special case when $F_1 = \emptyset$, we write

$$\begin{cases} \chi_{*,F_2} \\ \chi^{*,F_2} \end{cases};$$

in the special case when $F_2 = \{1,\ldots,\ell\}$, we write

$$\begin{cases} \chi_{F_1,*} \\ \chi^{F_1,*} \end{cases}.$$

Similar notational principles will be used in what follows without comment.

Put

$$\begin{cases} F_1^+(?) = \{i \in F_2 - F_1 : (\lambda_i^{F_1}, ?) > 0\} \\ F_2^-(?) = \{i \in F_2 - F_1 : (\lambda_{F_2}^i, ?) \leq 0\}. \end{cases}$$

Fixing an $H_0 \in V$, form now the following function on V,

$$\Gamma_{F_1}^{F_2}(H,H_0) = \sum_{\{F:F_1 \subset F \subset F_2\}} (-1)^{\#(F-F_1)} \chi_{F_1,F}(H) \cdot \chi^{F,F_2}(H-H_0).$$

Thanks to Lemma 2.6 in [2-(b)], the value of $\Gamma_{F_1}^{F_2}$ at H and H_0 depends only on their respective $V_{F_1} \cap V(F_2)$ - components

$$H(1,2) \quad \text{and} \quad H_0(1,2).$$

[Note: This definition of the Γ-function differs in sign from that of its analogue in Arthur [1-(a)]. The difference is not crucial but matters are somewhat simplified with the present convention.]

Lemma 2.1 *We have*

$$\Gamma_{F_1}^{F_2}(H,H_0) = (-1)^{\#(F_1^+(H))} \cdot \begin{cases} 1 & \text{if } F_1^+(H) = F_2^-(H - H_0) \\ 0 & \text{if } F_1^+(H) \neq F_2^-(H - H_0) \end{cases}.$$

Proof There is no loss of generality in supposing that $F_1 = \emptyset$, $F_2 = L$. This being the case, consider then the sum

$$\sum_{\{F: F \subset L\}} (-1)^{\#(F)} \chi_{*,F}(H) \cdot \chi^{F,*}(H - H_0) .$$

By definition,

$$\begin{cases} \chi_{*,F}(H) \neq 0 & \text{iff} \quad F \subset \emptyset^+(H) \\ \chi^{F,*}(H - H_0) \neq 0 & \text{iff} \quad F \supset L^-(H - H_0). \end{cases}$$

Our sum thus reduces to the sum over all F such that

$$L^-(H - H_0) \subset F \subset \emptyset^+(H)$$

of $(-1)^{\#(F)}$, and so must be zero unless $L^-(H - H\ ) = \emptyset^+(H)$, in which case it is

$$(-1)^{\#(\emptyset^+(H))} ,$$

as desired. //

Lemma 2.2 <u>Suppose that</u> $\Gamma_{F_1}^{F_2}(H, H_0) \neq 0$ -- <u>then</u>

$$\|H(1,2) - H_0(1,2)/2\| \leq \|H_0(1,2)\|/2.$$

Proof Again it can be assumed that $F_1 = \emptyset$, $F_2 = L$. Suppose that

$$F = \begin{cases} \emptyset^+(H) \\ L^-(H - H_0) . \end{cases}$$

Write

$$(H, H_0 - H) = \sum_{i=1}^{\ell} (\lambda_i, H) \cdot (\lambda^i, H_0 - H).$$

Then

$$\begin{cases} i \in F \Rightarrow (\lambda_i, H) \cdot (\lambda^i, H_0 - H) \geq 0 \\ i \notin F \Rightarrow (\lambda_i, H) \cdot (\lambda^i, H_0 - H) \geq 0. \end{cases}$$

Consequently, the nonvanishing of Γ_*^* forces the inequality

$$(H, H_0 - H) \geq 0.$$

But from this,

$$\begin{aligned} \|H - H_0/2\|^2 &= (H - H_0/2,\ H - H_0/2) \\ &= (H,H) - (H,H_0) + \|H_0\|^2/4 \\ &= (H, H - H_0) + \|H_0\|^2/4 \\ &\leq \|H_0\|^2/4, \end{aligned}$$

which is equivalent to our assertion. //

The rationale behind the introduction of the Γ-functions can be found in the following result.

Proposition 2.3 We have

$$\chi^{F_1, F_2}(H - H_0) = \sum_{\{F : F_1 \subset F \subset F_2\}} \chi^{F_1, F}(H) \cdot \Gamma_F^{F_2}(H, H_0).$$

The manipulations necessary for the proof will be facilitated if we consider first an auxillary function, namely

$$\sigma_{F_1}^{F_2, F_3}(H) = \sum_{\{F : F_2 \subset F \subset F_3\}} (-1)^{\#(F - F_2)} \chi_{F_1, F}(H) \cdot \chi^{F, F_3}(H).$$

The point here is this: In the special case when $F_3 = L$, this function is precisely the function figuring in Proposition 2.2 of [2-(b)]. Hence or otherwise

$$\sigma_F^{F,F_3}(H) = \begin{cases} 1 & \text{if } F = F_3 \\ 0 & \text{if } F \neq F_3 \end{cases} .$$

Proof of Proposition 2.3 We shall proceed by induction on $\#(F_2 - F_1)$. If $\#(F_2 - F_1) = 0$, then both sides of the purported equality are $\equiv 1$. In general, the difference

$$\Gamma_{F_1}^{F_2}(H,H_0) - \chi^{F_1,F_2}(H - H_0)$$

is equal to

$$\sum_{\{F:F_1 \not\subset F \subset F_2\}} (-1)^{\#(F-F_1)} \chi_{F_1,F}(H) \cdot \chi^{F,F_2}(H - H_0)$$

or still

$$\sum_{\{F,F':F_1 \not\subset F \subset F' \subset F_2\}} (-1)^{\#(F-F_1)} \chi_{F_1,F}(H) \cdot \chi^{F,F'}(H) \cdot \Gamma_{F'}^{F_2}(H,H_0)$$

or still

$$\sum_{\{F':F_1 \not\subset F' \subset F_2\}} \Gamma_{F'}^{F_2}(H,H_0) \cdot \left[\sum_{\{F:F_1 \not\subset F \subset F'\}} (-1)^{\#(F-F_1)} \chi_{F_1,F}(H) \cdot \chi^{F,F'}(H)\right]$$

or still

$$\sum_{\{F:F_1 \not\subset F \subset F_2\}} \Gamma_F^{F_2}(H,H_0) \cdot \left[\sigma_{F_1}^{F_1,F}(H) - \chi_{F_1,F_1}(H) \cdot \chi^{F_1,F}(H)\right]$$

or still

$$-\sum_{\{F:F_1 \not\subset F \subset F_2\}} \chi^{F_1,F}(H) \cdot \Gamma_F^{F_2}(H,H_0).$$

To finish the proof, we need only multiply through by -1 and then move $\Gamma_{F_1}^{F_2}(H,H_0)$ to the right hand side. //

The property of the Γ-functions embodied in the preceding proposition serves to characterize them. In precise terms:

Lemma 2.4 Suppose given a collection of functions

$$\daleth_{F_1}^{F_2}(H,H_0),$$

subject to

$$H,H_0 \in V, \quad \begin{cases} F_1 \subset L \\ F_2 \subset L \end{cases} \quad (F_1 \subset F_2),$$

with the property that for all values of these parameters

$$\chi^{F_1,F_2}(H-H_0) = \sum_{\{F:F_1 \subset F \subset F_2\}} \chi^{F_1,F}(H) \cdot \daleth_F^{F_2}(H,H_0).$$

Then

$$\daleth_{F_1}^{F_2}(H,H_0) = \Gamma_{F_1}^{F_2}(H,H_0).$$

[This follows by a straightforward induction on $\#(F_2 - F_1)$.]

In general, as can be seen by simple examples, a Γ-function need not be a characteristic function. However, if H_0 is suitably confined, then this will in fact be the case, at least up to a fixed sign.

<u>Proposition 2.5</u> <u>Suppose</u> <u>that</u> $H_0 \in \mathcal{C}^-$ -- <u>then</u>

$$\Gamma_{F_1}^{F_2}(H,H_0) = (-1)^{\#(F_2-F_1)} \cdot \begin{cases} 1 & \text{if } H(1,2) \in \mathcal{C}_{12} \cap (H_0(1,2) - \mathcal{D}_{12})^- \\ 0 & \text{if } H(1,2) \notin \mathcal{C}_{12} \cap (H_0(1,2) - \mathcal{D}_{12})^- . \end{cases}$$

[Note: The condition on H_0 can not be relaxed to $H_0 \in \mathcal{D}^-$.]

We need an elementary lemma.

<u>Lemma 2.6</u> <u>Let</u> F <u>be</u> <u>a</u> <u>subset</u> <u>of</u> F_2; <u>let</u> $i \in F$. <u>Suppose</u> <u>that</u> H <u>is</u> <u>an</u> <u>element</u> <u>of</u> V <u>such</u> <u>that</u>:

(i) $(\lambda_{F_2}^i, H) > 0$;

(ii) $(\lambda_{F_2}^j, H) \leq 0 \qquad \forall j \in F_2 - F$.

<u>Then</u>

$$(\lambda_F^i, H) \geq (\lambda_{F_2}^i, H).$$

<u>Proof</u> Since

$$\begin{aligned}(\lambda_F^i, H) &= (\lambda_{F_2}^i, P(F)H) \\ &= (\lambda_{F_2}^i, H) - (\lambda_{F_2}^i, P_F H),\end{aligned}$$

it will be enough to show that $(\lambda_{F_2}^i, P_F H) \leq 0$. To this end, first observe that

$$P(F_2)P_F H = \sum_{j \in F_2 - F} (\lambda_{F_2}^j, H)\lambda_j^F.$$

Consequently,

$$\begin{aligned}&(\lambda_{F_2}^i, P_F H) \\ &\quad = \sum_{j \in F_2 - F} (\lambda_{F_2}^i, \lambda_j^F) \cdot (\lambda_{F_2}^j, H).\end{aligned}$$

Because $j \in F_2 - F$, we have, by assumption, $(\lambda^j_{F_2}, H) \leq 0$. On the other hand,

$$\lambda^F_j \in \mathfrak{D}^-_{F_2} \Rightarrow (\lambda^i_{F_2}, \lambda^F_j) \geq 0.$$

These relations, when combined, suffice to complete the proof of the lemma. //

The proof of Proposition 2.5 depends in an essential way on the Combinatorial Lemma of Langlands (cf. Proposition 2.5 of [2-(b)]). The reason for this is easy to see. Indeed, apart from the factor $(-1)^{\#(F_2-F_1)}$, the putative expression for $\Gamma^{F_2}_{F_1}(H,H_0)$ is simply

$$\tau_{F_1,F_2}(\emptyset, H-H_0)\chi_{F_1,F_2}(H)$$

which, of course, is quite suggestive.

Proof of Proposition 2.5 Thanks to Lemma 2.4, the issue can be reduced to the equality of

$$\chi^{F_1,F_2}(H-H_0)$$

and of

$$\sum_{\{F:F_1 \subset F \subset F_2\}} \chi^{F_1,F}(H) \cdot (-1)^{\#(F_2-F)} \left[\tau_{F,F_2}(\emptyset, H-H_0)\chi_{F,F_2}(H)\right],$$

that is, of

$$\sum_{\{F:F_1 \subset F \subset F_2\}} (-1)^{\#(F_2-F)} \tau_{F_1,F}(F_2-F_1:H)\chi_{F,F_2}(H) \cdot \left[\tau_{F,F_2}(\emptyset, H-H_0)\right].$$

Put

$$F_0(H - H_0) = \{i \in F_2 - F_1 : (\lambda^i_{F_2}, H - H_0) > 0\}.$$

Then

$$\tau_{F,F_2}(\emptyset, H - H_0) = \begin{cases} 1 & \text{if } F_0(H - H_0) \subset F \subset F_2 \\ 0 & \text{otherwise.} \end{cases}$$

Thus the sum in question becomes

$$\sum_{\{F: F_0(H-H_0) \subset F \subset F_2\}} (-1)^{\#(F_2-F)} \tau_{F_1,F}(F_2 - F_1 : H) \chi_{F,F_1}(H).$$

Now by definition,

$$\tau_{F_1,F}(F_2 - F_1 : H)$$

is equal to the product of

$$\tau_{F_0(H-H_0),F}(F_2 - F_0(H - H_0) : H_0)$$

and

$$\begin{cases} 1 & \text{if } (\lambda_F^i, H) > 0 \qquad \forall i \in F_0(H - H_0) - F_1 \\ 0 & \text{otherwise.} \end{cases}$$

The $0 - 1$ factor in braces is actually $\equiv 1$. In fact,

$$\begin{cases} i \in F_0(H - H_0) - F_1 \Longrightarrow (\lambda_{F_2}^i, H - H_0) > 0 \\ j \in F_2 - F \Longrightarrow (\lambda_{F_2}^j, H - H_0) \le 0. \end{cases}$$

Therefore, in view of Lemma 2.6,

$$(\lambda_F^i, H - H_0) \ge (\lambda_{F_2}^i, H - H_0) > 0.$$

Since $H_0 \in \mathfrak{C}^-$,

$$(\lambda_F^i, H_0) \ge 0 \qquad \forall i \in F_0(H - H_0) - F.$$

So, for any i in $F_0(H - H_0) - F_1$,

$$(\lambda_F^i, H) = (\lambda_F^i, H - H_0) + (\lambda_F^i, H_0) > 0,$$

substantiating our contention. Accordingly, let us consider

$$\sum_{\{F:F_0(H-H_0)\subset F\subset F_2\}}(-1)^{\#(F_2-F)}\tau_{F_0(H-H_0),F}(F_2-F_0(H-H_0):H_0)\chi_{F,F_2}(H).$$

To evaluate this sum, we shall use the Combinatorial Lemma of Langlands. All that need be done is to adjust the signs. But, as

$$\#(F_2-F) = \#(F_2-F_0(H-H_0)) - \#(F-F_0(H-H_0)),$$

this is easy. We get, then

$$(-1)^{\#(F_2-F_0(H-H_0))} \times \begin{cases} 1 & \text{if } F_0(H-H_0) = F_2 \\ 0 & \text{otherwise,} \end{cases}$$

which is just $\chi^{F_1,F_2}(H-H_0)$. //

Here is a corollary. Supposing still that $H_0 \in \mathcal{C}^-$, if

$$F = F_1^+(H) = F_2^-(H-H_0),$$

then $F = F_2 - F_1$ (cf. Lemma 2.1).

Somewhat surprisingly, it is this observation which is at the basis for the proof of a variant on the above theme.

Proposition 2.7 *Suppose that* $H_0 \in -\mathcal{C}^-$ -- *then*

$$\Gamma_{F_1}^{F_2}(H,H_0) = \begin{cases} 1 & \text{if } H(1,2) \in -\mathcal{C}_{12}^- \cap (H_0(1,2) - \mathcal{D}_{12}) \\ 0 & \text{if } H(1,2) \notin -\mathcal{C}_{12}^- \cap (H_0(1,2) - \mathcal{D}_{12}). \end{cases}$$

[Note: The condition on H_0 can not be relaxed to $H_0 \in -\mathcal{D}^-$.]

A direct proof seems awkward. Maintaining the supposition that $H_0 \in -\mathfrak{C}^-$, we shall show instead that if

$$F = F_1^+(H) = F_2^-(H - H_0),$$

then $F = \emptyset$ (cf. Lemma 2.1). For this purpose, fix, as is clearly possible, an element $H^0 \in V$ such that

$$\begin{cases} (\lambda_{F_2}^i, H^0) < 0 & \forall i \in F \\ (\lambda_{F_1}^i, H^0) < 0 & \forall i \notin F. \end{cases}$$

On the basis of what has been said above, we have

$$\begin{cases} (\lambda_i^{F_1}, H) > 0 \quad , \quad (\lambda_{F_2}^i, H - H_0) \leq 0 & \forall i \in F \\ (\lambda_i^{F_1}, H) \leq 0 \quad , \quad (\lambda_{F_2}^i, H - H_0) > 0 & \forall i \notin F. \end{cases}$$

Now replace H by $H_t = H + tH^0$, $t > 0$ being taken small enough so as not to disturb the inequalities that are already strict -- then, of necessity, the remaining inequalities must be strict too. In other words,

$$\begin{cases} (\lambda_i^{F_1}, H_t) > 0 \quad , \quad (\lambda_{F_2}^i, H_t - H_0) < 0 & \forall i \in F \\ (\lambda_i^{F_1}, H_t) < 0 \quad , \quad (\lambda_{F_2}^i, H_t - H_0) > 0 & \forall i \notin F. \end{cases}$$

Finally, working with $-H_0 (\in \mathfrak{C}^-)$ and $-H_t$, we then infer that

$$\begin{aligned} F_1^+(-H_t) &= (F_2 - F_1) - F \\ &= F_2^-(-H_t - (-H_0)) = F_2 - F_1, \end{aligned}$$

i.e., $F = \emptyset$.

At this juncture, we shall shift the focus of our study from the geometrical to the analytical, the main objective being a consideration of the "Fourier transform"

$$\int_{V(F_2)/V(F_1)} \Gamma_{F_1}^{F_2}(H,H_0)e^{(?,H)}dH$$

of $\Gamma_{F_1}^{F_2}(?,H_0)$. Because

$$V(F_2)/V(F_1) \approx V_{F_1} \cap V(F_2),$$

it will not be restrictive to assume from the outset that $F_1 = \emptyset$, $F_2 = L$.

Consider, therefore

$$\Gamma_*^*(H,H_0) = \sum_{\{F:F\subset L\}} (-1)^{\#(F)} \chi_{*,F}(H)\cdot\chi^{F,*}(H-H_0).$$

Thanks to Lemma 2.2, $\Gamma_*^*(?,H_0)$ has compact support. But, unfortunately, the individual summands defining $\Gamma_*^*(?,H_0)$ do not. Accordingly, the evaluation of

$$\int_V \Gamma_*^*(H,H_0)e^{(\Lambda,H)}dH \qquad (\Lambda\in V+\sqrt{-1}\,V)$$

can not be done by simply integrating term by term without taking certain precautions.

Lemma 2.8 Fix an element $H_{00}\in -\mathfrak{C}^-$. Let $H_0\in H_{00}+\mathfrak{C}^-$ -- then, $\forall F$,

$$\chi_{*,F}(H)\cdot\chi^{F,*}(H-H_0) \neq 0 \Rightarrow H\in H_{00}+\mathfrak{D}.$$

Proof Fix an F. By definition,

$$\begin{cases} \chi_{*,F}(H) \neq 0 \iff \forall i \in F,\ (\lambda_i, H) > 0 \\ \chi^{F,*}(H - H_0) \neq 0 \iff \forall i \notin F,\ (\lambda_i, H - H_0) > 0. \end{cases}$$

In addition, the set

$$\{\lambda_i : i \in F\} \cup \{\lambda^i : i \notin F\}$$

is a basis of V, its dual being the set

$$\{\lambda_F^i : i \in F\} \cup \{\lambda_i^F : i \notin F\}.$$

This said, write

$$\begin{cases} H = \sum_{i \in F} c^i \lambda_F^i + \sum_{i \notin F} c_i \lambda_i^F \\ H_0 = \sum_{i \in F} c_0^i \lambda_F^i + \sum_{i \notin F} c_i^0 \lambda_i^F \\ H_{00} = \sum_{i \in F} c_{00}^i \lambda_F^i + \sum_{i \in F} c_i^{00} \lambda_i^F \quad . \end{cases}$$

Because $H_{00} \in -\mathcal{C}^-$, the c_{00}^i, c_i^{00} are all ≤ 0. On the other hand, as $H_0 - H_{00} \in \mathcal{C}^-$, the $c_0^i - c_{00}^i, c_i^0 - c_i^{00}$ are all ≥ 0. Finally, in view of our assumption that

$$\chi_{*,F}(H) \cdot \chi^{F,*}(H - H_0) \neq 0,$$

we have

$$\begin{cases} i \in F \implies c^i > 0 \geq c_{00}^i \\ i \notin F \implies c_i > c_i^0 \geq c_i^{00}. \end{cases}$$

Consequently, $\forall i$

$$\begin{cases} c^i - c_{00}^i > 0 \\ c_i - c_i^{00} > 0. \end{cases}$$

Thus $H - H_{00}$ is strictly positive on $\mathcal{C}^- - \{0\}$ and so $H - H_{00}$ does indeed belong to $\mathcal{D}$. //

It is easy to turn the preceding result around to produce the H_{00}, given the H_0. In fact, even more is true:

Lemma 2.9 Fix a compact subset $K_0 \subset V$ -- then $\exists H_{00} \in -\mathcal{C}^-$ such that

$$\forall H_0 \in K_0, \forall F:$$

$$\chi_{*,F}(H) \cdot \chi^{F,*}(H - H_0) \neq 0 \Rightarrow H \in H_{00} + \mathcal{D}.$$

Proof For $n = 1, 2, \ldots,$ put

$$H_n = -n\left(\sum_{i=1}^{\ell} \lambda^i\right).$$

Then

$$H_n + \mathcal{C} \subset H_{n+1} + \mathcal{C} \qquad \forall n.$$

Furthermore, $\exists N$ such that

$$n \geq N \Rightarrow K_0 \subset H_n + \mathcal{C}.$$

To complete the proof, therefore, we need only take $H_{00} = H_N$. //

Suppose now that H_0 is confined to a fixed compact subset $K_0 \subset V$ -- then Lemma 2.2 implies that $\Gamma_*^*(?, H_0)$ is compactly supported, the various supports $C(H_0)$ (say) all being contained in yet another compact set (depending on K_0).

Fix an $H_{00} \in -\mathcal{C}^-$ per Lemma 2.9. Let $\phi: V \to \underset{\sim}{C}$ be a measurable function which is integrable on $H_{00} + \mathcal{D}$ -- then

$$\int_V \Gamma_*^*(H,H_0)\phi(H)\,dH$$

is equal to

$$\sum_{\{F:F\subset L\}}(-1)^{\#(F)}\cdot\int_V \chi_{*,F}(H)\cdot\chi^{F,*}(H-H_0)\cdot\phi(H)\,dH.$$

Each of the integrals appearing here is supported on a subset of $H_{00}+\mathfrak{D}$. As it will suffice to study them separately, fix an F.

Write

$$\begin{cases} H = \sum_{i\in F} t^i\lambda_F^i + \sum_{i\notin F} t_i\lambda_i^F \\ H_0 = \sum_{i\in F} c_0^i\lambda_F^i + \sum_{i\notin F} c_i^0\lambda_i^F \end{cases}.$$

Then

$$\chi_{*,F}(H)\cdot\chi^{F,*}(H-H_0)\neq 0$$

iff

$$\begin{cases} i\in F \Rightarrow t^i>0 \\ i\notin F \Rightarrow t_i>c_i^0. \end{cases}$$

Thus, in terms of these parameters,

$$\int_V \chi_{*,F}(H)\cdot\chi^{F,*}(H-H_0)\cdot\phi(H)\,dH$$

becomes

$$C_F\cdot\underbrace{\int_0^\infty\cdots\int_0^\infty}_{i\in F}\ \underbrace{\int_{c_i^0}^\infty\cdots\int_{c_i^0}^\infty}_{i\notin F}[?]\,dt_F\,dt^F\ ,$$

where

$$[?] = \phi\Big(\sum_{i\in F} t^i\lambda_F^i + \sum_{i\notin F} t_i\lambda_i^F\Big),$$

and C_F is a certain positive constant which arises from the change in measure.

<u>Lemma 2.10</u> <u>We have</u>

$$C_F = |\det[(\lambda_i^F,\lambda_j^F):i,j\notin F]|^{1/2}/|\det[(\lambda_i,\lambda_j):i,j\in F]|^{1/2}.$$

[Let $\{e_1,\ldots,e_\ell\}$ be an orthonormal basis of V. Call T_F the linear transformation on V defined by the rule

$$H \mapsto \sum_{i\in F} (e_i,H)\lambda_F^i + \sum_{i\notin F} (e_i,H)\lambda_i^F.$$

Then it is clear that

$$C_F = |\det(T_F)|.$$

The matrix of T_F with respect to the e_i may be explicated by letting

$$\lambda(i:F) = \begin{cases} \lambda_F^i & \text{if } i\in F \\ \lambda_i^F & \text{if } i\notin F, \end{cases}$$

since, in this notation

$$[T_F]_{ij} = (e_i,\lambda(j:F)).$$

Bearing in mind that reversing the entries amounts to passing to the transpose of T_F, it follows that

$$C_F = |\det[(\lambda(i:F),\ \lambda(j:F))]|^{1/2}$$

which, when unraveled by the definitions, is the claimed value.]

It is not difficult to interpret the preceding expression for

C_F. Indeed, write $L(F)$ for the lattice in $V(F)$ generated by the $\lambda_i (i\in F)$ and write L_F for the lattice in V_F generated by the $\lambda_i^F (i\notin F)$ -- then, in these notations,

$$C_F = \mathrm{vol}(L_F)/\mathrm{vol}(L(F)).$$

Specialize ϕ to a function of the form

$$\phi(H) = e^{(\Lambda,H)} \qquad (H\in V),$$

where $\mathrm{Re}(\Lambda)\in -\mathcal{C}$ -- then obviously

$$\int_{H_{00}+\mathfrak{D}} |e^{(\Lambda,H)}|dH < +\infty.$$

So, in view of what has been said earlier,

$$\int_V \Gamma_*^*(H,H_0)e^{(\Lambda,H)}dH$$

can be written as the sum over all $F\subset L$ of $C_F\cdot(-1)^{\#(F)}$ times

$$\left[\prod_{i\in F}\int_0^\infty e^{t(\Lambda,\lambda_F^i)}dt\right]\cdot\left[\prod_{i\notin F}\int_{c_i^0}^\infty e^{t(\Lambda,\lambda_i^F)}dt\right],$$

that is, of $C_F\cdot(-1)^{\#(F)}$ times

$$\exp((\Lambda,\sum_{i\notin F}(\lambda^i,H_0)\lambda_i^F))$$

$$\times\ (1/\prod_{i\in F}(\Lambda,\lambda_F^i)\times\prod_{i\notin F}(\Lambda,\lambda_i^F)),$$

that is, of $C_F\cdot(-1)^{\#(F)}$ times

$$\frac{\exp((P_F\Lambda,H_0))}{\prod_{i\in F}(\Lambda,\lambda_F^i)\times\prod_{i\notin F}(\Lambda,\lambda_i^F)}.$$

Because

$$\int_V \Gamma_*^*(H,H_0)e^{(\Lambda,H)}dH$$

is an entire function of Λ,

$$\sum_{\{F:F\subset L\}} C_F\cdot(-1)^{\#(F)}\cdot\left[\frac{\exp((P_F\Lambda,H_0))}{\prod_{i\in F}(\Lambda,\lambda_F^i)\times\prod_{i\notin F}(\Lambda,\lambda_i^F)}\right]\quad(\mathrm{Re}(\Lambda)\in -\mathcal{C})$$

must extend to an entire function of Λ.

Proposition 2.11 <u>As a function of</u> H_0,

$$\int_V \Gamma_*^*(H,H_0)\,dH$$

<u>is a homogeneous polynomial of degree</u> ℓ.

Proof Take Λ in $-\mathcal{C}$ -- then, by the above, there exist constants $C_F(\Lambda)$ and $c_F(\Lambda:H_0)$ such that

$$\int_V \Gamma_*^*(H,H_0)e^{t(\Lambda,H)}dH$$
$$= t^{-\ell}\cdot\sum_{\{F:F\subset L\}} C_F(\Lambda)e^{tc_F(\Lambda:H_0)}$$

for all $t > 0$. Since the limit as $t\downarrow 0$ must exist and be independent of Λ,

$$\sum_{\{F:F\subset L\}} C_F(\Lambda)e^{tc_F(\Lambda:H_0)}$$

necessarily approaches zero faster than t^ℓ, thus its MacLaurin polynomial of order ℓ vanishes identically and so by L'Hôspital's rule,

$$\int_V \Gamma_*^*(H,H_0)\,dH = \frac{1}{\ell!}\cdot\sum_{\{F:F\subset L\}} C_F(\Lambda)(c_F(\Lambda:H_0))^\ell\ .$$

But the right hand side is evidently a homogeneous polynomial of degree ℓ in H_0. //

Let ϕ be a function defined on a subset $\mathrm{dom}(\phi)$ of $V + \sqrt{-1}\,V$ such that

$$\forall F \subset L, \quad P_F(\mathrm{dom}(\phi)) \subset \mathrm{dom}(\phi).$$

Formally, the A-transform A_ϕ of ϕ is given by

$$A_\phi(\Lambda) = \sum_{\{F:F\subset L\}} C_F\cdot(-1)^{\#F}\cdot\left[\frac{\phi(P_F\Lambda)}{\prod_{i\in F}(\Lambda,\lambda_F^i)\times\prod_{i\notin F}(\Lambda,\lambda_i^F)}\right].$$

[Note: Here it is understood that $\Lambda \in \mathrm{dom}(\phi)$ and, in addition, is ultraregular in the sense that Λ lies on none of the hyperplanes determined by the λ_F^i, λ_i^F.]

Proposition 2.12 <u>If</u> ϕ <u>is a</u> C^∞ <u>function on</u> $\sqrt{-1}\,V$, <u>then</u> A_ϕ <u>extends to a</u> C^∞ <u>function on</u> $\sqrt{-1}\,V$.

What does this result have to do with Γ-functions? The connection will emerge in the course of the proof of:

Lemma 2.13 <u>Let</u> $\phi \in \mathcal{C}(\sqrt{-1}\,V)$ -- <u>then</u> A_ϕ <u>extends to a</u> C^∞ <u>function on</u> $\sqrt{-1}\,V$.

Granted the lemma, the proposition readily follows. For in general,

$$\begin{array}{c}\phi(\Lambda) = \psi(\Lambda)\\ \Longrightarrow \\ A_\phi(\Lambda) = A_\psi(\Lambda).\end{array} \qquad (\|\Lambda\| < r)$$

If now ϕ be a C^∞ function on the ball $\sqrt{-1}B_r$ of radius r in $\sqrt{-1}V$, then, on any smaller such ball, ϕ agrees with an element of $\mathcal{C}(\sqrt{-1}V)$. Consequently, A_ϕ extends to a C^∞ function on $\sqrt{-1}B_r$.

To address Lemma 2.13, we shall need some preparation.

Write $A(V)$ for the set of all measurable functions $f:V \to \underset{\sim}{C}$ such that $\forall n \, \exists \, C_n$:

$$|f(H)| \le C_n \cdot (1 + \|H\|)^{-n} \qquad (H \in V).$$

Obviously,

$$\mathcal{C}(V) \subset A(V) \subset L^1(V).$$

Moreover, it is easy to check that if $f \in A(V)$, then

$$\forall n \, \exists \, M_n:$$

$$\int_{V-B_r} |f(H)| dH \le M_n \cdot (1 + r)^{-n} \qquad (\forall r > 0).$$

These facts presented, define the Γ-transform on $A(V)$ by the prescription

$$\Gamma_f(H) = \int_V \Gamma_*^*(H,H_0) f(H_0) dH_0 \qquad (f \in A(V)).$$

It is clear that $\forall f \in A(V)$, Γ_f exists as a function on V. What is less clear but still true is that $\Gamma(A(V)) \subset A(V)$. Indeed, if $\Gamma_*^*(H,H_0) \neq 0$, then $|\Gamma_*^*(H,H_0)| = 1$ and $\|H\| \le \|H_0\|$ (cf. Lemma 2.2), hence $\forall n$,

$$|\Gamma_f(H)| \leq \int_{V-B_{\|H\|}} |f(H_0)|dH_0 \leq M_n\cdot(1+\|H\|)^{-n},$$

i.e.,

$$f\in A(V) \Longrightarrow \Gamma_f\in A(V).$$

There is a canonical map

$$\wedge : A(V) \to C^\infty(\sqrt{-1}V),$$

namely the rule

$$\hat{f}(\Lambda) = \int_V f(H)e^{(\Lambda,H)}dH \qquad (\Lambda\in\sqrt{-1}V).$$

Of course,

$$\widehat{\mathcal{C}(V)} = \mathcal{C}(\sqrt{-1}V).$$

Furthermore,

$$f\in A(V) \Longrightarrow \hat{\Gamma}_f\in C^\infty(\sqrt{-1}V).$$

Proof of Lemma 2.13 No loss of generality is entailed if we suppose that $\phi = \hat{f}$, where $f\in\mathcal{C}(V)$. It will then be shown that

$$A_{\hat{f}} = \hat{\Gamma}_f$$

on the ultraregular points in $\sqrt{-1}V$ meaning, therefore, that the right hand side provides the desired extension of the left hand side to all of $\sqrt{-1}V$. Fix, accordingly, an ultraregular point Λ in $\sqrt{-1}V$ -- then we have

$$A_{\hat{f}}(\Lambda) = \sum_{\{F:F\subset L\}} c_F\cdot(-1)^{\#(F)}\cdot\left[\frac{\hat{f}(P_F\Lambda)}{\prod_{i\in F}(\Lambda,\lambda_F^i)\times\prod_{i\notin F}(\Lambda,\lambda_i^F)}\right]$$

$$= \int_V f(H_0)\cdot \sum_{\{F:F\subset L\}} c_F\cdot(-1)^{\#(F)}\cdot \left[\frac{\exp((P_F\Lambda,H_0))}{\prod_{i\in F}(\Lambda,\lambda_F^i)\times\prod_{i\notin F}(\Lambda,\lambda_i^F)}\right] dH_0$$

$$= \int_V f(H_0)\cdot \left\{\int_V \Gamma_*^*(H,H_0)e^{(\Lambda,H)}dH\right\}dH_0$$

$$= \int_V \left\{\int_V \Gamma_*^*(H,H_0)f(H_0)dH_0\right\}\cdot e^{(\Lambda,H)}dH$$

$$= \int_V \Gamma_f(H)e^{(\Lambda,H)}dH = \hat{\Gamma}_f(\Lambda),$$

the interchange in the order of integration being justified by Fubini's theorem. //

Proposition 2.14 *Suppose that ϕ is a holomorphic function on $B_{r'}\times\sqrt{-1}\,B_{r''}$ -- then A_ϕ is a holomorphic function on* $B_{r'}\times\sqrt{-1}\,B_{r''}$.

[The point is that A_ϕ is, a priori, meromorphic on $B_{r'}\times\sqrt{-1}\,B_{r''}$ and, at the same time, C^∞, thus locally bounded on $\sqrt{-1}\,B_{r''}$.]

§3. Geometry of Generalized Root Systems

The purpose of this § is to set up via a series of axioms a geometric framework of sufficient generality to accomodate the applications which will arise in the sequel. In brief, it is a question here of abstracting the familiar notion of "root system", stripping this concept to the bare essentials so as to be of maximum utility. We remark that "Weyl groups" play no role at all in these considerations.

Fix a finite dimensional inner product space

$$(V,(?,?))$$

of dimension ℓ, say. Let Φ be a finite subset of V subject to:

Axiom I $\lambda\in\Phi \Rightarrow \lambda \neq 0$;

Axiom II $\lambda\in\Phi \Rightarrow -\lambda\in\Phi$;

Axiom III $\mathrm{span}(\Phi) = V$.

We shall add other axioms later on, but there is no point in stating them now since they are not really needed initially.

Fix Φ per supra. Given $\lambda\in\Phi$, put

$$V_\lambda = \{H\in V: (\lambda,H) = 0\}.$$

Then, by a chamber $\mathfrak{C}$ of Φ, we understand any component of the

Φ-regular set

$$V_r = V - \bigcup_{\lambda \in \Phi} V_\lambda .$$

Evidently, chambers are open sets. For $\mathcal{C}$ a chamber of Φ, let

$$\begin{cases} \Phi_{\mathcal{C}}^{+} = \{\lambda \in \Phi : \forall H \in \mathcal{C}, \ (\lambda, H) > 0\} \\ \Phi_{\mathcal{C}}^{-} = \{\lambda \in \Phi : \forall H \in \mathcal{C}, \ (\lambda, H) < 0\}. \end{cases}$$

Obviously,

$$\Phi_{\mathcal{C}}^{+} \cap \Phi_{\mathcal{C}}^{-} = \emptyset, \ \Phi_{\mathcal{C}}^{+} = -\Phi_{\mathcal{C}}^{-} .$$

In addition, due to the connectedness of $\mathcal{C}$ and the fact that $\mathcal{C}$ meets no V_λ, we have

$$\Phi = \Phi_{\mathcal{C}}^{+} \cup \Phi_{\mathcal{C}}^{-} .$$

Lemma 3.1   Let $\mathcal{C}$ be a chamber of Φ -- then

$$\mathcal{C} = \{H \in V : (\lambda, H) > 0 \ \forall \lambda \in \Phi_{\mathcal{C}}^{+}\} .$$

[The left hand side is contained in the right hand side which, in turn, is a convex open subset of V_r, hence is contained in a unique component, namely $\mathcal{C}$.]

It is suggestive and convenient to refer to the elements of Φ as roots even though, of course, Φ itself may not be a root system. Given $\mathcal{C}$, the sets $\Phi_{\mathcal{C}}^{+}$ and $\Phi_{\mathcal{C}}^{-}$ are then the sets of positive and negative roots associated with $\mathcal{C}$ which, in view of Lemma 3.1, is consistent with the usual terminology. We shall also need an analogue of "simple root". Deferring this for the time being, as a start let us agree that a wall root of $\mathcal{C}$ is any $\lambda \in \Phi_{\mathcal{C}}^{+}$ for which $\mathcal{C}^{-} \cap V_{\lambda}$ has a nonempty interior in V_{λ}.

Maintaining the notation from §2 of [2-(b)], for S a subset of V denote by $\text{Pos}(S)$ the interior of

$$\{H \in V: (\sigma, H) > 0 \quad \forall \sigma \in S\}.$$

Observe that if S is compact, in particular finite, then necessarily

$$\text{Pos}(S) = \{H \in V: (\sigma, H) > 0 \quad \forall \sigma \in S\},$$

the set on the right being open in this case.

So, e.g., $\text{Pos}(\Phi_{\mathcal{C}}^{+}) = \mathcal{C}$. In general, a subset F of $\Phi_{\mathcal{C}}^{+}$ is said to generate $\mathcal{C}$ if $\text{Pos}(F) = \mathcal{C}$.

Lemma 3.2 *Let F be a subset of $\Phi_{\mathcal{C}}^{+}$; let λ be an element of $\Phi_{\mathcal{C}}^{+}$ such that*

$$\{r\lambda : r > 0\} \cap F = \emptyset.$$

Suppose that $F \cup \{\lambda\}$ generates $\mathcal{C}$ -- then F generates $\mathcal{C}$ iff λ is not a wall root of $\mathcal{C}$.

Proof $\Leftarrow$: Passing to the denial, let us assume that F does not generate $\mathcal{C}$ -- then we must prove that λ is a wall root of $\mathcal{C}$. By hypothesis, therefore,

$$\mathfrak{C} = \operatorname{Pos}(F \cup \{\lambda\}) \subset \operatorname{Pos}(F).$$

Since the first relation reads

$$\mathfrak{C} = \{H \in \operatorname{Pos}(F) : (\lambda, H) > 0\},$$

there exists an $H_0 \in \operatorname{Pos}(F)$ such that $(\lambda, H_0) < 0$, $\operatorname{Pos}(F)$ being open. Fix an open ball B_0 in $\mathfrak{C}$ -- then the intersection of the convex hull of $B_0 \cup \{H_0\}$ with V_λ is an open subset of V_λ so, to conclude that λ is a wall root of $\mathfrak{C}$, we have only to show that $\mathfrak{C}^- \cap V_\lambda$ contains it. For this purpose, take $H \in B_0$ and $t \in]0,1[$ with the property that

$$(\lambda, tH + (1-t)H_0) > 0.$$

Then we claim that

$$tH + (1-t)H_0 \in \mathfrak{C}$$

which, needless to say, leads immediately to our contention. However,

$$\left\{\begin{array}{l} H_0 \in \operatorname{Pos}(F) \\ H \in \mathfrak{C} \Longrightarrow H \in \operatorname{Pos}(F) \end{array}\right. \Longrightarrow tH + (1-t)H_0 \in \operatorname{Pos}(F).$$

Consequently,

$$tH + (1-t)H_0 \in \mathfrak{C},$$

as claimed.

$\Longrightarrow$: Let us assume that F generates $\mathfrak{C}$ but, to get a contradiction, that λ is a wall root of $\mathfrak{C}$. Because

$$\{r\lambda : r > 0\} \cap F = \emptyset \Rightarrow \underset{\sim}{R}\lambda \cap F = \emptyset,$$

the orthogonal projection $P_\lambda F$ of F onto V_λ consists of non-zero vectors. This said, fix, as is clearly possible, a point $H_0 \in V_\lambda$ and a $\delta_0 > 0$ such that

$$H \in V_\lambda \text{ and } \|H - H_0\| < \delta_0 \Rightarrow H \in \mathfrak{C}^-.$$

Take an arbitrary element $\lambda_0 \in F$ -- then

$$H \in V_\lambda \text{ and } \|H - H_0\| < \delta_0 \Rightarrow (\lambda_0, H) \geq 0.$$

When viewed as a map from V_λ to $\underset{\sim}{R}$, $P_\lambda \lambda_0$ is open. Therefore, it is actually true that

$$H \in V_\lambda \text{ and } \|H - H_0\| < \delta_0 \Rightarrow (\lambda_0, H) > 0.$$

As λ_0 can be any element of F, we have

$$H_0 \in \mathrm{Pos}(F) = \mathfrak{C} \Rightarrow (\lambda, H_0) > 0,$$

a contradiction.

Hence the lemma. //

When does a subset F of $\Phi_{\mathfrak{C}}^+$ generate $\mathfrak{C}$? Here is a simple criterion.

Proposition 3.3 *Let F be a subset of $\Phi_{\mathfrak{C}}^+$ -- then F generates $\mathfrak{C}$ iff for every wall root λ of $\mathfrak{C}$ there exists an $r > 0$ such that $r\lambda \in F$.*

Proof Suppose that F generates $\mathcal{C}$ -- then

$$\forall \lambda \in \Phi_{\mathcal{C}}^{+}, \quad \mathrm{Pos}(F \cup \{\lambda\}) = \mathcal{C}.$$

If

$$\{r\lambda : r > 0\} \cap F = \emptyset,$$

then, thanks to the preceding lemma, λ can not be a wall root of $\mathcal{C}$. Turning to the converse, begin with $\Phi_{\mathcal{C}}^{+}$ -- then, appealing once again to the lemma supra, we can say that the result F_0 of deleting one-by-one from $\Phi_{\mathcal{C}}^{+}$ those positive roots which are not wall roots of $\mathcal{C}$ must generate $\mathcal{C}$. Let now F be a subset of $\Phi_{\mathcal{C}}^{+}$ with the stated property -- then

$$\mathcal{C} = \mathrm{Pos}(F_0) \supset \mathrm{Pos}(F) \supset \mathcal{C},$$

i.e., $\mathrm{Pos}(F) = \mathcal{C}$, as desired. //

By definition, a simple root of $\mathcal{C}$ will be any wall root of $\mathcal{C}$ for which

$$r\lambda \in \Phi_{\mathcal{C}}^{+} \Rightarrow r \geq 1.$$

If $F_0(\mathcal{C})$ is the set of all simple roots of $\mathcal{C}$, then $F_0(\mathcal{C})$ generates $\mathcal{C}$ and, in fact, is a minimal generating set.

[Note: More picturesquely, the elements of $F_0(\mathcal{C})$ can be regarded as the "short" wall roots of $\mathcal{C}$; cf. infra.]

Proposition 3.4 $F_0(\mathcal{C})$ spans V.

This follows from:

Lemma 3.5 Suppose that $F \subset \Phi_{\mathcal{C}}^+$ generates $\mathcal{C}$ -- then $\text{span}(F) = V$.

Proof If $\text{span}(F) \neq V$, then there exists a nonzero $H_0 \in F^{\perp}$. Let $H \in \mathcal{C}$ -- then

$$H + tH_0 \in \text{Pos}(F) \quad \forall t \in \mathbf{R}.$$

Since Φ spans V, the same is true of $\Phi_{\mathcal{C}}^+$, so $(\lambda_0, H_0) \neq 0$ for some $\lambda_0 \in \Phi_{\mathcal{C}}^+$. Choose, as would be possible, a $t_0 \in \mathbf{R}$ such that

$$(\lambda_0, H + t_0H_0) \leq 0.$$

Then

$$H_0 + t_0H_0 \notin \mathcal{C} \Rightarrow \mathcal{C} \neq \text{Pos}(F),$$

contratrary to hypothesis. //

To make further progress, it will be necessary to add another axiom.

Axiom IV $\forall$ chamber $\mathcal{C}$ of Φ, $F_0(\mathcal{C})$ is linearly independent.

[Note: This axiom is not a consequence of the other three. For instance, there exist examples in $V = \mathbf{R}^3$ where $F_0(\mathcal{C})$ has arbitrarily large size.]

Let $\lambda \in \Phi$ -- then λ is said to be a short root of Φ if

$$r\lambda \in \Phi \Longrightarrow |r| \geq 1.$$

Write Φ_s for the set of such. Every line $\underset{\sim}{R}\lambda$ intersects Φ_s in exactly two points.

Take a $\lambda \in \Phi_s$ and consider the pair $(V_\lambda, \Phi_\lambda)$, where, if $P_\lambda : V \to V_\lambda$ is the orthogonal projection of V onto V_λ,

$$\Phi_\lambda = P_\lambda \Phi - \{0\}.$$

Do Axioms I - IV remain in force when (V, Φ) is replaced by $(V_\lambda, \Phi_\lambda)$? The answer is affirmative. Of course, the fact that Axioms I - III are inherited is virtually automatic, but, perhaps, this is not quite so clear of Axiom IV. In reality, as will be seen below, a good deal more can be said than might first be expected.

<u>Lemma 3.6</u> <u>Suppose that</u> $\lambda \in \Phi_s$ -- <u>then</u>

$$V_\lambda = \bigcup_{\{\mathcal{C} : \lambda \in \Phi_{\mathcal{C}}^+\}} \mathcal{C}^- \cap V_\lambda.$$

[Simply remark that for each H_0 in V_λ, one can certainly find a chamber $\mathcal{C}$ of Φ such that

$$H_0 \in \mathcal{C}^- \quad \text{and} \quad \forall H \in \mathcal{C}, \quad (\lambda, H) > 0.]$$

Given a λ in Φ_s, put

$$\mathfrak{C}_0(\lambda) = \{\mathcal{C} : \lambda \in F_0(\mathcal{C})\}.$$

Lemma 3.7 Suppose that $\lambda\in\Phi_s$. Let $\mathcal{C}\in\mathfrak{C}_0(\lambda)$ -- then

$$\mathcal{C}^- \cap V_\lambda = \{H\in V_\lambda : (\lambda_0,H) \geq 0 \quad \forall\lambda_0\in P_\lambda(F_0(\mathcal{C}))\}.$$

[On the grounds of the definitions, this is immediate.]

Given a λ in Φ_s and a $\mathcal{C}$ in $\mathfrak{C}_0(\lambda)$, write $\mathcal{C}_\lambda$ for the interior of $\mathcal{C}^-\cap V_\lambda$ (in V_λ).

Observe that:

(1) $\mathcal{C}_\lambda$ is a chamber of Φ_λ.

[The set

$$\{H\in V_\lambda : (\lambda_0,H) > 0 \quad \forall\lambda_0 \in P_\lambda(\Phi_{\mathcal{C}}^+ - \underset{\sim}{R}\lambda)\}$$

describes $\mathcal{C}_\lambda$ (cf. Lemma 3.7). Moreover, it is a convex open subset of

$$(V_\lambda)_r \ (= \Phi_\lambda \text{ - regular set})$$

and is contained in no larger connected subset thereof.]

(2) $P_\lambda(F_0(\mathcal{C}) - \{\lambda\})$, when shortened, is the set $F_0(\mathcal{C}_\lambda)$ of short wall roots of $\mathcal{C}_\lambda$.

[The set

$$\{H\in V_\lambda : (\lambda_0,H) > 0 \quad \forall\lambda_0 \in P_\lambda(F_0(\mathcal{C}) - \{\lambda\})\}$$

describes $\mathcal{C}_\lambda$ (cf. Lemma 3.7). Moreover,

$$\#(P_\lambda(F_0(\mathcal{C}) - \{\lambda\})) = \ell - 1,$$

so shortening merely serves to ensure that lengths are minimal (per Φ_λ!).]

We come now to the main result in this circle of ideas.

Proposition 3.8 Fix $\lambda \in \Phi_s$ -- then the map $\mathcal{C} \to \mathcal{C}_\lambda$ from $\mathfrak{C}_0(\lambda)$ to the chambers of Φ_λ is bijective.

Proof To establish injectivity, it suffices to show that for any H in any $\mathcal{C}_\lambda$, there is a small but positive t_H such that $H + t_H\lambda \in \mathcal{C}$. However,

$$\begin{cases} 0 < t << 1 \Rightarrow (\lambda_0, H + t\lambda) > 0 \quad \forall \lambda_0 \in F_0(\mathcal{C}) - \{\lambda\} \\ 0 < t \Rightarrow (\lambda, H + t\lambda) = t\|\lambda\|^2 > 0, \end{cases}$$

so t_H does in fact exist. As for the surjectivity, we need only prove that

$$V_\lambda = \bigcup_{\{\mathcal{C}\,:\,\mathcal{C} \in \mathfrak{C}_0(\lambda)\}} \mathcal{C}_\lambda^-,$$

i.e., that

$$V_\lambda = \bigcup_{\{\mathcal{C}\,:\,\mathcal{C} \in \mathfrak{C}_0(\lambda)\}} \mathcal{C}^- \cap V_\lambda.$$

Consider the difference

$$V_\lambda - \bigcup_{\{\mathcal{C}\,:\,\mathcal{C} \in \mathfrak{C}_0(\lambda)\}} \mathcal{C}^- \cap V_\lambda.$$

It is obviously open; at the same time, in view of Lemma 3.6, it is a finite union of nowhere dense sets, thus is nowhere dense, hence is empty. //

If, in the preceding discussion, λ is replaced by $-\lambda$, then still $V_\lambda = V_{-\lambda}$, $\Phi_\lambda = \Phi_{-\lambda}$, and all goes through as before. It is this point which is at the basis of the notion of "adjacency".

Two chambers $\mathcal{C}'$ and $\mathcal{C}''$ of Φ will be termed adjacent if there exists a $\lambda \in \Phi_s$ such that

$$\begin{cases} \mathcal{C}' \in \mathbf{C}_0(\lambda) \\ \mathcal{C}'' \in \mathbf{C}_0(-\lambda) \end{cases} \quad \text{and} \quad \mathcal{C}'_\lambda = \mathcal{C}''_{-\lambda}.$$

Under these conditions, the common closure of $\mathcal{C}'_\lambda = \mathcal{C}''_{-\lambda}$ is the intersection of the closures of $\mathcal{C}'$ and $\mathcal{C}''$. We then refer to $\mathcal{C}'_\lambda = \mathcal{C}''_{-\lambda}$ as the common wall of $\mathcal{C}'$ and $\mathcal{C}''$, λ (or $-\lambda$) being the wall root determining the wall.

Lemma 3.9 <u>Suppose that $\mathcal{C}'$ and $\mathcal{C}''$ are adjacent. Let</u> $H_0 \in \mathcal{C}'_\lambda = \mathcal{C}''_{-\lambda}$ -- <u>then there exists a</u> $\delta_0 > 0$ <u>such that</u>

$$H \notin V_\lambda \ \underline{\text{and}} \ \|H\| < \delta_0 \Rightarrow H + H_0 \in \begin{cases} \mathcal{C}' & \text{if } (\lambda, H) > 0 \\ \mathcal{C}'' & \text{if } (\lambda, H) < 0. \end{cases}$$

[Consider any $\delta_0 > 0$ for which

$$H \notin V_\lambda \ \text{and} \ \|H\| < \delta_0 \Rightarrow \begin{cases} (\lambda_0, H + H_0) > 0 & \forall \lambda_0 \in F_0(\mathcal{C}') - \{\lambda\} \\ (\lambda_0, H + H_0) > 0 & \forall \lambda_0 \in F_0(\mathcal{C}'') - \{-\lambda\}. \end{cases}$$

Because

$$(\lambda, H + H_0) = (\lambda, H),$$

such a δ_0 will work.]

Geometrically, this lemma means that one side of $\mathcal{C}'_\lambda = \mathcal{C}''_{-\lambda}$ is in $\mathcal{C}'$, the other in $\mathcal{C}''$; in addition, it says that

$$\mathcal{C}' \cup (\mathcal{C}'_\lambda = \mathcal{C}''_{-\lambda}) \cup \mathcal{C}''$$

is open in V.

Lemma 3.10 Let $\mathcal{C}'$ and $\mathcal{C}''$ be chambers of Φ -- then there exist chambers $\mathcal{C}_1, \ldots, \mathcal{C}_n$ of Φ such that $\mathcal{C}' = \mathcal{C}_1, \ldots, \mathcal{C}'' = \mathcal{C}_n$ and $\forall i$, $i=1,\ldots,n-1$, $\mathcal{C}_i$ and $\mathcal{C}_{i+1}$ are adjacent.

[The proof runs along familiar lines. Thus introduce the Φ-semiregular set

$$V_{sr} = V - \bigcup_{\lambda\in\Phi}(V_\lambda - (V_\lambda)_r).$$

Then $V_{sr} \supset V_r$ with

$$V_{sr} - V_r = \bigcup_{\lambda\in\Phi}(V_\lambda)_r.$$

Furthermore, V_{sr} is connected. Any two points in V_r can be joined by a polygonal path which stays in V_{sr}; such a path, if it crosses V_λ, must then cross V_λ in $(V_\lambda)_r$.]

Suppose that we define the distance between $\mathcal{C}'$ and $\mathcal{C}''$ to be the minimal n figuring in the lemma -- then this prescription is evidently a metric on the set of Φ-chambers, a fact which is useful for certain types of induction arguments.

Let F be a subset of Φ. Put

$$\begin{cases} V(F) = \mathrm{span}(F) \\ V_F = \bigcap_{\lambda \in F} V_\lambda . \end{cases}$$

Write $\Phi(F)$ for $V(F) \cap \Phi$ and denote by Φ_F the image of $\Phi - \Phi(F)$ under the orthogonal projection P_F of V onto V_F.

[Note: This notation is seemingly at odds with that of §2 in [2-(b)], as used already in §2 of this paper. The reconciliation will be made presently.]

Question: Do both $(V(F),\Phi(F))$ and (V_F,Φ_F) inherit Axioms I - IV? We have seen above that this is so of the pair (V_F,Φ_F) when F is a singleton, the extension to the general case then being handled by a simple iteration or, more precisely, by induction on $\dim(V(F))$, shifting, therefore, the onus to the pair $(V(F),\Phi(F))$ for which at least the validity of Axioms I - III is clear enough. That Axiom IV persists too is a consequence of the following lemma.

Lemma 3.11 *Let $\mathcal{C}(F)$ be a chamber of $\Phi(F)$ -- then there exists a chamber $\mathcal{C}$ of Φ such that $F_0(\mathcal{C}(F)) \subset F_0(\mathcal{C})$.*

Proof Fix a Φ_F-regular element H_F^0 in V_F. Choose a $\delta_F^0 > 0$ such that

$$H_F \in V_F \text{ and } \|H_F - H_F^0\| < \delta_F^0 \Longrightarrow H_F \in (V_F)_r.$$

Then there exists an $\varepsilon_F^0 > 0$ such that

$$\forall \lambda \in \Phi - \Phi(F):$$

$$H_F \in V_F \text{ and } \|H_F - H_F^0\| < \delta_F^0/2 \Rightarrow |(\lambda, H_F)| > \varepsilon_F^0,$$

and there exists a $\delta(F) > 0$ such that

$$\forall \lambda \in \Phi - \Phi(F):$$

$$H(F) \in V(F) \text{ and } \|H(F)\| < \delta(F) \Rightarrow |(\lambda, H(F))| < \varepsilon_F^0.$$

Having made these determinations, let S be the set of all points H in V of the form $H(F) + H_F$, where

$$\begin{cases} \|H(F)\| < \delta(F) \text{ and } H(F) \in \mathfrak{C}(F) \\ \|H_F - H_F^0\| < \delta_F^0/2 \text{ and } H_F \in V_F. \end{cases}$$

In view of the way in which matters have been arranged, for any $H(F) + H_F$ in S, we have

$$\begin{cases} \lambda \in \Phi(F) \Rightarrow |(\lambda, H(F) + H_F)| = |(\lambda, H(F))| > 0 \\ \lambda \in \Phi - \Phi(F) \Rightarrow |(\lambda, H(F) + H_F)| \geq \big|\, |(\lambda, H(F))| - |(\lambda, H_F)| \,\big| > 0. \end{cases}$$

Accordingly, S is a convex open subset of V_r, thus is contained in some chamber $\mathfrak{C}$ of Φ. We claim that $F_0(\mathfrak{C}(F)) \subset F_0(\mathfrak{C})$. Since lengths are not a problem, all we have to do is verify that $S^- \cap V_\lambda$ has a nonempty interior in V_λ for every $\lambda \in F_0(\mathfrak{C}(F))$. But, given such a λ, the subset S_λ of $S^- \cap V_\lambda$ consisting of the $H(F) + H_F$, where, additionally, $H(F)$ belongs to the interior of $\mathfrak{C}(F)^- \cap V(F)_\lambda$, is a nonempty open subset of V_λ. //

To summarize:

<u>Theorem 3.12</u> <u>Suppose that the pair</u> (V,Φ) <u>satisfies</u> Axioms I - IV -- <u>then</u>, $\forall F \subset \Phi$, <u>both pairs</u>

$$\begin{cases} (V(F),\Phi(F)) \\ \\ (V_F,\Phi_F) \end{cases}$$

<u>also satisfy</u> Axioms I - IV.

Let $\mathcal{C}$ be a chamber of Φ -- then $F_0(\mathcal{C})$ is a basis of Φ, consists of short roots, and every $\lambda \in \Phi$ is in

$$\begin{cases} \operatorname{span}_{\underset{\sim}{R}_{\geq 0}}(F_0(\mathcal{C})) \\ \quad\quad \text{or} \\ \operatorname{span}_{\underset{\sim}{R}_{\leq 0}}(F_0(\mathcal{C})). \end{cases}$$

Conversely, if F is a subset of Φ with these three properties, then $F = F_0(\mathcal{C})$, where

$$\mathcal{C} = \{H \in V: (\lambda,H) > 0 \ \forall \lambda \in F\}.$$

In general, a subset F of Φ is called a chamber set if there exists a chamber $\mathcal{C}$ of Φ such that $F \subset F_0(\mathcal{C})$.

<u>Lemma 3.13</u> <u>Let</u> $F \subset \Phi$ -- <u>then</u> F <u>is a chamber set iff</u> $F = F_0(\mathcal{C}(F))$ <u>for some chamber</u> $\mathcal{C}(F)$ <u>of</u> $\Phi(F)$.

[The necessity follows from the preceding observations (with (V,Φ) replaced by $(V(F),\Phi(F))$, while the sufficiency is a consequence of Lemma 3.11.]

Suppose that $F = \{\lambda\}$ ($\lambda \in \Phi_s$) -- then F is a chamber set. Extrapolating the notation of this special case, for an arbitrary chamber set F, put

$$\mathfrak{C}_0(F) = \{\mathcal{C} : F \subset F_0(\mathcal{C})\}.$$

Given a $\mathcal{C}$ in $\mathfrak{C}_0(F)$, we then write $\mathcal{C}_F$ for the interior of $\mathcal{C}^- \cap V_F$ (in V_F).

By induction on $\#(F)$, we have, as earlier:

(1) $\mathcal{C}_F$ is a chamber of Φ_F;

(2) $P_F(F_0(\mathcal{C}) - F)$, when shortened, is the set $F_0(\mathcal{C}_F)$ of short wall roots of $\mathcal{C}_F$.

Proposition 3.14 Fix a chamber set $F \subset \Phi$ -- then the map $\mathcal{C} \to \mathcal{C}_F$ from $\mathfrak{C}_0(F)$ to the chambers of Φ_F is bijective.

Proof If F is a singleton, then our statement is just Proposition 3.8. We may therefore assume that $\#(F) \geq 2$. Suppose that

$$\begin{cases} F \subset F_0(\mathcal{C}') \\ F \subset F_0(\mathcal{C}''), \end{cases}$$

with $\mathcal{C}'_F = \mathcal{C}''_F$ -- then the assertion of injectivity claims the equality of $\mathcal{C}'$ and $\mathcal{C}''$. But $\forall \lambda \in F$,

$$P_F = P_{P_\lambda(F - \{\lambda\})} \circ P_\lambda,$$

from which, by induction, we find that $P_\lambda(F_0(\mathcal{C}'))$, when shortened, is equal to $P_\lambda(F_0(\mathcal{C}''))$, when shortened, so, a fortiori, $\mathcal{C}' = \mathcal{C}''$. The assertion of surjectivity depends on an examination of the proof of Lemma 3.11. There, having fixed a Φ_F-regular element H_F^0 in $\mathcal{C}_F$, say, we constructed a chamber $\mathcal{C}$ of Φ whose closure $\mathcal{C}^-$ meets $\mathcal{C}_F$ in a nonempty open subset of V_F, meaning that $\mathcal{C}$ is the (unique) inverse image of $\mathcal{C}_F$ under the stated correspondence. //

Given a chamber $\mathcal{C}_F$ of Φ_F, let, in a deceptive notation,

$$\mathfrak{c}_F = \{\mathcal{C} : \mathcal{C}^- \cap V_F = \mathcal{C}_F^-\}.$$

Owing to Lemma 3.11 (or rather the proof thereof), one may assign to each chamber $\mathcal{C}(F)$ of $\Phi(F)$ a unique chamber $\mathcal{C} \in \mathfrak{c}_F$ with $F_0(\mathcal{C}(F)) \subset F_0(\mathcal{C})$, i.e., each $\mathcal{C}_F$ determines a map $\mathcal{C}(F) \to \mathcal{C}$ from the chambers of $\Phi(F)$ to $\mathfrak{c}_F$, characterized by the requirement that $F_0(\mathcal{C}(F)) \subset F_0(\mathcal{C})$. This map is actually a bijection. In fact, if

$$\left\{\begin{array}{l} F_0(\mathcal{C}'(F)) \\ \\ F_0(\mathcal{C}''(F)) \end{array}\right. \subset F_0(\mathcal{C}) \qquad (\mathcal{C} \in \mathfrak{c}_F),$$

but $\mathcal{C}'(F) \neq \mathcal{C}''(F)$, then

$$F_0(\mathcal{C}'(F)) \cap \operatorname{span}_{\mathbb{R}_{\leq 0}}(F_0(\mathcal{C}''(F))) \neq \emptyset,$$

an impossibility. On the other hand, take a $\mathcal{C}$ in $\mathfrak{C}_F$ -- then there necessarily exist

$$\begin{cases} H \in \mathcal{C} \\ H_F^0 \in \mathcal{C}_F \end{cases}$$

such that $H - H_F^0 \in V(F)_r$ implying, accordingly, that $F_0(\mathcal{C}(F)) \subset F_0(\mathcal{C})$, $\mathcal{C}(F)$ being defined by the condition

$$H - H_F^0 \in \mathcal{C}(F).$$

We have proved:

Proposition 3.15 Fix a chamber set $F \subset \Phi$ -- then, $\forall$ chamber $\mathcal{C}_F$ of Φ_F, the map $\mathcal{C}(F) \to \mathcal{C}$ from the chambers of $\Phi(F)$ to $\mathfrak{C}_F$ is bijective.

In passing, it should be noted that this correspondence preserves adjacency. More precisely, if

$$\begin{cases} \mathcal{C}'(F) \to \mathcal{C}' \\ \mathcal{C}''(F) \to \mathcal{C}'' \end{cases} \in \mathfrak{C}_F,$$

and if $\mathcal{C}'(F)$ and $\mathcal{C}''(F)$ are adjacent, then $\mathcal{C}'$ and $\mathcal{C}''$ are adjacent. Without pressing the point, observe that if

$$\begin{cases} \lambda \in F_0(\mathcal{C}'(F)) \subset F_0(\mathcal{C}') \\ -\lambda \in F_0(\mathcal{C}''(F)) \subset F_0(\mathcal{C}'') \end{cases}$$

with $\mathcal{C}'(F)_\lambda = \mathcal{C}''(F)_{-\lambda}$ as the common wall of $\mathcal{C}'(F)$ and $\mathcal{C}''(F)$, then the relation

$$V_\lambda = V(F)_\lambda \oplus V_F,$$

coupled with standard considerations, leads more or less immediately to our contention.

Consider now:

Axiom V $\forall$ chamber $\mathcal{C}$ of Φ,

$$\lambda',\lambda'' \in F_0(\mathcal{C}), \lambda' \neq \lambda'' \Rightarrow (\lambda',\lambda'') \leq 0.$$

It is this axiom which, in the presence of the other four, provides the link with the geometric situation envisioned in §2 of [2-(b)]. Indeed, to each chamber $\mathcal{C}$ of Φ there corresponds a basis $F_0(\mathcal{C}) = \{\lambda_1,\ldots,\lambda_\ell\}$ of V with the property that

$$(\lambda_i,\lambda_j) \leq 0 \qquad (i \neq j).$$

Note too that Axiom V is hereditary (in the obvious sense). Of course, we originally employed the $(V(F),V_F)$-mechanism only in this setting, F being, strictly speaking, a subset of $L = \{1,\ldots,\ell\}$. Actually, no conflict is present. For if F be a subset of Φ, first form $V(F)$ per supra, then select a chamber $\mathcal{C}(F)$ of $\Phi(F)$ and finally use Lemma 3.11 to produce a chamber $\mathcal{C}$ of Φ such that $F_0(\mathcal{C}(F)) \subset F_0(\mathcal{C})$. Plainly,

$$\begin{cases} V(F) = V(F_0(\mathcal{C}(F))) \\ V_F = V_{F_0(\mathcal{C}(F))}, \end{cases}$$

so that $(V(F), V_F)$ does in fact arise by picking a subset of L and applying the usual procedure, which implies, incidentally, that chamber sets suffice.

Simple examples show that Axiom IV $\not\Rightarrow$ Axiom V. On the other hand:

Lemma 3.16 Axiom V $\Rightarrow$ Axiom IV.

This statement is implied by:

Lemma 3.17 *Let $(V,(?,?))$ be a finite dimensional inner product space; let F be a finite subset of V such that*

(i) $\lambda', \lambda'' \in F,\ \lambda' \neq \lambda'' \Rightarrow (\lambda', \lambda'') \leq 0$;

(ii) $\exists H_0 \in V$ st: $\lambda \in F \Rightarrow (\lambda, H_0) > 0$.

Then F is linearly independent.

Proof The proof is via induction on $\dim(V)$, starting with the case when $\dim(V) = 1$, which is clear. If $V \neq \operatorname{span}(F)$, then $\dim(\operatorname{span}(F)) < \dim(V)$ and, after replacing H_0 by its orthogonal projection onto $\operatorname{span}(F)$, we can appeal to the induction hypothesis. Assume, therefore, that $V = \operatorname{span}(F)$. Fix a $\lambda_0 \in F$. Put

$$V_0 = \{H \in V : (\lambda_0, H) = 0\}$$

and denote by P_0 the associated orthogonal projection. Since it is a question of the equality $\#(F) = \dim(V)$, and since $P_0(F - \{\lambda_0\})$ spans V_0, to complete the proof it will be enough to verify that $P_0|F - \{\lambda_0\}$ is one-to-one and that

$$V_0, P_0(F - \{\lambda_0\}), P_0H_0$$

satisfy conditions (i) and (ii). Obviously, $F \cap \underset{\sim}{R}\lambda_0 = \{\lambda_0\}$. If now $P_0|F - \{\lambda_0\}$ failed to be injective, then there would exist a $\lambda \in F$ and a $t_0 \neq 0$ such that $\lambda + t_0\lambda_0 \in F$. In turn

$$0 \geq (\lambda, \lambda + t_0\lambda_0) > t_0(\lambda, \lambda_0) \Longrightarrow t_0 > 0$$

$$\Longrightarrow$$

$$0 \leq (\lambda + t_0\lambda_0, \lambda + t_0\lambda_0)$$

$$= (\lambda, \lambda + t_0\lambda_0) + t_0(\lambda_0, \lambda + t_0\lambda_0) \leq 0$$

$$\Longrightarrow$$

$$\lambda + t_0\lambda_0 = 0 \Longrightarrow t_0 = -1,$$

a contradiction. Next, let λ' and λ'' be distinct elements of $F - \{\lambda_0\}$. Write

$$\begin{cases} c' = -(\lambda', \lambda_0)/\|\lambda_0\|^2 \\ c'' = -(\lambda'', \lambda_0)/\|\lambda_0\|^2 \end{cases}.$$

Then we have

$$(P_0\lambda', P_0\lambda'')$$

$$= (\lambda' + c'\lambda_0, \lambda'' + c''\lambda_0)$$

$$= (\lambda', \lambda'') - \frac{(\lambda', \lambda_0)(\lambda'', \lambda_0)}{\|\lambda_0\|^2}$$

$$\leq (\lambda', \lambda'') \leq 0.$$

Finally, for any $\lambda \in F - \{\lambda_0\}$,

$$(P_0\lambda, P_0H_0)$$

$$= (P_0\lambda, H_0) = (\lambda, H_0) - \frac{(\lambda, \lambda_0)}{||\lambda_0||^2} \cdot (\lambda_0, H_0)$$

$$\geq (\lambda, H_0) > 0,$$

as desired. //

For us, a generalized root system (g.r.s.) is data as above subject to Axioms I-IV. A geometric g.r.s. is then a g.r.s. for which Axiom V is in force. In this connection, the adjective "geometric" is appended in order to emphasize that the various combinatorial tools developed in §2 of [2-(b)] are available for deployment.

We shall close this § with one last definition. Let (V, Φ) be a geometric g.r.s. -- then, attached to each chamber $\mathcal{C}$ of Φ, there is a notion of ultraregularity. That being, an element of $V + \sqrt{-1}\, V$ will be termed Φ-ultraregular if it is $\mathcal{C}$-ultraregular $\forall\, \mathcal{C}$.

§4. Detroit Families

The purpose of this § is to introduce the notion of a Detroit family, an artifice which allows one to deal with certain subtle questions of extensionality. This concept is due in principle to Arthur [1-(a)] although it figures implicitly in [2-(c)] as well. The present approach is an attempt on our part to systematize these earlier considerations.

Let (V,Φ) be a geometric g.r.s. -- then a family $\underset{\sim}{\phi} = \{\phi_{\mathcal{C}}\}$ of functions on subsets $\{\mathrm{dom}(\phi_{\mathcal{C}})\}$ of $V + \sqrt{-1}V$ parameterized by the chambers $\mathcal{C}$ of Φ is called a Detroit family provided that, whenever $\mathcal{C}'$ and $\mathcal{C}''$ are adjacent with common wall in V_λ,

$$\phi_{\mathcal{C}'}|V_\lambda + \sqrt{-1}V_\lambda = \phi_{\mathcal{C}''}|V_\lambda + \sqrt{-1}V_\lambda .$$

[Note: To avoid unnecessary fuss, we shall assume, as part of the definition, that

$$\forall\mathcal{C},\ \forall F \subset F_0(\mathcal{C}):$$

$$P_F(\mathrm{dom}(\phi_{\mathcal{C}})) \subset \mathrm{dom}(\phi_{\mathcal{C}}).]$$

The conditions of compatibility inherent in the definition of a Detroit family $\underset{\sim}{\phi} = \{\phi_{\mathcal{C}}\}$ are formulated in terms of the walls of adjacent chambers. Additional conditions follow automatically. Thus, suppose that F is a chamber set and that $\mathcal{C}'$ and $\mathcal{C}''$ are chambers with

$$(\mathcal{C}')^- \cap V_F = \mathcal{C}_F^- = (\mathcal{C}'')^- \cap V_F.$$

Claim:

$$\phi_{\mathcal{C}'}|V_F + \sqrt{-1}V_F = \phi_{\mathcal{C}''}|V_F + \sqrt{-1}V_F .$$

In fact, if

$$\begin{cases} \mathcal{C}'(F) \to \mathcal{C}' \\ \mathcal{C}''(F) \to \mathcal{C}'' \end{cases} \in \mathfrak{C}_F$$

and if $\mathcal{C}'(F)$ and $\mathcal{C}''(F)$ are linked by adjacent chambers, then $\mathcal{C}'$ and $\mathcal{C}''$ are linked by adjacent chambers (in $\mathfrak{C}_F$), so, on $V_F + \sqrt{-1}V_F$,

$$\phi_{\mathcal{C}'} = \cdots = \phi_{\mathcal{C}''} ,$$

the dots being the ϕ's attached to the links in the path leading from $\mathcal{C}'$ to $\mathcal{C}''$.

There are two lemmas of descent, the proofs of which are straightforward, hence can be omitted.

Lemma 4.1 *Let $\underset{\sim}{\phi} = \{\phi_{\mathcal{C}}\}$ be a Detroit family. Fix a chamber set* $F \subset \Phi$ -- *then the prescription*

$$\phi_{\mathcal{C}_F} = \phi_{\mathcal{C}}|V_F + \sqrt{-1}V_F \qquad (\mathcal{C} \in \mathfrak{C}_0(F))$$

determines a Detroit family $\underset{\sim}{\phi}_F$ (per (V_F, Φ_F)).

Lemma 4.2 *Let $\underset{\sim}{\phi} = \{\phi_{\mathcal{C}}\}$ be a Detroit family. Fix a chamber set* $F \subset \Phi$ -- *then, for every chamber $\mathcal{C}_F$ of Φ_F, the prescrip-*

tion

$$\phi_{\mathcal{C}(F)} = \phi_{\mathcal{C}}|V(F) + \sqrt{-1}V(F) \qquad (\mathcal{C} \in \mathcal{C}_F)$$

determines a Detroit family $\underset{\sim}{\phi}(F)$ (*per* $(V(F),\Phi(F))$).

[Note: This procedure actually gives rise to a collection of Detroit families, one for each choice of the chamber $\mathcal{C}_F$ of Φ_F.]

Detroit families make an appearance in geometric problems centering, roughly speaking, on the nature of the volume of certain convex hulls. It would be out of place to go into detail now, but the reader who is familiar with the theory of weighted orbital integrals will realize what we have in mind. Accordingly, we shall be content to settle for a definition and a simple statement.

Let (V,Φ) be a geometric g.r.s. -- then a set $\{H_{\mathcal{C}}\}$ of points in V parameterized by the chambers $\mathcal{C}$ of Φ is said to be a Φ-orthogonal set provided that, whenever $\mathcal{C}'$ and $\mathcal{C}''$ are adjacent with common wall in V_λ,

$$H_{\mathcal{C}'} - H_{\mathcal{C}''} \in \underset{\sim}{R}\lambda.$$

Lemma 4.3 *The points* $H_{\mathcal{C}}$ *form a* Φ-*orthogonal set iff the functions*

$$\phi_{\mathcal{C}}(?) = \exp((?,H_{\mathcal{C}}))$$

constitute a Detroit family.

Fix, henceforth, a geometric g.r.s. (V,Φ) of dimension ℓ, say. To prepare for the main order of business it will be convenient to introduce some notation. Let F be a finite subset of V (not just Φ). Put

$$\Theta_F(?) = \frac{|\det[(\lambda_i,\lambda_j):i,j\in F]|^{1/2}}{\prod_{\lambda\in F}(?,\lambda)} .$$

Of course, if F is linearly dependent then $\Theta_F \equiv 0$. In addition, if F_t is obtained from F by a replacement $\lambda \to t_\lambda\lambda(t_\lambda > 0)$, then still

$$\Theta_F = \Theta_{F_t} .$$

Let $\mathfrak{C}$ be a chamber of Φ -- then, since the definition of A-transform presupposes a choice for $\mathfrak{C}$, we shall write $A_{\mathfrak{C},\phi}$ in place of A_ϕ, so

$$A_{\mathfrak{C},\phi}(\Lambda) = \sum_{\{F:F\subset F_0(\mathfrak{C})\}} C_F\cdot(-1)^{\#(F)}\cdot\left[\frac{\phi(P_F\Lambda)}{\prod_{i\in F}(\Lambda,\lambda_F^i)\times\prod_{i\notin F}(\Lambda,\lambda_i^F)}\right].$$

To recast this expression in terms of the Θ's, fix an $F\subset F_0(\mathfrak{C})$ -- then, taking into account Lemma 2.10, we can say that

$$\frac{C_F}{\prod_{i\in F}(\Lambda,\lambda_F^i)\times\prod_{i\notin F}(\Lambda,\lambda_i^F)}$$

is equal to

$$\Theta_{F^*}(\Lambda)\Theta_{F_0(\mathfrak{C}_F)}(P_F\Lambda).$$

Here and in what follows, F^* is the basis dual to F in $V(F)$, F any chamber set. Consequently,

$$A_{\mathfrak{C},\phi}(\Lambda) = \sum_{\{F:F\subset F_0(\mathfrak{C})\}} (-1)^{\#(F)}\cdot\Theta_{F^*}(\Lambda)\Theta_{F_0(\mathfrak{C}_F)}(P_F\Lambda)\cdot\phi(P_F\Lambda).$$

Given a Detroit family $\underset{\sim}{\phi} = \{\phi_{\mathfrak{C}}\}$, put

$$Ш_{\underset{\sim}{\phi}} = \sum_{\mathfrak{C}} \Theta_{F_0(\mathfrak{C})}\cdot\phi_{\mathfrak{C}} \quad .$$

$Ш_{\underset{\sim}{\phi}}$ is defined on the Φ-ultraregular points in $\bigcap_{\mathfrak{C}}\mathrm{dom}(\phi_{\mathfrak{C}})$, as is

$$A_{\underset{\sim}{\phi}} = \sum_{\mathfrak{C}} A_{\mathfrak{C},\phi_{\mathfrak{C}}} ,$$

and:

Theorem 4.4 $\quad Ш_{\underset{\sim}{\phi}} = A_{\underset{\sim}{\phi}}$.

The assertion is one of pointwise equality. To see what the basic idea is, start with the right hand side -- then

$$\begin{aligned} &A_{\underset{\sim}{\phi}}(\Lambda) \\ &= \sum_{\mathfrak{C}}\sum_{\{F:F\subset F_0(\mathfrak{C})\}} (-1)^{\#(F)}\cdot\Theta_{F^*}(\Lambda)\Theta_{F_0(\mathfrak{C}_F)}(P_F\Lambda)\cdot\phi_{\mathfrak{C}}(P_F\Lambda) \\ &= \sum_{F} (-1)^{\#(F)}\cdot\Theta_{F^*}(\Lambda)\cdot\left[\sum_{\{\mathfrak{C}:F\subset F_0(\mathfrak{C})\}} \Theta_{F_0(\mathfrak{C}_F)}(P_F\Lambda)\cdot\phi_{\mathfrak{C}}(P_F\Lambda)\right] , \end{aligned}$$

the outer sum $\sum_F$ being over the chamber sets of Φ. But, by definition,

$$\mathfrak{C}_0(F) = \{\mathfrak{C}:F\subset F_0(\mathfrak{C})\},$$

thus, in view of Lemma 4.1, we have

$$\text{Ш}_{\underset{\sim}{\phi}_F}(P_F\Lambda) = \sum_{\{\mathcal{C}:F\subset F_0(\mathcal{C})\}} \boldsymbol{\Theta}_{F_0(\mathcal{C}_F)}(P_F\Lambda)\cdot\phi_{\mathcal{C}}(P_F\Lambda).$$

Accordingly,

$$A_{\underset{\sim}{\phi}}(\Lambda) = \sum_F (-1)^{\#(F)}\cdot\boldsymbol{\Theta}_{F*}(\Lambda)\cdot\text{Ш}_{\underset{\sim}{\phi}_F}(P_F\Lambda).$$

Take now $\sum\limits_F$ and rewrite it as a double sum

$$\sum_{V_0}\sum_{\{F:V(F)=V_0\}}.$$

Fix a V_0, together with a chamber set F_0 for which $V_0 = V(F_0)$ and call $P_0:V\to V_0$ the corresponding orthogonal projection -- then

$$\sum_{\{F:V(F)=V_0\}}(-1)^{\#(F)}\cdot\boldsymbol{\Theta}_{F*}(\Lambda)\cdot\text{Ш}_{\underset{\sim}{\phi}_F}(P_F\Lambda)$$

$$= \pm\left[\sum_{\mathcal{C}_0}\boldsymbol{\Theta}_{F_0(\mathcal{C}_0)*}(P_0\Lambda)\right]\cdot\text{Ш}_{\underset{\sim}{\phi}_{F_0}}(P_{F_0}\Lambda).$$

The extreme possibility is when $F_0 = \emptyset$. In this case, we pick up $\text{Ш}_{\underset{\sim}{\phi}}$ alone. To eliminate all the others, we need only establish that

$$F_0 \neq \emptyset \Longrightarrow \sum_{\mathcal{C}_0}\boldsymbol{\Theta}_{F_0(\mathcal{C}_0)*}\circ P_0 \equiv 0.$$

Let us argue by induction on $\dim(V_0)$, supposing the assertion to be true for all nonempty F_0 of cardinality $\leq \ell - 1$, so

$$A_{\underset{\sim}{\phi}} = \text{Ш}_{\underset{\sim}{\phi}} + (-1)^{\ell}\cdot\left[\sum_{\mathcal{C}}\boldsymbol{\Theta}_{F_0(\mathcal{C})*}\right]\cdot C_0,$$

C_0 a constant. Set $\phi_{\mathcal{C}} \equiv 1\ \forall\,\mathcal{C}$ -- then, in this situation,

$$\begin{cases} \underset{\sim}{A}_{\phi} & \text{is entire (cf. Proposition 2.14)} \\ \underset{\sim}{Ш}_{\phi} & \text{is entire (cf. Proposition 4.5),} \end{cases}$$

implying, therefore, that

$$\sum_{\mathfrak{C}} \Theta_{F_0}(\mathfrak{C})^*$$

is entire. Let Λ be an ultraregular point -- then

$$\sum_{\mathfrak{C}} \Theta_{F_0}(\mathfrak{C})^*(t\Lambda) = t^{-\ell} \cdot \sum_{\mathfrak{C}} \Theta_{F_0}(\mathfrak{C})^*(\Lambda) \qquad (t > 0).$$

Since the left hand side must stay finite as $t \downarrow 0$, the right hand side must vanish, as desired.

This completes the proof of Theorem 4.4, modulo:

Proposition 4.5 *Let $\underset{\sim}{\phi} = \{\phi_{\mathfrak{C}}\}$ be a Detroit family. Let D be a connected open subset of $V + \sqrt{-1}V$ on which all the $\phi_{\mathfrak{C}}$ are defined and holomorphic -- then $Ш_{\underset{\sim}{\phi}}$ extends to a holomorphic function on D.*

[Note: $Ш_{\underset{\sim}{\phi}}$ is only, a priori, a meromorphic function on D.]

Proof The singular set of $Ш_{\underset{\sim}{\phi}}$ is either $\emptyset$ or of pure dimension $\ell - 1$. On the other hand, the singularities of $Ш_{\underset{\sim}{\phi}}$ in D, if they exist at all, must lie along hyperplanes. Such terms come in pairs, say at $\mathfrak{C}'$ and $\mathfrak{C}''$, where $\mathfrak{C}'$ and $\mathfrak{C}''$ are adjacent with common wall in V_λ:

$$\Theta_{F_0}(\mathfrak{C}')(\Lambda) \cdot \phi_{\mathfrak{C}'}(\Lambda) + \Theta_{F_0}(\mathfrak{C}'')(\Lambda) \cdot \phi_{\mathfrak{C}''}(\Lambda),$$

i.e.,

$$\frac{1}{(\Lambda,\lambda)}\cdot E(\Lambda),$$

where, on $V_\lambda \cap D$,

$$E(\Lambda) = \|\lambda\|\cdot\Big[\Theta_{F_0(\mathcal{C}'_\lambda)}(\Lambda)\cdot\phi_{\mathcal{C}'}(\Lambda) - \Theta_{F_0(\mathcal{C}''_{-\lambda})}(\Lambda)\cdot\phi_{\mathcal{C}''}(\Lambda)\Big] .$$

But the term in brackets is evidently zero on $V_\lambda \cap D$, so, by L'Hôspital's rule, the singularity along V_λ is actually removable. //

There is also a C^∞ version of this result, namely:

Proposition 4.6 Let $\underset{\sim}{\phi} = \{\phi_{\mathcal{C}}\}$ be a Detroit family. Suppose that each $\phi_{\mathcal{C}}$ is a C^∞ function on $\sqrt{-1}V$ -- then $Ш_{\underset{\sim}{\phi}}$ extends to a C^∞ function on $\sqrt{-1}V$.

[Thanks to Proposition 2.12, each $A_{\mathcal{C},\phi_{\mathcal{C}}}$ is a C^∞ function on $\sqrt{-1}V$, thus the same holds for

$$A_{\underset{\sim}{\phi}} = \sum_{\mathcal{C}} A_{\mathcal{C},\phi_{\mathcal{C}}}$$

or still, for $Ш_{\underset{\sim}{\phi}}$ (cf. Theorem 4.4).]

§5. Differentiation in the Truncation Parameter

The purpose of this § is to study the differential dependence of a truncated inner product on the truncation parameter, the main result being a proposition of transition. These considerations are in turn based on an interesting property of the truncation operator. This property was not mentioned in [2-(b)], primarily because it was not needed there, but also because it depends on the theory of Γ-functions, the development of which at that juncture would have represented a digression.

It will be presupposed that the reader is familiar with [2-(b)]. Let, therefore, G be a reductive Lie group, Γ a lattice in G, both subject to the usual conditions. We recall that the parameter space $\underset{\sim}{a}$ for the theory of truncation was constructed from a set of representatives for the Γ-conjugacy classes of maximal Γ-cuspidal split parabolic subgroups of G. Inside $\underset{\sim}{a}$ was a certain nonempty subset $\underset{\sim}{a}_Q$, cofinal in $\underset{\sim}{a}$ (per <). Since $\underset{\sim}{a}$ is choice dependent, some complications will eventually arise if we do not stop to take a few precautions now. The easiest way out is to simply alter our conventions a little, recasting the definitions so as to fit our current needs.

Thus let $\mathcal{C}_\Gamma$ be the set of all Γ-cuspidal split parabolic

subgroups of G. Put

$$\mathrm{Lie}(\Gamma) = \prod_{P \in \mathcal{C}_\Gamma} a_P ,$$

a_P being the Lie algebra of the special split component A_P of

(P,S). Evidently, $\mathrm{Lie}(\Gamma)$ is a vector space (over $\underset{\sim}{R}$). Write

$\underset{\sim}{a}(\Gamma)$ for the subset of $\mathrm{Lie}(\Gamma)$ consisting of all ΠH_P such that

$$(P'',S'';A_{P''}'') \succeq (P',S';A_{P'}')$$

$$\Longrightarrow H_{P''} = \mathrm{Proj}(H_{P'}).$$

Denote by $\underset{\sim}{a}$ the subset of $\underset{\sim}{a}(\Gamma)$ comprised of those ΠH_P for

which

$$H_{\gamma P \gamma^{-1}} = I_\Gamma(\gamma P \gamma^{-1}:P)H_P \qquad (\gamma \in \Gamma).$$

To justify this notation, observe that the present $\underset{\sim}{"a"}$ is in

fact canonically bijective with the $\underset{\sim}{"a"}$ mentioned above, namely,

$$\Pi H_P \in \underset{\sim}{a} \; - \; \Pi H_{P_m^{\max}} \in \underset{\sim}{a}.$$

Of course, the point is that the new description of $\underset{\sim}{a}$ is, in an

obvious sense, invariant. Call $\underset{\sim}{a_0}$ the subset of $\underset{\sim}{a}(\Gamma)$ made up

of the ΠH_P satisfying

$$H_{\gamma P \gamma^{-1}} = I(\gamma P \gamma^{-1}|A_{\gamma P \gamma^{-1}}:P|A_P)H_P \qquad (\gamma \in \Gamma).$$

Then $\underset{\sim}{a_0}$ is a vector space and $\underset{\sim}{a}$ is an affine space attached

to $\underset{\sim}{a_0}$. Accordingly, in the customary terminology, $\underset{\sim}{a_0}$ may be

regarded as the translation space of $\underset{\sim}{a}$, acting, as it does, faithfully and transitively on $\underset{\sim}{a}$:

$$\begin{cases} \underset{\sim}{H} \in \underset{\sim}{a} \\ \underset{\sim}{H}_0 \in \underset{\sim}{a}_0 \end{cases} \Rightarrow \underset{\sim}{H} + \underset{\sim}{H}_0 \in \underset{\sim}{a}.$$

Given $\underset{\sim}{H}_1, \underset{\sim}{H}_2 \in \underset{\sim}{a}$, it will also be necessary to reassess the meanings of

$$\begin{cases} \underset{\sim}{H}_1 < \underset{\sim}{H}_2 \\ \underset{\sim}{H}_1 \lessdot \underset{\sim}{H}_2. \end{cases}$$

Put

$$\underset{\sim}{\mathcal{C}}_0 = \{\Pi H_P \in \underset{\sim}{a}_0 : \forall P, H_P \in \mathcal{C}_P(a_P)\}.$$

Then here

$$\underset{\sim}{H}_1 < \underset{\sim}{H}_2 \Leftrightarrow \underset{\sim}{H}_2 - \underset{\sim}{H}_1 \in \underset{\sim}{\mathcal{C}}_0.$$

To define $\lessdot$, first introduce $\underset{\sim}{a}_0(\cdot)$, the vector subspace of $\underset{\sim}{a}_0$ whose elements are the ΠH_P such that

$$H_{xPx^{-1}} = I(xPx^{-1}|A_{xPx^{-1}}:P|A_P)H_P \qquad (x \in G).$$

If, as usual, (P_0, S_0) is some fixed Γ-percuspidal split parabolic subgroup of G with special split component A_0, let

$$\underset{\sim}{\mathcal{C}}_0(\cdot) = \{\Pi H_P \in \underset{\sim}{a}_0(\cdot) : H_{P_0} \in \mathcal{C}_{P_0}(a_0)\}.$$

Then here

$$\underset{\sim}{H}_1 \lessdot \underset{\sim}{H}_2 \Leftrightarrow \underset{\sim}{H}_2 - \underset{\sim}{H}_1 \in \underset{\sim}{\mathcal{C}}_0(\cdot).$$

Needless to say,

$$\underset{\sim}{H}_1 \lessdot \underset{\sim}{H}_2 \Rightarrow \underset{\sim}{H}_1 < \underset{\sim}{H}_2,$$

the converse being true if Γ has one cusp, but not in general.

The revised definition of $\underset{\sim}{a}_Q$ offers no difficulty, hence will be omitted. It should be mentioned, however, that $\underset{\sim}{a}_Q$ can be thought of as open, a point which was not stressed in [2-(b)], although the inequalities required for a detailed verification can be found there. In what follows, we shall tacitly agree that when $\underset{\sim}{H} \in \underset{\sim}{a}_Q$ is given, only those $\underset{\sim}{H}_0 \in \underset{\sim}{a}_0$ will be considered such that still

$$\underset{\sim}{H} + \underset{\sim}{H}_0 \in \underset{\sim}{a}_Q.$$

Let us also remind ourselves that

$$\forall \underset{\sim}{H}_1, \forall \underset{\sim}{H}_2:$$

$$\underset{\sim}{H}_1 \ll \underset{\sim}{H}_2, \ \underset{\sim}{H}_2 \in \underset{\sim}{a}_Q \implies \underset{\sim}{H}_1 \in \underset{\sim}{a}_Q.$$

If (P,S) is a Γ-cuspidal split parabolic subgroup of G, then, adhering to the usual practice, its special split component shall be denoted by A rather than A_P, the sub-P supra having been used as an index only. This said, let $P = M \cdot A \cdot N$ be the corresponding Langlands decomposition -- then there is a natural map

$$I_M : \underset{\sim}{a} \to \underset{\sim}{a}_M,$$

$\underset{\sim}{a}_M$ playing the same role per (M, Γ_M) as is played by $\underset{\sim}{a}$ per (G, Γ). Bear in mind too that I_M has a cofinal image and is order preserving (relative to $<$), with

$$I_M(\underset{\sim}{a}_Q) \subset (\underset{\sim}{a}_M)_Q .$$

Notationally, it will be convenient to write $\underset{\sim}{H}(P)$ for the component H_P of $\underset{\sim}{H} = \Pi H_P$ indexed by P.

Let f be a complex valued locally bounded (measurable) function on G/Γ -- then, for any $\underset{\sim}{H} \in \underset{\sim}{a}$, the truncation operator $Q^{\underset{\sim}{H}}$, when applied to f, is given by

$$Q^{\underset{\sim}{H}} f(x)$$

$$= \sum_{P \in \mathcal{C}_\Gamma} (-1)^{\mathrm{rank}(P)} \chi_{P,A:\mathfrak{D}}(\underset{\sim}{H}(P) - H_{P|A}(x)) \cdot f^P(x).$$

Our objective now will be to see what happens when $\underset{\sim}{H}$ is replaced by a translate $\underset{\sim}{H} + \underset{\sim}{H}_0$, where $\underset{\sim}{H}_0 \in \underset{\sim}{a}_0$. Of course, initially

$$Q^{\underset{\sim}{H} + \underset{\sim}{H}_0} f(x)$$

$$= \sum_{P \in \mathcal{C}_\Gamma} (-1)^{\mathrm{rank}(P)} \chi_{P,A:\mathfrak{D}}(\underset{\sim}{H}(P) + \underset{\sim}{H}_0(P) - H_{P|A}(x)) \cdot f^P(x).$$

Owing to Proposition 2.3, we have

$$\chi_{P,A:\mathfrak{D}}(\underset{\sim}{H}(P) + \underset{\sim}{H}_0(P) - H_{P|A}(x))$$

$$= \sum_{\{F: F \subset \Sigma_P^0(\mathfrak{g},\mathfrak{a})\}} \chi^{*,F}(\underset{\sim}{H}(P) - H_{P|A}(x)) \cdot \Gamma_F^*(\underset{\sim}{H}(P) - H_{P|A}(x), -\underset{\sim}{H}_0(P)).$$

In this context, let us agree to write Γ_P in place of Γ_*^*. If, as in [2-(b)], $\mathrm{Dom}_\Gamma(P)$ stands for the set of all Γ-cuspidal split parabolic subgroups of G which are dominated predecessors of (P,S), then, employing a standard trick, replace

$$\sum_{P\in\mathcal{C}_\Gamma}\sum_{\{F:F\subset\Sigma_P^0(\mathfrak{g},\mathfrak{a})\}}$$

by

$$\sum_{P\in\mathcal{C}_\Gamma}\ \sum_{P'\in\mathrm{Dom}_\Gamma(P)} .$$

Taking into account the definition of the partial truncation operator $Q_P^{\underset{\sim}{H}}$ (see §8 of [2-(b)]), after a short calculation we end up with

$$Q^{\underset{\sim}{H}+\underset{\sim}{H}_0}f(x)$$

$$= \sum_{P\in\mathcal{C}_\Gamma}(-1)^{\mathrm{rank}(P)}Q_P^{\underset{\sim}{H}}f(x)\cdot\Gamma_P(\underset{\sim}{H}(P) - H_{P|A}(x), -\underset{\sim}{H}_0(P))$$

or still, in view of the commutativity of partial truncation with (right) Γ-translation and the behavior of I_Γ under Γ-conjugation,

$$Q^{\underset{\sim}{H}+\underset{\sim}{H}_0}f(x)$$

$$= \sum_{i=1}^{r}\ \sum_{\gamma_i\in\Gamma/\Gamma\cap P_i}(-1)^{\mathrm{rank}(P_i)}Q_{P_i}^{\underset{\sim}{H}}f(x\gamma_i)\cdot\Gamma_{P_i}(\underset{\sim}{H}(P_i)-H_{P_i|A_i}(x\gamma_i),-\underset{\sim}{H}_0(P_i)).$$

Here,

$$\{(P_i,S_i):1 \le i \le r\}$$

is a set of representatives for the Γ-conjugacy classes of Γ-cuspidal split parabolic subgroups of G.

So:

<u>Lemma 5.1</u> Let $\underline{H}\in\underline{a}$, $\underline{H}_0\in\underline{a}_0$ -- then, for any complex valued locally bounded (measurable) function f on G/Γ,

$$Q^{\underline{H}+\underline{H}_0}f(x)$$

$$=\sum_{i=1}^{r}\sum_{\gamma_i\in\Gamma/\Gamma\cap P_i}(-1)^{\mathrm{rank}(P_i)}Q_{P_i}^{\underline{H}}f(x\gamma_i)\cdot\Gamma_{P_i}(\underline{H}(P_i)-H_{P_i|A_i}(x\gamma_i),-\underline{H}_0(P_i)).$$

Suppose that

$$f\in\bigcup_r S_r^\infty(G/\Gamma).$$

Then, as is known (cf. §7 of [2-(b)]),

$$Q^{\underline{H}}f\in R(G/\Gamma)\qquad(\underline{H}\in\underline{a}_Q).$$

Let

$$g\in\bigcup_r S_r^\infty(G/\Gamma).$$

Then, $\forall\underline{H}\in\underline{a}_Q$, the inner product

$$(Q^{\underline{H}+\underline{H}_0}f,g)$$

$$=\int_{G/\Gamma}Q^{\underline{H}+\underline{H}_0}f(x)\overline{g(x)}d_G(x)$$

exists and formally, at least, is equal to

$$\sum_{i=1}^{r}(-1)^{\mathrm{rank}(P_i)}\int_{G/\Gamma\cap P_i}Q_{P_i}^{\underline{H}}f(x)\cdot\Gamma_{P_i}(\underline{H}(P_i)-H_{P_i|A_i}(x),-\underline{H}_0(P_i))\cdot\overline{g(x)}d_G(x).$$

To justify this procedure, it is clearly enough to establish:

<u>Lemma 5.2</u> Let (P,S) be a Γ-cuspidal split parabolic subgroup of G with special split component A -- then, for all $\underline{H}\in\underline{a}_Q$ and

$$\forall f,g\in\bigcup_r S_r^\infty(G/\Gamma),$$

the integral

$$\int_{G/\Gamma\cap P} Q_P^{\underset{\sim}{H}} f(x)\cdot\Gamma_P(\underset{\sim}{H}(P) - H_{P|A}(x), -\underset{\sim}{H}_0(P))\cdot\overline{g(x)}\,d_G(x)$$

is absolutely convergent.

Proof The integral in question can be rewritten as an integral

$$\int_K\int_{M/\Gamma_M}\int_A [\ldots]a^{2\rho}d_K(k)d_M(m)d_A(a).$$

If f_P and g_P are assigned their usual meanings (cf. TES, p. 79), then, thanks to the relation between partial truncation on G and complete truncation on M (see §8 of [2-(b)]), this integral can be put in the form

$$\int_A \Gamma_P(\underset{\sim}{H}(P) - \log a, -\underset{\sim}{H}_0(P))[?]d_A(a),$$

where

$$[?]$$

$$= (1\times Q^{I_M(\underset{\sim}{H})} f_P(?:?:a), g_P(?:?:a)),$$

the inner product being per $K\times M/\Gamma_M$. Because $\underset{\sim}{H}_0$ is fixed, the integral

$$\int_A \Gamma_P(\underset{\sim}{H}(P) - \log a, -\underset{\sim}{H}_0(P))d_A(a)$$

is compactly supported. On the other hand, [?], qua a function of $a\in A$, is obviously bounded on compacta. //

Keeping to the preceding assumptions and notations, we claim that the dependence on $\underset{\sim}{H}_0$ in

$$(Q^{\underset{\sim}{H}+\underset{\sim}{H}_0}f,g)$$

is actually C^∞. Indeed, since it has just been shown that this expression is, up to sign, the sum over i of the

$$\int_{G/\Gamma\cap P_i} Q_{P_i}^{H} f(x)\cdot\Gamma_{P_i}(\underset{\sim}{H}(P_i) - H_{P_i|A_i}(x),\ -\underset{\sim}{H}_0(P_i))\cdot\overline{g(x)}\,d_G(x),$$

we need only verify:

<u>Lemma 5.3</u> <u>Let</u> (P,S) <u>be a</u> Γ-<u>cuspidal split parabolic subgroup of</u> G <u>with special split component</u> A -- <u>then, for all</u> $\underset{\sim}{H}\in\underset{\sim}{a}_Q$ <u>and</u>

$$\forall f,g\in\bigcup_r S_r^\infty(G/\Gamma),$$

<u>the absolutely convergent integral</u>

$$\int_{G/\Gamma\cap P} Q_P^{H} f(x)\cdot\Gamma_P(\underset{\sim}{H}(P) - H_{P|A}(x),\ -\underset{\sim}{H}_0(P))\cdot\overline{g(x)}\,d_G(x)$$

<u>is a</u> C^∞ <u>function of</u> $\underset{\sim}{H}_0(P)$.

<u>Proof</u> It is not difficult to see that the maps

$$\begin{cases} H \mapsto f_P(?:?:\exp(H)) \\ H \mapsto g_P(?:?:\exp(H)) \end{cases}$$

from

$$a \text{ to } \bigcup_r S_r^\infty(K\times M/\Gamma_M)$$

are each strongly C^∞. Consequently,

$$\phi(H) \equiv (1\times Q^{I_M(\underset{\sim}{H})} f_P(?:?:\exp(H)),\ g_P(?:?:\exp(H)))$$

is a C^∞ function of H. That being, write

$$\int_A \Gamma_P(\underset{\sim}{H}(P) - \log a,\ -\underset{\sim}{H}_0(P))[?]d_A(a)$$

as

$$\int_{\mathfrak{a}} \Gamma_P(H, -H_0(P))\phi(H(P) - H)d_{\mathfrak{a}}(H).$$

There is no loss of generality in supposing that $H_0(P)$ is restricted to a ball of finite radius centered at the origin. With this understanding, let ψ be a corresponding C^∞ cut-off function -- then

$$\int_{\mathfrak{a}} \Gamma_P(H, -H_0(P))\phi(H(P) - H)d_{\mathfrak{a}}(H)$$
$$= \int_{\mathfrak{a}} \Gamma_P(H, -H_0(P))\phi(H(P) - H)\psi(H)d_{\mathfrak{a}}(H).$$

The latter integral can be evaluated exactly as in the discussion following Lemma 2.9. The result is certainly a C^∞ function of $H_0(P)$. //

Having ensured ourselves that

$$(Q^{H+H_0}f, g)$$

is a C^∞ function of H_0, let us try to compute

$$\frac{d}{dH}(Q^H f, g) = \lim_{H_0 \to 0} \frac{d}{dH_0}(Q^{H+H_0}f, g).$$

Of course, as it stands, this expression is meaningless, but, as we shall see, when matters are appropriately arranged, the differentiation can be given a sensible interpretation, the final result being the proposition of transition mentioned at the beginning.

Consider again

$$\int_{G/\Gamma\cap P} Q_P^H f(x)\cdot\Gamma_P(H(P) - H_{P|A}(x), -H_0(P))\cdot\overline{g(x)}d_G(x)$$

or still, as above,

$$\int_{a} \Gamma_P(H, -\underset{\sim}{H}_0(P))\phi(\underset{\sim}{H}(P) - H)d_a(H),$$

ϕ being

$$(1 \times Q^{I_M(\underset{\sim}{H})} f_P(?:?:\exp(?)), g_P(?:?:\exp(?))).$$

It will be assumed that $\underset{\sim}{H}_0 \in -\mathcal{C}_0$, so that $-\underset{\sim}{H}_0(P) \in \mathcal{C}_P(a_P)$. Fix now a realization

$$\underset{\sim}{a} \sim \bigoplus_m a_m^{max} .$$

Given an index m, freeze all the components of $\underset{\sim}{H}_0$ save for the mth, it being reserved for the direction of differentiation. If P_m^{max} dominates no Γ-conjugate of P, then $\int_a \Gamma_P \ldots$ is constant along P_m^{max}, hence, after differentiating, contributes nothing. Suppose, therefore, that

$$(P_m^{max}, S_m^{max}; A_m^{max}) \succeq (P,S;A).$$

If $\Sigma_P^o(\mathfrak{g},a) = \{\lambda_1,\ldots,\lambda_\ell\}$, then

$$a_m^{max} = \bigcap_{i>1} \mathrm{Ker}(\lambda_i), \text{ say.}$$

Write H_m for H_{λ_1} and call H_m^{max} the projection of H_m onto a_m^{max}. In these notations, the first stage in our evaluation consists in calculating the limit as $\underset{\sim}{H}_0$ approaches $\underset{\sim}{0}$ of

$$\lim_{t \downarrow 0} \frac{1}{t} \cdot \int_a [\Gamma_P(H, -\underset{\sim}{H}_0(P) - t \frac{H_m}{\|H_m^{max}\|}) - \Gamma_P(H, -\underset{\sim}{H}_0(P))]\phi(\underset{\sim}{H}(P) - H)d_a(H).$$

It turns out that the double limit is zero unless $P = P_m^{max}$, in which case the result is simply

$$\phi(\underset{\sim}{H}(P)).$$

[Note: To clarify the appearance of

$$\frac{H_m}{\|H_m^{max}\|},$$

we remark that its projection onto a_m^{max} is a (positive) unit vector.]

Proceeding to the details, consider instead

$$\frac{1}{\|H_m^{max}\|}$$

times

$$\lim_{t\downarrow 0} \frac{1}{t}\cdot\int_{a}[\Gamma_P(H,-\underset{\sim}{H}_0(P)+tH_m)-\Gamma_P(H,-\underset{\sim}{H}_0(P))]\phi(\underset{\sim}{H}(P)-H)d_a(H).$$

For t sufficiently small,

$$-\underset{\sim}{H}_0(P)+tH_m \in \mathfrak{C}_P(a).$$

Accordingly, thanks to Proposition 2.5,

$$\Gamma_P(?,-\underset{\sim}{H}_0(P)+tH_m)-\Gamma_P(?,-\underset{\sim}{H}_0(P))$$

is equal to

$$(-1)^{\ell}\cdot[\chi_{\mathfrak{C}_P(a)\cap(-\underset{\sim}{H}_0(P)+tH_m-\mathfrak{D}_P(a)^-)}-\chi_{\mathfrak{C}_P(a)\cap(-\underset{\sim}{H}_0(P)-\mathfrak{D}_P(a)^-)}],$$

the χ's being the characteristic functions of the indicated sets.

Because

$$-\underset{\sim}{H}_0(P)+tH_m-\mathfrak{D}_P(a)^- \supset -\underset{\sim}{H}_0(P)-\mathfrak{D}_P(a)^-,$$

we need only determine the product of

$$\frac{(-1)^{\ell+1}}{\|H_m^{max}\|}$$

with

$$\lim_{t\downarrow 0} \frac{1}{t}\cdot\int_{\mathcal{C}_P(a)\cap S_t(\underset{\sim}{H}_0(P))} \phi(\underset{\sim}{H}(P)-H)d_a(H)$$

$$= \lim_{t\downarrow 0} \frac{1}{t}\cdot\int_{S_t(\underset{\sim}{H}_0(P))} \phi(\underset{\sim}{H}(P)-H)\chi_{P,A:\mathcal{C}}(H)d_a(H),$$

where $S_t(\underset{\sim}{H}_0(P))$ is the difference

$$(-\underset{\sim}{H}_0(P) + tH_m - \mathfrak{D}_P(a)^-) - (-\underset{\sim}{H}_0(P) - \mathfrak{D}_P(a)^-).$$

Using the definitions, this difference can be represented as the intersection of

$$\{H : t \geq (H + \underset{\sim}{H}_0(P), \lambda^1) > 0\}$$

with

$$\bigcap_{i>1}\{H : (H + \underset{\sim}{H}_0(P), \lambda^i) \leq 0\}.$$

Put

$$S(\underset{\sim}{H}_0(P):\tau) = \{H : (H + \underset{\sim}{H}_0(P), \lambda^1) = \tau\}$$

$$\cap \bigcap_{i>1}\{H : (H + \underset{\sim}{H}_0(P), \lambda^i) \leq 0\}.$$

Then

$$\bigcup_{0<\tau\leq t} S(\underset{\sim}{H}_0(P):\tau)$$

fibers our region of integration, so the integral

$$\int_{S_t(\underset{\sim}{H}_0(P))} \phi(\underset{\sim}{H}(P)-H)\chi_{P,A:\mathcal{C}}(H)d_a(H)$$

admits the iteration

$$\int_0^t \left[\int_{S(\underset{\sim}{H}_0(P):\tau)} \phi(\underset{\sim}{H}(P)-H)\chi_{P,A:\mathfrak{C}}(H)\,dH \right] d\tau.$$

Multiply up by $1/t$ and let $t \downarrow 0$ -- then we get an $(\ell-1)$-dimensional integral

$$\int_{S(\underset{\sim}{H}_0(P):0)} \phi(\underset{\sim}{H}(P)-H)\chi_{P,A:\mathfrak{C}}(H)\,dH.$$

The dependence of this integral on $\underset{\sim}{H}_0$ is continuous. If $\ell > 1$, then, as $\underset{\sim}{H}_0(P)$ approaches the origin along a straight line, the area of

$$S(\underset{\sim}{H}_0(P):0) \cap \mathfrak{C}_P(a)$$

is proportional to $\|\underset{\sim}{H}_0(P)\|^{\ell-1}$, thus tends to zero. But if $\ell = 1$, so that $P = P_m^{max}$, then

$$\begin{cases} (-1)^{\ell+1} = 1 \\ \|H_m^{max}\| = \|H_m\|, \end{cases}$$

and $\lim_{t \downarrow 0} \ldots$ collapses to

$$\frac{1}{\|H_m\|} \cdot [\|H_m\| \cdot \phi(\underset{\sim}{H}(P)-\underset{\sim}{H}_0(P))]$$

giving

$$\phi(\underset{\sim}{H}(P))$$

for $\underset{\sim}{H}_0 = \underset{\sim}{0}$.

In other words, employing a suggestive terminology, the derivative of

$$(Q^{\underset{\sim}{H}} f, g)$$

in the direction corresponding to $P = P_m^{max}$ is

$$-(1 \times Q^{I_M(\underset{\sim}{H})} f_P(?:?:\exp(\underset{\sim}{H}(P))),\ g_P(?:?:\exp(\underset{\sim}{H}(P)))),$$

there being no contribution from the remaining terms.

To formulate a general result, which is somewhat more involved due to a number of technical complications, it will be best to set up our problem relative to (P_0, S_0) -- then

$$\underset{\sim}{a}_0(\cdot) \sim a_0$$

and $\underset{\sim}{H}_0$ can be confined to $\underset{\sim}{a}_0(\cdot)$. Let F be a subset of $\Sigma_{P_0}^{o}(\mathfrak{g}, a_0)$. Attached to F is a differential operator $\underset{\sim}{D}_F$ defined on C^∞ functions $\underset{\sim}{f}_0 : \underset{\sim}{a}_0(\cdot) \to \underset{\sim}{C}$ by the prescription

$$\underset{\sim}{D}_F \underset{\sim}{f}_0 \Big|_{\underset{\sim}{H}_0} = \frac{\partial^{\#(ꟻ)}}{\ldots \partial t_i \ldots} \underset{\sim}{f}_0\Big(\underset{\sim}{H}_0(P_0) + \sum_{\lambda_i \in ꟻ} t_i H_{\lambda_i}\Big)\Big|_{t_i = 0\ (\forall i)}.$$

Here we have written ꟻ for the complement of F. Observe that $\underset{\sim}{D}_F$ is a directional derivative <u>unnormalized</u> by unit vectors. Set

$$\underset{\sim}{D}(F : \underset{\sim}{f}_0) = \underset{\sim}{D}_F \underset{\sim}{f}_0 \Big|_{\underset{\sim}{0}}.$$

Let $\mathcal{C}$ be an association class of Γ-cuspidal split parabolic subgroups of G; let

$$\mathcal{C} = \coprod_i \mathcal{C}_i$$

be a decomposition of $\mathcal{C}$ into G-conjugacy classes $\mathcal{C}_i$. Given $\mathcal{C}_i$, let $P_{i\mu}(1 \le \mu \le r_i)$ be a set of representatives for $\Gamma \backslash \mathcal{C}_i$. Suppose now that F is a subset of $\Sigma_{P_0}^{o}(\mathfrak{g}, a_0)$ with the property that

$$(P_0,S_0)_F \in \mathcal{C}_i.$$

In terms of this data, the main result of the present § is:

Proposition 5.4 Let $H \in \mathfrak{a}_Q$ -- then

$$\forall f,g \in \bigcup_{\Gamma} S_{\Gamma}^{\infty}(G/\Gamma),$$

we have

$$D(F:(Q^{H+?}f,g)) = (-1)^{\text{rank}(\mathcal{C})}|\det[(\lambda_i^F,\lambda_j^F):i,j\notin F]|^{1/2}$$

$$\times \sum_{\mu=1}^{r_i}(1\times Q^{I_{M_{i\mu}}(H)}f_{P_{i\mu}}(?:?:\exp(H(P_{i\mu}))),\ g_{P_{i\mu}}(?:?:\exp(H(P_{i\mu})))).$$

The proof is by induction on rank $(\mathcal{C})$, the case when rank$(\mathcal{C})$ = 1 being the gist of the spadework supra.

[Note: To account for the fact that no lengths appeared earlier, bear in mind that the derivative in the direction corresponding to $P = P_m^{max}$ was normalized.]

Assuming the result to be true for all association classes of rank ℓ or less, let $\mathcal{C}$ be an association class of rank = ℓ and let $\mathcal{C}'$ be an association class of rank = $\ell+1$, with

$$\begin{cases}(P_0,S_0)_F \in \mathcal{C}_i \\ (P_0,S_0)_{F'} \in \mathcal{C}'_{i'}\end{cases} \qquad (\mathcal{C}_i \succeq \mathcal{C}'_{i'},\ \text{say}).$$

To push the induction forward, choose the representatives $P'_{i\mu:\mu'}$ for $\Gamma\backslash\mathcal{C}'_{i'}$ compatibly with the representatives $P_{i\mu}$ for $\Gamma\backslash\mathcal{C}_i$

(cf. TES, pp. 164-166). If $F = F'\cup\{\lambda_0\}$, then

$$\underset{\sim}{D}_{F'} = \underset{\sim}{D}_{\{\lambda_0\}}\circ\underset{\sim}{D}_F$$

and

$$\underset{\sim}{D}(F':(Q^{\underset{\sim}{H}+?}f,g))$$

$$= \underset{\sim}{D}(\{\lambda_0\}:\underset{\sim}{D}_F(Q^{\underset{\sim}{H}+?}f,g))$$

$$= (-1)^{\operatorname{rank}(\mathfrak{C})}|\det[(\lambda_i^F,\lambda_j^F):i,j\notin F]|^{1/2}$$

$$\times \sum_{\mu=1}^{r_i} \underset{\sim}{D}(\{\lambda_0\}:(1\times Q^{I_{M_{i\mu}}}(\underset{\sim}{H})+?\ \ldots,\ldots))$$

$$= (-1)^{\operatorname{rank}(\mathfrak{C}')}|\det[(\lambda_i^F,\lambda_j^F):i,j\notin F]|^{1/2}\cdot\|\lambda_0^{F'}\|$$

$$\times \sum_{\mu=1}^{r_i}\sum_{\mu'=1}^{r_\mu}$$

$$(1\times Q^{I_{M'_{i\mu:\mu'}}}(\underset{\sim}{H}) f_{P'_{i\mu:\mu'}}(?:?:\exp(\underset{\sim}{H}(P'_{i\mu:\mu'}))),g_{P'_{i\mu:\mu'}}(?:?:\exp(\underset{\sim}{H}(P'_{i\mu:\mu'})))),$$

which finishes the induction modulo the relation

$$|\det[(\lambda_i^{F'},\lambda_j^{F'}):i,j\notin F']|^{1/2}$$

$$= |\det[(\lambda_i^F,\lambda_j^F):i,j\notin F]^{1/2}\cdot\|\lambda_0^{F'}\|.$$

Because this equality can be interpreted in the setting of a geometric g.r.s., we see that its validity is a consequence of the following generality.

Lemma 5.5 Let (V,Φ) be a geometric g.r.s.; let $\mathfrak{C}$ be a chamber of Φ -- then

$\forall F \subset F_0(\mathfrak{C})$:

$$|\det[(\lambda_i,\lambda_j):i,j\in L]|^{1/2}$$

$$= |\det[(\lambda_i,\lambda_j):i,j\in F]|^{1/2}\cdot|\det[(P_F\lambda_i,P_F\lambda_j):i,j\notin F]|^{1/2}.$$

[This is a variant on Lemma 2.10. One can, e.g., start by showing directly that

$$|\det[(\lambda_i,\lambda_j):i,j\in L]|^{1/2}$$

$$= \|\lambda_k\|\cdot|\det[(P_{\{k\}}\lambda_i,P_{\{k\}}\lambda_j):i,j\in L-\{k\}]|^{1/2}$$

and then proceed by induction on $\#(F)$.]

We want now to consider an application of the foregoing differentiation procedure which will be useful later on. To this end, let, as usual, $\mathcal{A}(G/\Gamma)$ be the space of automorphic forms on G/Γ -- then, of course,

$$\mathcal{A}(G/\Gamma) \subset \bigcup_r S_r^{\infty}(G/\Gamma).$$

Proposition 5.6 *Suppose that* $f,g\in\mathcal{A}(G/\Gamma)$ -- *then the inner product*

$$(Q^{\underset{\sim}{H}}f,g) \qquad (\underset{\sim}{H}\in\underset{\sim}{\mathfrak{a}}_Q)$$

is an exponential polynomial on $\underset{\sim}{\mathfrak{a}}_Q$.

Before giving the proof, we had best explain the meaning of the term "exponential polynomial on $\underset{\sim}{\mathfrak{a}}_Q$".

To begin with, let V be a finite dimensional vector space

over $\underset{\sim}{R}$ -- then by an exponential polynomial on an open subset U of V we understand a finite linear combination of functions of the form

$$p_\Lambda(H)\cdot e^{\langle H,\Lambda\rangle},$$

where

$$\begin{cases} p_\Lambda \in \underset{\sim}{C}[V] \\ \Lambda \in \operatorname{Hom}_{\underset{\sim}{R}}(V,\underset{\sim}{C}). \end{cases}$$

If, more generally, $V = V_0 + H_0$ is a finite dimensional affine space over $\underset{\sim}{R}$, then a function f on an open subset U of V is termed an exponential polynomial provided $f(? + H_0)$ is an exponential polynomial on $U_0 (\equiv U - H_0) \hookrightarrow V_0$. Naturally, this agreement is independent of the choice of H_0.

So, as $\underset{\sim}{a}_Q$ is an open subset of $\underset{\sim}{a}$, the assertion that

$$(\underset{\sim}{Q}^H f, g)$$

is an exponential polynomial on $\underset{\sim}{a}_Q$ is at least meaningful. To prove it, we need a simple lemma.

<u>Lemma 5.7</u>   <u>Let</u> f <u>be a</u> C^1 <u>function on an open convex subset</u> U of $\underset{\sim}{R}^n$. <u>Suppose that</u>

$$\forall k, \ \frac{\partial f}{\partial x_k} \text{ is an exponential polynomial on } U.$$

<u>Then</u> f <u>is an exponential polynomial on</u> U.

<u>Proof</u> We may assume that $n > 1$ and use induction. Put

$$F(x_1,\ldots,x_n) = \int_0^{x_n} \frac{\partial f}{\partial x_n}(x_1,\ldots,x_{n-1},t)dt.$$

In this connection, observe that $\frac{\partial f}{\partial x_n}$, being an exponential polynomial, is defined on all of $\underset{\sim}{R}^n$. On U itself,

$$\frac{\partial F}{\partial x_n} = \frac{\partial f}{\partial x_n} .$$

Accordingly, if

$$g(x_1,\ldots,x_n) = F(x_1,\ldots,x_n) - f(x_1,\ldots,x_n),$$

then $\frac{\partial g}{\partial x_n} = 0$. Furthermore, since F is an exponential polynomial,

$$\frac{\partial g}{\partial x_k} = \frac{\partial F}{\partial x_k} - \frac{\partial f}{\partial x_k}$$

is an exponential polynomial on U. Let

$$\pi : \underset{\sim}{R}^n \to \underset{\sim}{R}^{n-1}$$

be the canonical projection -- then g is constant on the fibers over $\pi(U)$, hence does not depend on x_n. One may thus define a function $g_\pi (= g \circ \pi^{-1})$ on $\pi(U)$ satisfying there the relations

$$\frac{\partial g_\pi}{\partial x_k}(x_1,\ldots,x_{n-1}) = \frac{\partial g}{\partial x_k}(x_1,\ldots,x_n).$$

By induction, g_π is an exponential polynomial. The same is therefore true of g and, consequently, of f. //

<u>Proof of Proposition 5.6</u> Since $\underset{\sim}{a}_Q$ is an open convex subset of $\underset{\sim}{a}$, we need only verify that the conditions of the preceding lemma are in force here. Fix, as was done earlier, a realization

$$\underset{\sim}{a} \sim \bigoplus_m a_m^{max} \quad .$$

Given an index m, the claim is that the derivative of

$$(Q^{\underset{\sim}{H}} f, g)$$

in the direction corresponding to $P = P_m^{max}$ is an exponential polynomial. But this derivative is precisely the negative of

$$(1 \times Q^{I_M(\underset{\sim}{H})} f_P(?:?:\exp(\underset{\sim}{H}(P))),\ g_P(?:?:\exp(\underset{\sim}{H}(P)))).$$

Bearing in mind that $f,g \in A(G/\Gamma)$, one can say that there exist finitely many elements

$$\begin{cases} \Lambda_i,\ \Lambda_j \in \check{\underset{\sim}{a}} + \sqrt{-1}\check{\underset{\sim}{a}} \\ u_i,\ u_j \in \underset{\sim}{C}[a], \end{cases}$$

along with automorphic forms

$$\Phi_i,\ \Phi_j \quad \text{on} \quad K \times M/\Gamma_M ,$$

such that

$$\begin{cases} f_P(kma) = \sum_i a^{\Lambda_i} p_{u_i}(H)\Phi_i(km) \\ g_P(kma) = \sum_j a^{\Lambda_j} p_{u_j}(H)\Phi_j(km) \quad . \end{cases} \qquad (H = \log a)$$

Using these decompositions, the inner product

$$(1 \times Q^{I_M(\underset{\sim}{H})} f_P(?:?:\exp(\underset{\sim}{H}(P))),\ g_P(?:?:\exp(\underset{\sim}{H}(P))))$$

can be written as a sum over i and j of the

$$p_i(\underset{\sim}{H}(P))\overline{p_j(\underset{\sim}{H}(P))} \cdot \exp(\langle \underset{\sim}{H}(P), \Lambda_i + \bar{\Lambda}_j \rangle) \cdot (1 \times Q^{I_M(\underset{\sim}{H})} \Phi_i, \Phi_j).$$

Proceeding inductively, the relation

$$I_M(\underset{\sim}{a}_Q) \subset (\underset{\sim}{a}_M)_Q$$

allows one to infer that

$$(1 \times Q^{I_M(\underset{\sim}{H})}\Phi_i,\Phi_j)$$

is an exponential polynomial in $I_M(\underset{\sim}{H})$. On the other hand, $I_M(\underset{\sim}{H})$ is independent of $\underset{\sim}{H}(P)$ so, to draw the required conclusion, it suffices to note that in coordinates, any function of the form

$$p(x_n)e^{Cx_n}\cdot[\text{exp. poly. in } (x_1,\ldots,x_{n-1})]$$

is an exponential polynomial per $(x_1,\ldots,x_n)$. //

§6. Exponents

The purpose of this § is to introduce the notion of "exponent", there being two variations on this theme, the one involving truncation, the other involving constant terms. A central conclusion is that the latter completely determine the former, a fact which, in the cases of interest to us, means that their positions can be ascertained with certainty.

Let $\check{\underset{\sim}{a}} + \sqrt{-1}\check{\underset{\sim}{a}}$ be the affine dual of $\underset{\sim}{a}$ -- then, $\forall\Gamma$-cuspidal $P = M\cdot A\cdot N$, there exists a canonical map

$$\begin{cases} \check{a} + \sqrt{-1}\check{a} \rightarrow \check{\underset{\sim}{a}} + \sqrt{-1}\check{\underset{\sim}{a}} \\ \Lambda \mapsto \underset{\sim}{\Lambda}_P, \end{cases}$$

namely

$$\langle \underset{\sim}{H}, \underset{\sim}{\Lambda}_P \rangle = \langle \underset{\sim}{H}(P), \Lambda \rangle \qquad (\underset{\sim}{H} \in \underset{\sim}{a}).$$

An affine functional $\underset{\sim}{\Lambda}$ on $\underset{\sim}{a}$ is said to be P-pure if it is of the form $\underset{\sim}{\Lambda} = \underset{\sim}{\Lambda}_P$ $(\exists\Lambda)$. P-purity is evidently a function of the Γ-conjugacy class of P.

Let $f, g \in A(G/\Gamma)$ -- then, according to Proposition 5.6, the inner product

$$(Q^{\underset{\sim}{H}} f, g)$$

is an exponential polynomial on $\underset{\sim}{a}_Q$. The affine functionals $\underset{\sim}{\Lambda}$

thereby determined are essentially unique. They will be called the truncation exponents of the pair (f,g), the set of such being denoted by

$$\underset{\sim}{E}(f,g).$$

A natural question then suggests itself: Is it possible to predict the nature of the elements in the set $\underset{\sim}{E}(f,g)$ from, e.g., the nature of the f_P and g_P? As will be shown below, the answer is affirmative.

To formulate a precise statement, let us first agree on what we shall understand by the "normal form" of a constant term. Thus let $f \in A(G/\Gamma)$ -- then, given any Γ-cuspidal split parabolic subgroup (P,S) of G with special split component A, there exist finitely many elements

$$\begin{cases} \Lambda_i \in \check{\mathfrak{a}} + \sqrt{-1}\check{\mathfrak{a}} \\ u_i \in \underset{\sim}{C}[\mathfrak{a}], \end{cases}$$

along with automorphic forms

$$\Phi_i \quad \text{on} \quad K \times M/\Gamma_M,$$

with the property that

$$f_P(kma) = \sum_i a^{\Lambda_i} p_{u_i}(H)\Phi_i(km) \qquad (H = \log a),$$

which, after regrouping, can be written as

$$f_P(kma) = \sum_i a^{\Lambda_i}\left(\sum_j p_{ij}(H)\Phi_{ij}(km)\right) \qquad (H=\log a),$$

where the Λ_i are distinct, and the p_{ij}, Φ_{ij} are linearly independent. This is "the" normal form of f_P.

Lemma 6.1 There exist a finite set of points $a_k \in A$ and a finite set of constants $C_k \in \mathbb{C}$ such that on $K \times M$

$$\Phi_{ij} = \sum_k C_k \cdot (f_P \circ R_{a_k}),$$

R_{a_k} being the right translation operator.

[Let $\{\phi_k\}$ be an enumeration of the $\{p_{ij} e^{\Lambda_i}\}$. Because the $p_{ij} e^{\Lambda_i}$ comprise a linearly independent set of exponential polynomials, there exists a finite set of points $a_k \in A$ for which the matrix

$$[\phi_{k'}(a_{k''})]$$

is invertible. A little linear algebra then leads easily to the assertion.]

Let again $f \in A(G/\Gamma)$ -- then, per the normal form of f_P, we put

$$E_P(f) = \{\Lambda_i\},$$

the constant term exponents of f along P.

Given now $f, g \in A(G/\Gamma)$, write

$$\underset{\sim}{E}_P(f,g)$$

for the set

$$\{\underset{\sim}{\Lambda}_P' + \bar{\underset{\sim}{\Lambda}}_P'':\Lambda'\in E_P(f),\ \Lambda''\in E_P(g)\}.$$

The elements of $\underset{\sim}{E}_P(f,g)$ will be referred to as the constant term exponents of the pair (f,g) along P.

Theorem 6.2 Suppose that $f,g\in A(G/\Gamma)$ -- then

$$\underset{\sim}{E}(f,g) \subset \bigcup_{P\in \mathcal{C}_\Gamma} \underset{\sim}{E}_P(f,g).$$

[Note: The union on the right is finite.]

Proof Fix a $\underset{\sim}{\Lambda}\in\underset{\sim}{E}(f,g)$. Obviously,

$$\underset{\sim}{0}\in\underset{\sim}{E}(f,g) \Longrightarrow \underset{\sim}{0}\in\underset{\sim}{E}_G(f,g) = \{\underset{\sim}{0}\}.$$

Assume, therefore, that $\underset{\sim}{\Lambda} \neq \underset{\sim}{0}$ -- then, relative to a realization

$$\underset{\sim}{a} \sim \bigoplus_m a_m^{max},$$

$\exists$ an index m such that the derivative $\partial_m \underset{\sim}{\Lambda}$ associated with m does not vanish. Taking $P = P_m^{max}$, consider the derivative of

$$(Q^{\underset{\sim}{H}}f,g)$$

in the direction corresponding to $P = P_m^{max}$, that is (cf. §5), consider the sum over i and j of

$$\exp(\langle \underset{\sim}{H}(P),\Lambda_i + \bar{\Lambda}_j\rangle)$$

times the sum over k and ℓ of

$$p_{ik}(\underset{\sim}{H}(P))\overline{p_{j\ell}(\underset{\sim}{H}(P))}\cdot(1\times Q^{I_M(\underset{\sim}{H})}\Phi_{ik},\Phi_{j\ell}).$$

Choose i and j such that the action of $\underset{\sim}{\Lambda}$ on $\underset{\sim}{H}(P)$ is $\Lambda_i + \bar{\Lambda}_j$ -- then the action of $\underset{\sim}{\Lambda}$ on the rest of $\underset{\sim}{a}$ must be given by a truncation exponent of a pair $(\Phi_{ik},\Phi_{j\ell})$. But, proceeding inductively, the truncation exponents of any pair are controlled by the conclusion of our theorem. In other words, there exists a dominated predecessor P' of P, indices i_0,k_0 and j_0,ℓ_0, and elements

$$\begin{cases} '\Lambda_f \in E_{'P}(\Phi_{i_0k_0}) \\ '\Lambda_g \in E_{'P}(\Phi_{j_0\ell_0}) \end{cases}$$

such that

$$\langle \underset{\sim}{H},\underset{\sim}{\Lambda}\rangle = \langle \underset{\sim}{H}(P'),\Lambda_{i_0} + '\Lambda_f + \bar{\Lambda}_{j_0} + '\bar{\Lambda}_g\rangle .$$

Accordingly, it need only be shown that

$$\begin{cases} \Lambda_{i_0} + '\Lambda_f \in E_{P'}(f) \\ \Lambda_{j_0} + '\Lambda_g \in E_{P'}(g). \end{cases}$$

Since the situation is symmetric, we shall deal explicitly with $\Lambda_{i_0} + '\Lambda_f$. Utilizing the decomposition $M = K_M\cdot 'P$, write, in a suggestive notation,

$$(\Phi_{ik})_{'P}(k_M'm'a) = \sum_{'i} ('a)^{\Lambda_{'i}}(\sum_{'k} p_{'i'k}('H)\Phi_{'i'k}(k_M'm))\;('H = \log\, 'a).$$

Then

$$f_{P'}(m'a')$$

$$= f_{P'}('m('aa))$$

$$= \sum_i a^{\Lambda_i}(\sum_k p_{ik}(H)(\Phi_{ik})_{'P}('m'a))$$

$$= \sum_{i,'i} (a')^{\Lambda_i + \Lambda_{'i}}(\sum_{k,'k} p_{ik}(H)p_{'i'k}('H)\Phi_{'i'k}('m)).$$

Because

$$'\Lambda_f \in E_{'P}(\Phi_{i_0k_0}),$$

$'\Lambda_f$ must appear as a $\Lambda_{'i}$. That

$$\Lambda_{i_0} + '\Lambda_f \in E_{P'}(f)$$

is then seen to follow from Lemma 6.1. //

It is a corollary that every truncation exponent Λ of the pair (f,g) is P-pure for some P.

To tie down the truncation exponents even more, we shall need a preliminary result which is of interest in its own right.

Proposition 6.3 *Let $\mathcal{C}$ be an association class of Γ-cuspidal split parabolic subgroups of G. Suppose that f is an automorphic form on G/Γ such that*

$$\forall P \notin \mathcal{C},$$

$$f_P \sim 0,$$

but

$$\forall P \in \mathcal{C},\ \forall \Lambda_i \in E_P(f),$$

$$\mathrm{Re}(\Lambda_i) \in \mathcal{D}_P(\check{\mathfrak{a}})^-.$$

Then

$\forall P \in \mathcal{C}_\Gamma$, $\forall \Lambda_i \in E_P(f)$,

$$\mathrm{Re}(\Lambda_i) \in \mathfrak{D}_P(\check{\mathfrak{a}})^-.$$

This is the "principle of permanence" for exponents. Its proof depends on an elementary lemma, the thrust of which is basically familiar, if not explicitly known.

Given an association class $\mathcal{C}$, recall that a Γ-cuspidal P_0 dominates $\mathcal{C}$ iff

$$\mathrm{Dom}_{\mathcal{C}}(P_0) = \{P \in \mathcal{C} : P \preceq P_0\}$$

is nonempty.

Lemma 6.4 Let $\mathcal{C}$ be an association class of Γ-cuspidal split parabolic subgroups of G. Suppose that f is an automorphic form on G/Γ such that

$\forall P \notin \mathcal{C}$,

$$f_P \sim 0.$$

Fix a P_0 dominating $\mathcal{C}$ -- then

$\forall i,j$, $\forall P \preceq P_0$,

$$(\Phi_{ij})_{P^\dagger} \sim 0$$

unless

$$P \in \mathrm{Dom}_{\mathcal{C}}(P_0).$$

[The indices i,j referring to P_0, Lemma 6.1 implies that

$$(\Phi_{ij})_{P^\dagger} = \sum_k C_k \cdot (f_P \circ R_{a_k}),$$

from which our assertion follows immediately.]

Proof of Proposition 6.3 If P dominates no element of $\mathcal{C}$, then actually $f_P \equiv 0$ (by Langlands lemma; cf. TES, p. 82). Consequently, we may assume that P dominates $\mathcal{C}$. Write, as above,

$$f_P(kma) = \sum_i a^{\Lambda_i}\Big(\sum_j p_{ij}(H)\Phi_{ij}(km)\Big) \qquad (H = \log a).$$

In particular, the Λ_i are all distinct and the contention is that their real parts lie in $\mathcal{D}_P(\check{\mathfrak{a}})^-$. To get a contradiction, suppose not -- then, for some index i_0, say, $\mathrm{Re}(\Lambda_{i_0}) \notin \mathcal{D}_P(\check{\mathfrak{a}})^-$. This in turn implies that

$$\forall P' \preceq P, \ \forall {}'\Lambda \in {}'\check{\mathfrak{a}} + \sqrt{-1}\,{}'\check{\mathfrak{a}},$$

$$\mathrm{Re}(\Lambda_{i_0} + {}'\Lambda) \notin \mathcal{D}_{P'}(\check{\mathfrak{a}}')^-.$$

Owing to Lemma 6.4 (and, of course, Langlands lemma again), we can find a $P' \in \mathrm{Dom}_{\mathcal{C}}(P)$ and an index j_0 for which

$$(\Phi_{i_0 j_0})_{{}'P} \neq 0.$$

Then, as in the proof of Theorem 6.2,

$$f_{P'}(m'a')$$

$$= \sum_{i,{}'i} (a')^{\Lambda_i + \Lambda_{'i}}\Big(\sum_{j,{}'j} p_{ij}(H)p_{'i'j}({}'H)\Phi_{'i'j}({}'m)\Big),$$

an impossibility, since

$$\mathrm{Re}(\Lambda_{i_0} + \Lambda_{'i}) \notin \mathfrak{D}_{P'}(\check{a}')^-$$

for all $\Lambda_{'i}$. //

Denote by $\underset{\sim}{\mathrm{J}}_C$ that subset of the affine dual of $\underset{\sim}{a}$ comprised of those $\underset{\sim}{\Lambda}$ for which

$$\underset{\sim}{H}_1 < \underset{\sim}{H}_2 \Rightarrow \mathrm{Re}(\langle \underset{\sim}{H}_1, \underset{\sim}{\Lambda}\rangle) \leq \mathrm{Re}(\langle \underset{\sim}{H}_2, \underset{\sim}{\Lambda}\rangle).$$

Given a Γ-cuspidal P,

$$\underset{\sim}{\Lambda}_P \in \underset{\sim}{\mathrm{J}}_C \text{ iff } \mathrm{Re}(\Lambda) \in \mathfrak{D}_P(\check{a})^-,$$

as can be easily verified by employing ♦.

Theorem 6.5 Let $\mathcal{C}$ be an association class of Γ-cuspidal split parabolic subgroups of G. Suppose that f, g are automorphic forms on G/Γ such that

$$\begin{cases} \forall P \notin \mathcal{C}, \\ \quad f_P \sim 0 \end{cases} \qquad \begin{cases} \forall P \notin \mathcal{C}, \\ \quad g_P \sim 0, \end{cases}$$

but

$$\begin{cases} \forall P \in \mathcal{C}, \forall \Lambda_i \in E_P(f) \\ \quad \mathrm{Re}(\Lambda_i) \in \mathfrak{D}_P(\check{a})^- \end{cases} \qquad \begin{cases} \forall P \in \mathcal{C}, \forall \Lambda_j \in E_P(g), \\ \quad \mathrm{Re}(\Lambda_j) \in \mathfrak{D}_P(\check{a})^-. \end{cases}$$

Then

$$\underset{\sim}{E}(f,g) \subset \underset{\sim}{\mathrm{J}}_C .$$

[To prove this, one simply combines Theorem 6.2 with the preceding proposition. It should also be noted that if the association classes were distinct, then $\underset{\sim}{E}(f,g)$ would be empty.]

In passing, let us formally isolate for later reference an important deduction, the proof of which the reader will agree is implicit in the foregoing.

Scholium Let $\mathfrak{C}$ be an association class of Γ-cuspidal split parabolic subgroups of G. Suppose that f is an automorphic form on G/Γ such that

$$\forall P \notin \mathfrak{C},$$

$$f_P \sim 0 .$$

Then

$$\forall P \in \mathfrak{C}_\Gamma, \quad \forall \Lambda_i \in E_P(f),$$

$$\exists P_i \in \mathrm{Dom}_{\mathfrak{C}}(P), \quad \exists \Lambda_i(i) \in E_{P_i}(f)$$

for which

$$\mathrm{pro}(\Lambda_i(i)) = \Lambda_i .$$

[Here, pro stands for the ambient projection. Needless to say, $E_P(f)$ is empty unless P dominates $\mathfrak{C}$.]

We shall terminate this § with some remarks about square integrable automorphic forms.

Lemma 6.6 Let $f \in \mathcal{A}(G/\Gamma) \cap L^2(G/\Gamma)$ -- then $\forall\Gamma$-cuspidal $P \neq G$,

$$E_P(f) \subset \mathcal{D}_P(\check{\mathfrak{a}}).$$

[In fact, $f \in L^2_{dis}(G/\Gamma)$, thus, for basic reasons, is a finite linear combination of certain Eisenstein system functions

$$E(\mathfrak{X}:G|\{1\}:P_{i_0}|A_{i_0}:T_{i_0}:X(P_{i_0},A_{i_0}):?).$$

Our assertion therefore follows from conditions E-S:I and Geom:III in TES.]

There is also a converse result.

Lemma 6.7 Let $f \in \mathcal{A}(G/\Gamma)$. If $\forall\Gamma$-cuspidal $P \neq G$ and if $\forall\Lambda_i \in E_P(f)$,

$$\text{Re}(\Lambda_i) \in \mathcal{D}_P(\check{\mathfrak{a}}),$$

then $f \in L^2(G/\Gamma)$.

[Note: As an automatic consequence, $\text{Im}(\Lambda_i)$ must necessarily be zero $\forall i$.]

The proof depends on a

Sublemma Let $f \in S_r^\infty(G/\Gamma)$. Suppose that $(Q^H f, f)$ stays bounded as $H \to -\infty$ in $\mathfrak{a}_Q$ -- then $f \in L^2(G/\Gamma)$.

Deferring the proof of this momentarily, let us return to Lemma 6.7. Owing to Theorem 6.2 (and the hypotheses), all the terms in the expression

$$\sum_{\Lambda \in E(f,f)} P_\Lambda(H) \cdot e^{\langle H, \Lambda \rangle}$$

of $(Q^H f, f)$ as an exponential polynomial on $\underline{a}_Q$ have to approach

0 as $H \to -\infty$ except, perhaps, for the one corresponding to $\underline{0}$.

However, $\underline{0}$ does not survive differentiation in any direction so

P_0 is a constant. Accordingly, $(Q^H f, f)$ tends to a finite limit

as $H \to -\infty$, hence, by the criterion supra, f is indeed L^2.

Proof of Sublemma   Fix an $M > 0$ with the property that

$$M \geq (Q^H f, f) = \|Q^H f\|^2 \qquad \forall H \in \underline{a}_Q.$$

Let C be a compact subset of G/Γ, χ_C its characteristic func-

tion -- then, C being arbitrary, to force f into $L^2(G/\Gamma)$,

it need only be shown that

$$\|\chi_C f\| \leq \sqrt{M} \ .$$

To this end, take H sufficiently negative -- then

$$Q^H \cdot (\chi_C \cdot ?) = \chi_C \cdot ?,$$

which implies that

$$\forall g \in L^2(C),$$

$$(Q^H f, g) = (f, Q^H g) = (f, g),$$

i.e., $f = Q^H f$ a.e. on C. Consequently,

$$(Q^H f, \chi_C f) = (f, \chi_C f)$$

$$\Rightarrow \qquad = \|\chi_C f\|^2 \leq \|Q^H f\| \cdot \|\chi_C f\|$$

$$\|\chi_C f\| \leq \sqrt{M} \ ,$$

as contended. //

One last point should be noted. Suppose that f and g are square integrable automorphic forms on G/Γ. Write

$$(Q^{H}f,g) = \sum_{\Lambda\in E(f,g)} p_{\Lambda}(H)\cdot e^{\langle H,\Lambda\rangle}. \qquad (H\in a_Q).$$

Then we claim that

$$p_0 = (f,g),$$

a constant. Here is a simple way to see this. Bearing in mind Lemma 6.6, one can say that

$$p_0(H) - (f,g),$$

i.e.,

$$(Q^{H}f,g) - (f,g) - \sum_{\substack{\Lambda\in E(f,g)\\ \Lambda\neq 0}} p_{\Lambda}(H)\cdot e^{\langle H,\Lambda\rangle}$$

goes to 0 as H goes to $-\infty$. For any $H\in a_Q$,

$$H + tH_{\rho}\in a_Q \qquad (t\leq 0).$$

Consequently,

$$\lim_{t\to-\infty} [p_0(H+tH_{\rho}) - (f,g)] = 0$$

$$\Rightarrow p_0(H+tH_{\rho}) = (f,g) \qquad \forall t$$

$$\Rightarrow p_0(H) = (f,g).$$

Because a_Q is open, it is now clear that

$$p_0 = (f,g).$$

§7. Residues

The purpose of this § is to take the inner product formula obtained in §7 of [2-(c)] and examine the form it assumes upon application of the residue taking process. The results of the preceding § will then eventually be brought to bear to study the nature of the exponents arising thereby.

Let us begin by recalling the point of departure for all this. Thus fix an association class $\mathcal{C}_0$ of Γ-cuspidal split parabolic subgroups of G. Let $\mathcal{C}_{i_0}$ be a G-conjugacy class in $\mathcal{C}_0$, $\mathfrak{X}$ an equivariant system of admissible affine subspaces attached to $\mathcal{C}_{i_0}$. Suppose given an Eisenstein system $\{E,\nabla\}$ belonging to $\mathfrak{X}$. There is then attached to each

$$(P_{i_0},S_{i_0};A_{i_0})\in\mathcal{C}_{i_0}$$

an E-function

$$E(\mathfrak{X}:G|\{1\}:P_{i_0}|A_{i_0}:T_{i_0}:\Lambda_{i_0}:?).$$

Let $\mathcal{C}_{j_0}$ be another G-conjugacy class in $\mathcal{C}_0$ -- then with any

$$(P_{j_0},S_{j_0};A_{j_0})\in\mathcal{C}_{j_0}$$

there is associated an Eisenstein series

$$E(P_{j_0}|A_{j_0}:\Phi_{j_0}:\Lambda_{j_0}:?).$$

Here, of course,

$$\begin{cases} T_{i_0} \in \mathrm{Hom}(S_{\mathfrak{X}(P_{i_0}, A_{i_0})}, E_{cus}(\delta, \mathcal{O}_{i_0})) \\ \Phi_{j_0} \in E_{cus}(\delta, \mathcal{O}_{j_0}), \end{cases}$$

$\mathcal{O}_{i_0}$ and $\mathcal{O}_{j_0}$ belonging to the same orbit type, $\underset{\sim}{\mathcal{O}}_0$, say. Confining $\underset{\sim}{H}$ to $\underset{\sim}{a}_Q$, on general grounds, one knows that the inner product

$$(Q^{\underset{\sim}{H}} E(\mathfrak{X}:G|\{1\}:P_{i_0}|A_{i_0}:T_{i_0}:\Lambda_{i_0}:?), E(P_{j_0}|A_{j_0}:\Phi_{j_0}:\Lambda_{j_0}:?))$$

is a meromorphic function of $(\Lambda_{i_0}, \bar{\Lambda}_{j_0})$. The thrust of Theorem 7.3 in [2-(c)] is to provide an explicit expression for it. Indeed, in the notation of that theorem, we can say that our inner product is equal to the sum

$$(-1)^{\ell_0} \cdot \mathrm{vol}(\mathfrak{C}_0) \cdot \sum_{k_0=1}^{r_0} \sum_{\xi_0=1}^{r_{k_0}} \sum_{w_{k_0\xi_0}:j_0 \in W(A_{k_0\xi_0}, A_{j_0})} \sum_{w_{k_0\xi_0}:i_0 \in W_{\{G\}}(\mathfrak{X}; A_{k_0\xi_0}, A_{i_0})}$$

of the

$$\nabla(\mathfrak{X}:G|(\{1\},\{1\}):P_{k_0\xi_0}|A_{k_0\xi_0}:P_{i_0}|A_{i_0}:w_{k_0\xi_0}:i_0:\Lambda_{i_0})T_{i_0}$$

paired with the

$$d(\exp\langle \underset{\sim}{H}(P_{k_0\xi_0}), w_{k_0\xi_0}:j_0\Lambda_{j_0} - ?\rangle)(1/\prod_{\lambda_{k_0\xi_0}} (w_{k_0\xi_0}:j_0\Lambda_{j_0} - ?, \lambda_{k_0\xi_0})))(-w_{k_0\xi_0}:i_0\bar{\Lambda}_{i_0})$$
$$\otimes\, c_{cus}(P_{k_0\xi_0}|A_{k_0\xi_0}:P_{j_0}|A_{j_0}:w_{k_0\xi_0}:j_0:\Lambda_{j_0})\Phi_{j_0}.$$

Actually, the proof of this result makes use of the E-function

$$E(\mathfrak{X}:G|\{1\}:P_{i_0}|A_{i_0}:T_{i_0}:\Lambda_{i_0}:?)$$

only through the form of its constant term along $P_{k_0\xi_0}$, as is reflected by the inner sum over

$$W_{\{G\}}(\mathfrak{X};A_{k_0\xi_0},A_{i_0}).$$

A more general statement is therefore valid in that the E-function can be replaced by an arbitrary automorphic form $f\in A(G/\Gamma)$. The point is simply this. There exist finitely many elements

$$\begin{cases} \Lambda_h \in \check{a}_{k_0\xi_0} + \sqrt{-1}\check{a}_{k_0\xi_0} \\ u_h \in \underset{\sim}{C}[a_{k_0\xi_0}], \end{cases}$$

along with automorphic forms

$$\Phi_h \quad \text{on} \quad K \times M_{k_0\xi_0}/\Gamma_{M_{k_0\xi_0}},$$

in which, prior to normalization,

$$f_{P_{k_0\xi_0}} (= f_{K \times P_{k_0\xi_0}}):$$

$$f_{P_{k_0\xi_0}}(kma) = \sum_h a^{\Lambda_h} p_{u_h}(H)\Phi_h(km) \quad (H = \log a).$$

Accordingly, the inner product

$$(\underset{\sim}{Q}^H f,\ E(P_{j_0}|A_{j_0}:\Phi_{j_0}:\Lambda_{j_0}:?))$$

is equal to the sum

$$(-1)^{\ell_0}\cdot \mathrm{vol}(\mathfrak{C}_0)\cdot \sum_{k_0=1}^{r_0} \sum_{\xi_0=1}^{r_{k_0}} \sum_{w_{k_0\xi_0:j_0}\in W(A_{k_0\xi_0},A_{j_0})} \sum_h$$

of the

$$u_h \otimes \Phi_h$$

paired with the

$$d(\exp(\langle \underset{\sim}{H}(P_{k_0\xi_0}), w_{k_0\xi_0:j_0}\Lambda_{j_0} - ?\rangle)(1/\prod_{\lambda_{k_0\xi_0}}(w_{k_0\xi_0:j_0}\Lambda_{j_0} - ?, \lambda_{k_0\xi_0})))(-\bar{\Lambda}_h)$$

$$\otimes\ c_{cus}(P_{k_0\xi_0}|A_{k_0\xi_0}:P_{j_0}|A_{j_0}:w_{k_0\xi_0:j_0}:\Lambda_{j_0})\Phi_{j_0}$$

or still, of the

$$\Phi_h$$

paired with the

$$d(\exp(\langle \underset{\sim}{H}(P_{k_0\xi_0}), w_{k_0\xi_0:j_0}\Lambda_{j_0} - ?\rangle)(1/\prod_{\lambda_{k_0\xi_0}}(w_{k_0\xi_0:j_0}\Lambda_{j_0} - ?, \lambda_{k_0\xi_0})))(-\bar{\Lambda}_h)(u_h^*)$$

$$\cdot\ c_{cus}(P_{k_0\xi_0}|A_{k_0\xi_0}:P_{j_0}|A_{j_0}:w_{k_0\xi_0:j_0}:\Lambda_{j_0})\Phi_{j_0}\ .$$

Naturally, the latter pairing is nothing more than the usual inner product on

$$K \times M_{k_0\xi_0}/\Gamma_{M_{k_0\xi_0}},$$

$d(\ldots)(u_h^*)$ being a scalar. To simplify this expression, put

$$\phi_{k_0\xi_0}(H:\Lambda',\Lambda'') = \frac{e^{\langle H, \Lambda' - \Lambda''\rangle}}{\prod_{\lambda_{k_0\xi_0}}(\Lambda' - \Lambda'', \lambda_{k_0\xi_0})}\ .$$

Then

$$(\Phi_h, d(\ldots)(u_h^*)\cdot c_{cus}(\ldots)\Phi_{j_0})$$

is the same as

$$(\Phi_h, c_{cus}(P_{k_0\xi_0}|A_{k_0\xi_0}:P_{j_0}|A_{j_0}:w_{k_0\xi_0:j_0}:\Lambda_{j_0})\Phi_{j_0})$$

times the complex conjugate of

$$d[\phi_{k_0\xi_0}(\underset{\sim}{H}(P_{k_0\xi_0}):w_{k_0\xi_0:j_0}\Lambda_{j_0}, ?)](-\ \bar{\Lambda}_h)(u_h^*),$$

i.e., is the same as

$$(\Phi_h, c_{cus}(P_{k_0\xi_0}|A_{k_0\xi_0}:P_{j_0}|A_{j_0}:w_{k_0\xi_0:j_0}:\Lambda_{j_0})\Phi_{j_0})$$

times

$$d(\phi_{k_0\xi_0}(\underset{\sim}{H}(P_{k_0\xi_0}):?,-w_{k_0\xi_0:j_0}\bar{\Lambda}_{j_0})](\Lambda_h)(u_h)$$

To recapitulate:

Theorem 7.1 Let $f\in A(G/\Gamma)$ -- then the inner product

$$(Q^{\underset{\sim}{H}}f, E(P_{j_0}|A_{j_0}:\Phi_{j_0}:\Lambda_{j_0}:?))$$

is equal to the sum

$$(-1)^{\ell_0}\cdot\operatorname{vol}(\mathfrak{C}_0)\cdot\sum_{k_0=1}^{r_0}\sum_{\xi_0=1}^{r_{k_0}}\sum_{w_{k_0\xi_0:j_0}\in W(A_{k_0\xi_0},A_{j_0})}\sum_h$$

of the

$$(\Phi_h, c_{cus}(P_{k_0\xi_0}|A_{k_0\xi_0}:P_{j_0}|A_{j_0}:w_{k_0\xi_0:j_0}:\Lambda_{j_0})\Phi_{j_0})$$

times

$$d[\phi_{k_0\xi_0}(\underset{\sim}{H}(P_{k_0\xi_0}):?,-w_{k_0\xi_0:j_0}\bar{\Lambda}_{j_0})](\Lambda_h)(u_h).$$

[In this connection, bear in mind that $\underset{\sim}{H}\in\underset{\sim}{a}_Q$.]

The differential

$$d[\phi_{k_0\xi_0}(\underset{\sim}{H}(P_{k_0\xi_0}):?,-w_{k_0\xi_0:j_0}\bar{\Lambda}_{j_0})](\Lambda_h)(u_h)$$

can be regarded as

$$\partial_{u_h}[\phi_{k_0\xi_0}(\underset{\sim}{H}(P_{k_0\xi_0}):?,-w_{k_0\xi_0:j_0}\bar{\Lambda}_{j_0})](\Lambda_h),$$

a certain derivative of

$$\frac{e^{\langle \underset{\sim}{H}(P_{k_0\xi_0}), ? + w_{k_0\xi_0}:j_0\bar{\Lambda}_{j_0}\rangle}}{\prod_{\lambda_{k_0\xi_0}}(? + w_{k_0\xi_0}:j_0\bar{\Lambda}_{j_0},\lambda_{k_0\xi_0})}$$

evaluated at Λ_h. It is not difficult to be explicit about its form. In fact, agreeing to suppress the indices, we have:

<u>Lemma 7.2</u> $\forall u,\Lambda',\Lambda''$:

$$H \mapsto \partial_u[\phi(H:?,\Lambda'')](\Lambda')$$

<u>is given by</u>

$$H \mapsto U(H:\Lambda',\Lambda'') \cdot e^{\langle H,\Lambda' - \Lambda''\rangle},$$

<u>where</u>

$$U(H:\Lambda',\Lambda'') \in \underset{\sim}{C}[a,(\Lambda' - \Lambda'',\lambda_1)^{-1},\ldots,(\Lambda' - \Lambda'',\lambda_\ell)^{-1}].$$

[Here, as usual, $\lambda_1,\ldots,\lambda_\ell$ is an enumeration of $\Sigma_p^0(\mathfrak{g},\mathfrak{a})$.]

<u>Proof</u> The issue is the ∂_u-stability of

$$\underset{\sim}{C}[a,(? - \Lambda'',\lambda_1)^{-1},\ldots,(? - \Lambda'',\lambda_\ell)^{-1}] \cdot e^{\langle H,? - \Lambda''\rangle}.$$

To establish this, decompose u as a sum of products of degree one -- then, with $u = \Lambda$, say, it need only be shown that

$$\partial_\Lambda \left\{\prod_{i=1}^{L} \langle H,\Lambda_i\rangle \cdot \prod_{i=1}^{\ell} (? - \Lambda'',\lambda_i)^{-n_i}\right\} \cdot e^{\langle H,? - \Lambda''\rangle}$$

has the correct form, or still, that

$$\frac{d}{dt}\left\{\prod_{i=1}^{L}\langle H,\Lambda_i\rangle \cdot \prod_{i=1}^{\ell} (\Lambda' + t\Lambda - \Lambda'',\lambda_i)^{-n_i}\right\} \cdot e^{\langle H,\Lambda' + t\Lambda - \Lambda''\rangle}\Big|_{t=0}$$

has the correct form which, however, follows immediately by an elementary computation. //

Returning to Theorem 7.1, set

$$T_h(H:\Lambda) = e^{-\langle H,\Lambda\rangle} \cdot \partial_{u_h}\left(\frac{e^{\langle H,?\rangle}}{\prod_{\lambda_{k_0\xi_0}}(?,\lambda_{k_0\xi_0})}\right)(\Lambda).$$

Then the inner product

$$(Q_{\underset{\sim}{H}}f,\ E(P_{j_0}|A_{j_0}:\Phi_{j_0}:\Lambda_{j_0}:?))$$

is equal to the sum

$$(-1)^{\ell_0}\cdot\mathrm{vol}(\mathfrak{C}_0)\cdot\sum_{k_0=1}^{r_0}\ \sum_{\xi_0=1}^{r_{k_0}}\ \sum_{w_{k_0\xi_0:j_0}\in W(A_{k_0\xi_0},A_{j_0})}\ \sum_h$$

of the

$$T_h(\underset{\sim}{H}(P_{k_0\xi_0}):\Lambda_h+w_{k_0\xi_0:j_0}\bar{\Lambda}_{j_0})\cdot\exp(\langle\underset{\sim}{H}(P_{k_0\xi_0}),\Lambda_h+w_{k_0\xi_0:j_0}\bar{\Lambda}_{j_0}\rangle)$$
$$\times\ (\Phi_h, c_{cus}(P_{k_0\xi_0}|A_{k_0\xi_0}:P_{j_0}|A_{j_0}:w_{k_0\xi_0:j_0}:\Lambda_{j_0})\Phi_{j_0}),$$

an expression which, for the purposes at hand, is slightly more convenient than the one appearing in the statement of the theorem per se. Now the dependence of our function on Λ_{j_0} is conjugate meromorphic, a minor but irritating nuisance. We shall therefore rectify the situation by taking complex conjugates throughout, meaning, then, that the issue will be that of the inner product

$$(Q_{\underset{\sim}{H}}E(P_{j_0}|A_{j_0}:\Phi_{j_0}:\Lambda_{j_0}:?),f)$$

or still, the sum

$$(-1)^{\ell_0}\cdot \mathrm{vol}(\mathfrak{C}_0)\cdot \sum_{k_0=1}^{r_0}\sum_{\xi_0=1}^{r_{k_0}}\sum_{w_{k_0\xi_0:j_0}\in W(A_{k_0\xi_0},A_{j_0})}\sum_h$$

of the

$$\frac{}{T_h(\underset{\sim}{H}(P_{k_0\xi_0}):\Lambda_h + w_{k_0\xi_0:j_0}\Lambda_{j_0})}\cdot \exp(\langle \underset{\sim}{H}(P_{k_0\xi_0}), w_{k_0\xi_0:j_0}\Lambda_{j_0} + \bar{\Lambda}_h\rangle)$$

$$\times\ (c_{cus}(P_{k_0\xi_0}|A_{k_0\xi_0}:P_{j_0}|A_{j_0}:w_{k_0\xi_0:j_0}:\Lambda_{j_0})\Phi_{j_0},\Phi_h).$$

Taking into account Lemma 7.2, if we combine $T_h(\ldots)$ with $(c_{cus}(\ldots),\Phi_h)$, then the resulting product $Q_h(H:w_{k_0\xi_0:j_0}:\Lambda_{j_0})$ is in

$$\mathfrak{m}_{j_0}[H],$$

$\mathfrak{m}_{j_0}$ being the algebra of meromorphic functions on $\check{\mathfrak{a}}_{j_0} + \sqrt{-1}\check{\mathfrak{a}}_{j_0}$ whose singularities lie along hyperplanes. Consequently, the inner product

$$(Q^{\underset{\sim}{H}}E(P_{j_0}|A_{j_0}:\Phi_{j_0}:\Lambda_{j_0}:?),f)$$

is equal to the sum

$$(-1)^{\ell_0}\cdot \mathrm{vol}(\mathfrak{C}_0)\cdot \sum_{k_0=1}^{r_0}\sum_{\xi_0=1}^{r_{k_0}}\sum_{w_{k_0\xi_0:j_0}\in W(A_{k_0\xi_0},A_{j_0})}\sum_h$$

of the

$$Q_h(\underset{\sim}{H}(P_{k_0\xi_0}):w_{k_0\xi_0:j_0}:\Lambda_{j_0})\cdot \exp(\langle \underset{\sim}{H}(P_{k_0\xi_0}), w_{k_0\xi_0:j_0}\Lambda_{j_0} + \bar{\Lambda}_h\rangle).$$

It is at this juncture that residues enter the picture.

To facilitate the transition, first change the notation, replacing f by g and j_0 by i_0 -- then the discussion supra provides a formula for the inner product

$$(Q_{\underset{\sim}{H}} E(P_{i_0}|A_{i_0}:\Phi_{i_0}:\Lambda_{i_0}:?),g)$$

in terms of a sum of certain meromorphic functions

$$Q_h(\underset{\sim}{H}(P_{k_0\xi_0}):w_{k_0\xi_0}:i_0:\Lambda_{i_0})$$

multiplied by exponentials

$$\exp(\langle \underset{\sim}{H}(P_{k_0\xi_0}), w_{k_0\xi_0}:i_0\Lambda_{i_0} + \bar{\Lambda}_h\rangle).$$

Observe that the definitions imply that the dependence of each Q_h on Φ_{i_0} is linear.

We now ask: What happens if

$$E(P_{i_0}|A_{i_0}:\Phi_{i_0}:\Lambda_{i_0}:?)$$

is replaced by an arbitrary Eisenstein system function f, say? To be precise, let, as before, $\mathfrak{C}_{i_0}$ be a G-conjugacy class in $\mathfrak{C}_0$, $\mathfrak{X}$ an equivariant system of admissible affine subspaces attached to $\mathfrak{C}_{i_0}$. Suppose given an Eisenstein system $\{E,\nabla\}$ belonging to $\mathfrak{X}$ -- then, in all cases of interest, an arbitrary E-function

$$E(\mathfrak{X}:G|\{1\}:P_{i_0}|A_{i_0}:T_{i_0}:\Lambda_{i_0}:?)$$

can be written as a finite linear combination of iterated residues of cuspidal Eisenstein series, a main conclusion of TES (cf. Theorem 5.12 of that work). Accordingly, consider, without loss of generality, an iterated residue f of

$$E(P_{i_0}|A_{i_0}:\Phi_{i_0}(\Lambda_{i_0}):\Lambda_{i_0}:?).$$

Specifically, therefore, f is a meromorphic function on $\mathfrak{X}(P_{i_0}, A_{i_0})$ which, in fact, is in

$$\mathfrak{m}_{i_0, \mathfrak{X}},$$

the algebra of meromorphic functions on $\mathfrak{X}(P_{i_0}, A_{i_0})$ whose singularities lie along hyperplanes.

Proposition 7.3 *The inner product*

$$(Q^{\underset{\sim}{H}} f\big|_{\Lambda_{i_0}}, g) \qquad (\underset{\sim}{H} \in \underset{\sim}{a}_Q)$$

is equal to the sum

$$(-1)^{\ell_0} \cdot \mathrm{vol}(\mathfrak{C}_0) \cdot \sum_{k_0=1}^{r_0} \sum_{\xi_0=1}^{r_{k_0}} \sum\nolimits_{w_{k_0\xi_0:i_0} \in W(A_{k_0\xi_0}, A_{i_0})} \sum_h$$

of the

$$Q_h(\mathfrak{X}:\underset{\sim}{H}(P_{k_0\xi_0}):w_{k_0\xi_0:i_0}:\Lambda_{i_0}) \cdot \exp(\langle \underset{\sim}{H}(P_{k_0\xi_0}), w_{k_0\xi_0:i_0}\Lambda_{i_0} + \bar{\Lambda}_h \rangle),$$

where

$$Q_h(\mathfrak{X}:H:w_{k_0\xi_0:i_0}:\Lambda_{i_0}) \in \mathfrak{m}_{i_0,\mathfrak{X}}[H].$$

[Note: The notation is deceptive in that Q_h also depends (linearly) on f.]

Since the residues are iterated, it can be assumed that $\mathfrak{X}$ is of codimension one. To treat this situation, let us agree to suppress the indices and let H be a hyperplane in $\check{\mathfrak{a}} + \sqrt{-1}\check{\mathfrak{a}}$ defined over $\underset{\sim}{R}$. Denoting by $\mathfrak{m}_H$ the algebra of meromorphic functions on H whose singularities lie along hyperplanes, we have:

__Lemma 7.4__ $\forall h,w$:

$$e^{-\langle H,?\rangle}\cdot \mathrm{res}_{\mathfrak{H}}(Q_h(H:w:?)e^{\langle H,?\rangle}) \in \mathfrak{m}_{\mathfrak{H}}[H].$$

__Proof__ Pick a real unit normal $\Lambda_\perp$ to $\mathfrak{H}$ -- then, on $\mathfrak{H}$, the Laurent expansion gives

$$Q_h(H:w:\Lambda + \zeta\Lambda_\perp)e^{\langle H,\Lambda + \zeta\Lambda_\perp\rangle}$$

$$= e^{\langle H,\Lambda\rangle}\cdot\left[e^{\langle H,\zeta\Lambda_\perp\rangle}\cdot\sum_{n=-N}^{\infty} q_n(h:H:w:\Lambda)\zeta^n\right],$$

where, $\forall n$,

$$q_n(h:H:w:?) \in \mathfrak{m}_{\mathfrak{H}}[H].$$

To within a constant, our residue

$$\mathrm{res}_{\mathfrak{H}}(Q_h(H:w:?)e^{\langle H,?\rangle})\Big|_{\Lambda\in\mathfrak{H}}$$

is therefore seen to be

$$e^{\langle H,\Lambda\rangle}\cdot\frac{d^{N-1}}{d\zeta^{N-1}}\left[e^{\langle H,\zeta\Lambda_\perp\rangle}\cdot(\zeta^N Q_h(H:w:\Lambda + \zeta\Lambda_\perp))\right]_{\zeta=0}.$$

Thanks to what has been said above and the Leibnitz rule, the derivative in question qua a function of Λ, belongs to $\mathfrak{m}_{\mathfrak{H}}[H]$. Hence the lemma. //

The inner product formula provided by Proposition 7.3 is qualitative in nature. It should be contrasted with Theorem 7.1 where, so to speak, we assume more and therefore get more. To be as explicit in the residual situation, some new principles will need to be introduced; cf. §8. Nevertheless, it is important to observe that even with what we have now, one can at least assert that

$$\underset{\sim}{E}(f|_{\Lambda_{i_0}}, g)$$

is contained in

$$\bigsqcup_{k_0} \bigsqcup_{\xi_0} \bigsqcup_{W(A_{k_0\xi_0}, A_{i_0})} \bigsqcup_{h} \{(\underset{\sim}{w_{k_0\xi_0}: i_0}\Lambda_{i_0} + \bar{\Lambda}_h) P_{k_0\xi_0}\} .$$

[Note: The set of Λ_h depends, of course, on the choice of k_0 and ξ_0.]

On general grounds, some of the exponents above are actually superfluous; they will be eliminated in due course.

§8. Control and Isolation

The purpose of this § is to consider the concept of "control", applying it to isolate the various terms which contribute to $(Q^H f,g)$ $(f,g \in A(G/\Gamma))$; cf. Theorem 8.3. We then specialize the choice of f and g to E-functions of the sort used to construct residual Eisenstein series, leading thereby to the main results, which we term the first and second theorems of control. It should be noted that the level one considerations in [2-(c)] are subsumed by what is to be found here, the present study being valid at any level.

We shall begin by reexamining the differential operator

$$D(F:?)$$

figuring in Proposition 5.4. There, the role of F, vis-à-vis the given association class $\mathfrak{C}$, is to single out a specific G-conjugacy class $\mathfrak{C}_i$. Furthermore, by definition (see §9 of [2-(b)]), the factor

$$|\det[(\lambda_i^F,\lambda_j^F):i,j\notin F]|^{1/2}$$

is nothing more nor less than

$$\mathrm{vol}(\mathfrak{C}).$$

From now on, therefore, we shall agree to write

$\underset{\sim}{D}(\mathcal{C}_i:?)$ in place of $\underset{\sim}{D}(F:?)$.

Suppose that $\underset{\sim}{\Lambda}$ is a P-pure affine functional on $\underset{\sim}{a}$, so that $\underset{\sim}{\Lambda} = \underset{\sim}{\Lambda}_P$ $(\exists\Lambda)$. Put

$$F_\Lambda = \{\lambda_i \in \Sigma_P^o(\mathfrak{g},\mathfrak{a}):(\lambda_i,\Lambda) = 0\}$$

Call $\mathcal{C}_i(\underset{\sim}{\Lambda})$ the G-conjugacy class of Γ-cuspidal split parabolic subgroups of G determined by P_{F_Λ}. Claim: $\mathcal{C}_i(\underset{\sim}{\Lambda})$ is uniquely determined. In other words, if

$$\underset{\sim}{\Lambda} \text{ is } \begin{cases} P' & \text{pure} \\ P'' & \text{pure,} \end{cases} \quad \text{(maximal w.r.t. this property)}$$

then the

G-conjugacy class per P' = G-conjugacy class per P".

To see this, simply remark that

$$\{\mathcal{C}_i:\underset{\sim}{D}(\mathcal{C}_i:e^{\langle \underset{\sim}{H}+?,\underset{\sim}{\Lambda}\rangle}) \not\equiv 0\}$$

is, alternatively,

$$\{\mathcal{C}_i:\mathcal{C}_i \succeq \mathcal{C}_i(\underset{\sim}{\Lambda})\}.$$

To help motivate the considerations which are to follow, it seems best to establish first a simple lemma of prediction.

Let $f,g \in \mathcal{A}(G/\Gamma)$ -- then an association class $\mathcal{C}$ is said to control

$$(Q^{\underset{\sim}{H}}f,g)$$

if

$$\forall \underset{\sim}{\Lambda} \in \underset{\sim}{E}(f,g), \quad \mathcal{C}_i(\underset{\sim}{\Lambda}) \subset \mathcal{C} .$$

Under these circumstances, to reflect the decomposition

$$\mathcal{C} = \coprod_i \mathcal{C}_i ,$$

set

$$\underset{\sim}{E}_i(f,g) = \{\underset{\sim}{\Lambda} \in \underset{\sim}{E}(f,g) : \mathcal{C}_i(\underset{\sim}{\Lambda}) \subset \mathcal{C}_i\}.$$

Then we have

$$\underset{\sim}{E}(f,g) = \coprod_i \underset{\sim}{E}_i(f,g).$$

Lemma 8.1 *Let* $f,g \in \mathcal{A}(G/\Gamma)$. *Suppose that* $\mathcal{C}$ *controls*

$$(Q^{\underset{\sim}{H}} f,g).$$

If $\forall i$ *and* $\forall \underset{\sim}{\Lambda} \in \underset{\sim}{E}_i(f,g)$, *there is a polynomial* $p_{\underset{\sim}{\Lambda}}$ *on* $\underset{\sim}{a}$ *such that*

$$\underset{\sim}{D}(\mathcal{C}_i : (Q^{\underset{\sim}{H} + ?} f,g)) = \sum_{\underset{\sim}{\Lambda} \in \underset{\sim}{E}_i(f,g)} \underset{\sim}{D}(\mathcal{C}_i : p_{\underset{\sim}{\Lambda}}(\underset{\sim}{H} + ?) \cdot e^{\langle \underset{\sim}{H} + ?, \underset{\sim}{\Lambda} \rangle}),$$

then

$$(Q^{\underset{\sim}{H}} f,g) = \sum_{\underset{\sim}{\Lambda} \in \underset{\sim}{E}(f,g)} p_{\underset{\sim}{\Lambda}}(\underset{\sim}{H}) \cdot e^{\langle \underset{\sim}{H}, \underset{\sim}{\Lambda} \rangle} .$$

[Note: Here, $\underset{\sim}{H} \in \underset{\sim}{a}_Q$ (cf. Proposition 5.6).]

Proof On general grounds, there are polynomials $q_{\underset{\sim}{\Lambda}}$ on $\underset{\sim}{a}$ for which

$$(Q^{\underset{\sim}{H}} f,g) = \sum_{\underset{\sim}{\Lambda} \in \underset{\sim}{E}(f,g)} q_{\underset{\sim}{\Lambda}}(\underset{\sim}{H}) \cdot e^{\langle \underset{\sim}{H}, \underset{\sim}{\Lambda} \rangle} ,$$

our contention thus being that

$$p_{\underset{\sim}{\Lambda}} = q_{\underset{\sim}{\Lambda}} \qquad \forall \underset{\sim}{\Lambda} \in \underset{\sim}{E}(f,g).$$

To this end, fix an index i -- then, from the hypothesis,

$$\sum_{\underset{\sim}{\Lambda}\in \underset{\sim}{E}_i(f,g)} \underset{\sim}{D}(\mathfrak{C}_i : p_{\underset{\sim}{\Lambda}}(\underset{\sim}{H} + ?)\cdot e^{\langle \underset{\sim}{H} + ?, \underset{\sim}{\Lambda}\rangle})$$

$$= \sum_{\underset{\sim}{\Lambda}\in \underset{\sim}{E}_i(f,g)} \underset{\sim}{D}(\mathfrak{C}_i : q_{\underset{\sim}{\Lambda}}(\underset{\sim}{H} + ?)\cdot e^{\langle \underset{\sim}{H} + ?, \underset{\sim}{\Lambda}\rangle})$$

$$+ \sum_{\underset{\sim}{\Lambda}\in \underset{\sim}{E}(f,g) - \underset{\sim}{E}_i(f,g)} \underset{\sim}{D}(\mathfrak{C}_i : q_{\underset{\sim}{\Lambda}}(\underset{\sim}{H} + ?)\cdot e^{\langle \underset{\sim}{H} + ?, \underset{\sim}{\Lambda}\rangle})$$

or still, upon rewriting,

$$\sum_{\underset{\sim}{\Lambda}\in \underset{\sim}{E}_i(f,g)} \underset{\sim}{D}(\mathfrak{C}_i : (p_{\underset{\sim}{\Lambda}} - q_{\underset{\sim}{\Lambda}})(\underset{\sim}{H} + ?)\cdot e^{\langle \underset{\sim}{H} + ?, \underset{\sim}{\Lambda}\rangle})$$

$$= \sum_{\underset{\sim}{\Lambda}\in \underset{\sim}{E}(f,g) - \underset{\sim}{E}_i(f,g)} \underset{\sim}{D}(\mathfrak{C}_i : q_{\underset{\sim}{\Lambda}}(\underset{\sim}{H} + ?)\cdot e^{\langle \underset{\sim}{H} + ?, \underset{\sim}{\Lambda}\rangle}).$$

Because the exponents in the last expression are distinct, it follows by comparison that both sides have all terms $\equiv 0$. In particular: $\forall \underset{\sim}{\Lambda}\in \underset{\sim}{E}_i(f,g)$,

$$\underset{\sim}{D}(\mathfrak{C}_i : (p_{\underset{\sim}{\Lambda}} - q_{\underset{\sim}{\Lambda}})(\underset{\sim}{H} + ?)\cdot e^{\langle \underset{\sim}{H} + ?, \underset{\sim}{\Lambda}\rangle}) \equiv 0 \Rightarrow p_{\underset{\sim}{\Lambda}} = q_{\underset{\sim}{\Lambda}},$$

since $\mathfrak{C}_i(\underset{\sim}{\Lambda}) = \mathfrak{C}_i$. Hence the lemma. //

Application We have seen in §7 that if

$$E(P_{i_0} | A_{i_0} : \Phi_{i_0} : \Lambda_{i_0} : ?)$$

is a cuspidal Eisenstein series and if g is an automorphic form, then the inner product

$$(Q^{\underset{\sim}{H}} E(P_{i_0} | A_{i_0} : \Phi_{i_0} : \Lambda_{i_0} : ?), g)$$

is equal to the sum

$$(-1)^{\ell_0}\cdot \mathrm{vol}(\mathfrak{C}_0)\cdot \sum_{k_0=1}^{r_0} \sum_{\xi_0=1}^{r_{k_0}} \sum_{w_{k_0\xi_0}:i_0 \in W(A_{k_0\xi_0}, A_{i_0})} \sum_h$$

of the

$$\overline{T_h(\underset{\sim}{H}(P_{k_0\xi_0}):\Lambda_h + w_{k_0\xi_0:i_0}\bar{\Lambda}_{i_0})\cdot\exp(\langle\underset{\sim}{H}(P_{k_0\xi_0}),w_{k_0\xi_0}\Lambda_{i_0} + \bar{\Lambda}_h\rangle)}$$

$$\times\ (c_{cus}(P_{k_0\xi_0}|A_{k_0\xi_0}:P_{i_0}|A_{i_0}:w_{k_0\xi_0:i_0}:\Lambda_{i_0})\Phi_{i_0},\Phi_h).$$

We want now to indicate a different way of viewing this result in terms of the lemma of prediction supra. By taking Λ_{i_0} sufficiently general, it can and will be assumed that

$$\forall k_0,\ \mathcal{C}_i((\underset{\sim}{w_{k_0\xi_0:i_0}\Lambda_{i_0}} + \bar{\Lambda}_h)_{P_{k_0\xi_0}}) = \mathcal{C}_{k_0}.$$

The inner product formula itself then implies that $\mathcal{C}_0$ controls

$$(Q^{\underset{\sim}{H}}E(P_{i_0}|A_{i_0}:\Phi_{i_0}:\Lambda_{i_0}:?),g).$$

We remark that this inference is <u>not</u> an a priori consequence of Theorem 6.2. On the other hand, if it is assumed that $\mathcal{C}_0$ controls

$$(Q^{\underset{\sim}{H}}E(P_{i_0}|A_{i_0}:\Phi_{i_0}:\Lambda_{i_0}:?),g),$$

then we claim that the inner product formula can be obtained via an application of Lemma 8.1. Indeed, thanks to Proposition 5.4,

$$\underset{\sim}{D}(\mathcal{C}_{k_0}:(Q^{\underset{\sim}{H}+?}E(P_{i_0}|A_{i_0}:\Phi_{i_0}:\Lambda_{i_0}:?),g))$$

is equal to the sum

$$(-1)^{\ell_0}\cdot\mathrm{vol}(\mathcal{C}_0)\cdot\sum_{\xi_0=1}^{r_{k_0}}\ \sum_{w_{k_0\xi_0:i_0}\in W(A_{k_0\xi_0},A_{i_0})}\ \sum_h$$

of the

$$\overline{P_{u_h}(\underset{\sim}{H}(P_{k_0\xi_0}))}\cdot\exp(\langle\underset{\sim}{H}(P_{k_0\xi_0}),w_{k_0\xi_0}\Lambda_{i_0} + \bar{\Lambda}_h\rangle)$$

$$\times\ (c_{cus}(P_{k_0\xi_0}|A_{k_0\xi_0}:P_{i_0}|A_{i_0}:w_{k_0\xi_0:i_0}:\Lambda_{i_0})\Phi_{i_0},\Phi_h).$$

In passing, note that the partial truncation operator can be suppressed, it being here a question of cusp forms. Accordingly, to complete our discussion, we need only show that

$$\underset{\sim}{D}(\mathbf{C}_{k_0}:T_h(\underset{\sim}{H}+?,?)\cdot e^{\langle \underset{\sim}{H}+?,?\rangle})$$

$$= p_{u_h}(\underset{\sim}{H}(P_{k_0\xi_0}))\cdot \exp(\langle \underset{\sim}{H}(P_{k_0\xi_0}),?\rangle).$$

But, using the definitions, this follows by an easy calculation.

The preceding observations make it plain that control is tantamount to the inner product formula at the cuspidal level. Unfortunately, there does not seem to be a direct way to gain control ab initio! However, at the residual level, it will be possible to achieve partial control from first principles, viz. Proposition 7.3. Before embarking on this, we shall take up a result which serves to place the later developments in perspective.

An elementary generality will be needed.

Lemma 8.2 Suppose that $\underset{\sim}{\Lambda}$ is a P-pure affine functional on $\underset{\sim}{a}$ to which there are attached polynomials $p_{\underset{\sim}{\Lambda}}$ and $q_{\underset{\sim}{\Lambda}}$. Let

$$\begin{cases} E \\ F \end{cases}$$

be exponential polynomials on $\mathfrak{a}$ *such that*

$$\begin{cases} p_{\Lambda}\cdot e^{\langle ?,\Lambda\rangle} \in E \\ q_{\Lambda}\cdot e^{\langle ?,\Lambda\rangle} \in F \ . \end{cases}$$

Then

$$p_{\Lambda} = q_{\Lambda}$$

provided the term corresponding to Λ *in*

$$D(\mathcal{C}_i(\Lambda):E)$$

is equal to the term corresponding to Λ *in*

$$D(\mathcal{C}_i(\Lambda):F) \ .$$

[The verification is immediate.]

Let us also agree to some notation.

Suppose that $f(\Lambda':\Lambda'')$ is a meromorphic function of $(\Lambda':\Lambda'')$ on $\mathbb{C}^{\ell}\times\mathbb{C}^{\ell}$ (or on a suitable open subset thereof). Given polynomials p',p'', write

$$\partial'_{p'}\ \bar{\partial}''_{p''}(f)$$

for

$$\overline{\partial''_{p''}\{\overline{\partial'_{p'}f(\Lambda':\bar{\Lambda}'')}\}}.$$

Note that the condition

$$f(\bar{\Lambda}',\bar{\Lambda}'') = \overline{f(\Lambda',\Lambda'')},$$

allows one to say that

$$\partial_{p'}^{\prime},\ \bar{\partial}_{p''}^{\prime\prime}(f)$$

is, alternatively,

$$\partial_{p'}^{\prime}\overline{\{\partial_{p''}^{\prime\prime} f\,(\bar{\Lambda}',\Lambda'')\}}.$$

Let $f,g\in A(G/\Gamma)$ -- then by

$$p(f,g:H),$$

we shall understand the polynomial associated with the truncation exponent 0, so that

$$(Q^{H}f,g) = p(f,g:H) + \sum_{\substack{\Lambda\in E(f,g)\\ \Lambda\neq 0}} p_{\Lambda}(H)\cdot e^{\langle H,\Lambda\rangle}$$

on a_Q. Relative to a set

$$\{(P_i,S_i): 1\le i\le r\}$$

of representatives for the Γ-conjugacy classes of Γ-cuspidal split parabolic subgroups of G, put

$$\begin{cases} f_{P_i}(?m_i a_i) = \sum_k^{(i)} a_i^{\Lambda_k}\Big(\sum_\mu^{(i,k)} p_{k\mu}(H_i)\Phi_{k\mu}(?m_i)\Big) \\ g_{P_i}(?m_i a_i) = \sum_\ell^{(i)} a_i^{\Lambda_\ell}\Big(\sum_\nu^{(i,\ell)} p_{\ell\nu}(H_i)\Phi_{\ell\nu}(?m_i)\Big) \end{cases} \qquad (H_i = \log a_i).$$

Because

$$\Phi_{k\mu},\Phi_{\ell\nu}$$

are automorphic forms on $K\times M_i/\Gamma_{M_i}$, the symbol

$$p(\Phi_{k\mu},\Phi_{\ell\nu}:I_{M_i}(H)) \qquad (H\in a_Q)$$

is meaningful. Finally, for each i, denote by

$$\sum_k^{[i]}\ \sum_\ell^{[i]}$$

a sum over those pairs (k,ℓ) such that $\Lambda_k + \bar{\Lambda}_\ell$ is orthogonal to no element λ_i of $\Sigma^o_{P_i}(\mathfrak{g},\mathfrak{a}_i)$.

Theorem 8.3 Suppose that $f,g\in\mathcal{A}(G/\Gamma)$ -- then the inner product

$$(Q^{H}f,g) \quad (H\in\mathfrak{a}_Q)$$

is equal to the sum

$$\sum_{i=1}^{r}(-1)^{\operatorname{rank}(P_i)}\operatorname{vol}(\mathfrak{C}_{P_i})\cdot\sum_k{}^{[i]}\sum_\ell{}^{[i]}\sum_\mu{}^{(i,k)}\sum_\nu{}^{(i,\ell)}$$

of the

$$\partial'_{p_{k\mu}}\bar{\partial}''_{p_{\ell\nu}}(e_i(H))\Big|_{(\Lambda_k,\Lambda_\ell)}\cdot p(\Phi_{k\mu},\Phi_{\ell\nu}:I_{M_i}(H)),$$

where

$$e_i(H:\Lambda':\Lambda'') = \frac{e^{\langle H(P_i),\Lambda'+\Lambda''\rangle}}{\prod_{\lambda_i}(\Lambda'+\Lambda'',\lambda_i)}.$$

[Here, $\mathfrak{C}_{P_i}$ is the association class to which P_i belongs.]

Proof Each $\Lambda\in E(f,g)$ is P-pure (cf. §6), hence $\mathfrak{C}_i(\Lambda)$ exists. This said, to establish the equality in question, we shall utilize Lemma 8.2 (permissible, both sides being exponential polynomials). So fix a $\Lambda\in E(f,g)$. Upon applying $D(\mathfrak{C}_i(\Lambda):\ldots)$ to each and taking into account Proposition 5.4, the issue is thus seen to be that of the equality of the term corresponding to Λ in

$$\sum_{i:P_i\in\mathfrak{C}_i(\Lambda)}\cdot\ \cdot\ \cdot e^{\langle H(P_i),\Lambda_k+\bar{\Lambda}_\ell\rangle}$$

$$\times\ p_{k\mu}(H(P_i))\overline{p_{\ell\nu}(H(P_i))}\cdot(1\times Q^{I_{M_i}(H)}\Phi_{k\mu},\Phi_{\ell\nu})$$

and of the term corresponding to $\underset{\sim}{\Lambda}$ in

$$\sum_{i:P_i \in \mathfrak{C}_i(\underset{\sim}{\Lambda})} \cdot \cdot \cdot \partial'_{p_{k\mu}} \bar{\partial}''_{p_{\ell\nu}} (e^{\langle \underset{\sim}{H}(P_i), ?' + ?'' \rangle}) \Big|_{(\Lambda_k, \Lambda_\ell)}$$

$$\times \; p(\Phi_{k\mu}, \Phi_{\ell\nu} : I_{M_i}(\underset{\sim}{H})).$$

Naturally, not all the terms are relevant. To isolate those that are, set

$$S_{\underset{\sim}{\Lambda}} = \{(i:(k,\ell)) : \underset{\sim}{\Lambda} \leftrightarrow \Lambda_k + \bar{\Lambda}_\ell\}.$$

Since the $\underset{\sim}{\Lambda}$-terms ignore $I_{M_i}(\underset{\sim}{H})$ in the exponent, we are thereby reduced to showing that

$$\sum_{(i:(k,\ell)) \in S_{\underset{\sim}{\Lambda}}} \sum_{\mu,\nu} e^{\langle \underset{\sim}{H}(P_i), \Lambda_k + \bar{\Lambda}_\ell \rangle}$$

$$\times \; p_{k\mu}(\underset{\sim}{H}(P_i)) \overline{p_{\ell\nu}(\underset{\sim}{H}(P_i))} \cdot p(\Phi_{k\mu}, \Phi_{\ell\nu} : I_{M_i}(\underset{\sim}{H}))$$

is the same as

$$\sum_{(i:(k,\ell))\; S_{\underset{\sim}{\Lambda}}} \sum_{\mu,\nu} \partial'_{p_{k\mu}} \bar{\partial}''_{p_{\ell\nu}} (e^{\langle \underset{\sim}{H}(P_i), ?' + ?'' \rangle}) \Big|_{(\Lambda_k, \Lambda_\ell)}$$

$$\times \; p(\Phi_{k\mu}, \Phi_{\ell\nu} : I_{M_i}(\underset{\sim}{H}))$$

or still, that

$$p_{k\mu}(\underset{\sim}{H}(P_i)) \overline{p_{\ell\nu}(\underset{\sim}{H}(P_i))} \cdot e^{\langle \underset{\sim}{H}(P_i), \Lambda_k + \bar{\Lambda}_\ell \rangle}$$

is the same as

$$\partial'_{p_{k\mu}} \bar{\partial}''_{p_{\ell\nu}} (e^{\langle \underset{\sim}{H}(P_i), ?' + ?'' \rangle}) \Big|_{(\Lambda_k, \Lambda_\ell)} ,$$

which, however, is clear. //

This, then, is in a certain sense an explicit inner product

formula, at least in abstracto. Of course, the main point is the nature of the

$$p(f,g:\underset{\sim}{H}).$$

For example, as was observed at the end of §6, if $f,g \in A(G/\Gamma) \cap L^2(G/\Gamma)$, then

$$p(f,g:\underset{\sim}{H}) = (f,g).$$

Even so, one cannot automatically conclude that

$$p(\Phi_{k\mu},\Phi_{\ell\nu}:I_{M_i}(\underset{\sim}{H})) = (\Phi_{k\mu},\Phi_{\ell\nu}),$$

there being no guarantee that

$$\Phi_{k\mu},\Phi_{\ell\nu} \in L^2(K \times M/\Gamma_{M_i}).$$

At the opposite extreme, if f and g are taken to be cuspidal Eisenstein series, then we already know what $(Q^{\underset{\sim}{H}}f,g)$ has to be, thus our theorem gives us less than what we might hope to have. In this situation, a great deal of cancellation must occur, but we have no real idea as to what the internal mechanism might be. It is therefore conceivable that a systematic study in general of the properties of

$$p(f,g:\underset{\sim}{H})$$

could very well be profitable in that a deeper understanding of $(Q^{\underset{\sim}{H}}f,g)$ would no doubt result.

Here is one simple illustration, itself of some interest.

Application Let Γ be a rank-1 lattice in G. Suppose that $f,g\in A(G/\Gamma)\cap L^2_{res}(G/\Gamma)$ -- then, as is known (see [2-(a)]),

$$\begin{cases} f_{P_i}(?m_i a_i) = \sum_k^{(i)} a_i^{\Lambda_k}\Phi_{ik}(?m_i) \\ g_{P_i}(?m_i a_i) = \sum_\ell^{(i)} a_i^{\Lambda_\ell}\Phi_{i\ell}(?m_i), \end{cases}$$

Λ_k and Λ_ℓ being, in the case at hand, nonzero and real, with

$$(\Lambda_k+\Lambda_\ell,\lambda_i)\neq 0.$$

Our theorem thus tells us that

$$(Q^H f,g)$$

$$=(f,g)\cdot\mathrm{vol}(\mathcal{C})\cdot\sum_{i=1}^{r}\sum_k^{(i)}\sum_\ell^{(i)}\frac{e^{\langle H(P_i),\Lambda_k+\Lambda_\ell\rangle}}{(\Lambda_k+\Lambda_\ell,\lambda_i)}\cdot(\Phi_{ik},\Phi_{i\ell}),$$

a relation which is at the basis for the proof of the so-called Arthur conjecture (detailed in the next paper in this series).

We turn now to the main theorems of control.

Fix anew an association class $\mathcal{C}_0$ of Γ-cuspidal split parabolic subgroups of G. Let $\mathcal{C}_{i_0}$ be a G-conjugacy class in $\mathcal{C}_0$, $\mathfrak{X}$ an equivariant system of admissible affine subspaces attached to $\mathcal{C}_{i_0}$. Take $\mathfrak{X}$ principal (cf. TES, p. 156) -- then, in the usual way, $\mathfrak{X}$ determines an association class $\mathcal{C}(\mathrm{Dis}(\mathfrak{X}))$, containing a G-conjugacy class $\mathcal{C}_k(\mathrm{Dis}(\mathfrak{X}))$ such that

$$\mathfrak{C}_k(\mathrm{Dis}(\mathfrak{X})) \succeq \mathfrak{C}_{i_0}.$$

Suppose given an Eisenstein system $\{E,\nabla\}$ belonging to $\mathfrak{X}$ of "residue type", as well as an automorphic form g. We shall agree to say that the parameter Λ_{i_0} of the E-function

$$f|_{\Lambda_{i_0}} = E(\mathfrak{X}:G|\{1\}:P_{i_0}|A_{i_0}:T_{i_0}:\Lambda_{i_0}:?)$$

is in general position with respect to g if, in the notation of Proposition 7.3, the following condition is met:

$$\begin{cases} \forall k_0, \forall \xi_0, \forall w_{k_0\xi_0:i_0}, \forall h, \\ \forall \lambda_{k_0\xi_0} \in \Sigma^0_{P_{k_0\xi_0}}(g, a_{k_0\xi_0}) \text{ st} \end{cases}$$

$$\Longrightarrow \begin{array}{l} (w_{k_0\xi_0:i_0}(\mathfrak{X}(P_{i_0},A_{i_0})) + \bar{\Lambda}_h, \lambda_{k_0\xi_0}) \not\equiv 0 \\ (w_{k_0\xi_0:i_0}\Lambda_{i_0} + \bar{\Lambda}_h, \lambda_{k_0\xi_0}) \neq 0. \end{array}$$

Theorem 8.4 (**First Theorem of Control**) Fix a $\underset{\sim}{\Lambda} \in \underset{\sim}{E}(f|_{\Lambda_{i_0}}, g)$ then, under the assumption that Λ_{i_0} is g-general,

$$\mathfrak{C}(\mathrm{Dis}(\mathfrak{X})) \succeq \mathfrak{C}(\underset{\sim}{\Lambda}) \succeq \mathfrak{C}_0.$$

We remark that in view of our notational principles, $\mathfrak{C}(\underset{\sim}{\Lambda})$ stands for the association class corresponding to $\mathfrak{C}_i(\underset{\sim}{\Lambda})$, the thrust of the assertion being therefore one of association as opposed to conjugacy.

For sake of definiteness, write, as is possible (cf. Proposition 7.3),

$$\boldsymbol{\Lambda} = \underline{(w_{k_0\xi_0:i_0}\Lambda_{i_0} + \bar{\Lambda}_h)}_{P_{k_0\xi_0}}.$$

Then

$$\mathcal{C}_i(\boldsymbol{\Lambda}) \succeq \mathcal{C}_{k_0},$$

hence, a fortiori,

$$\mathcal{C}(\boldsymbol{\Lambda}) \succeq \mathcal{C}_0.$$

The other domination, however, is not so immediate. Put

$w = w_{k_0\xi_0:i_0}$. To visualize the set-up, consider

$$\begin{array}{cccccccccc}
 & A_{Dis(\mathfrak{X})} & \hookrightarrow & A_{i_0} & \xrightarrow{\ \ w\ \ } & & A_{k_0\xi_0} & \hookleftarrow & A_{\boldsymbol{\Lambda}} & \\
\mathcal{C}(Dis(\mathfrak{X})) & \Big| & & \Big| & & & \Big| & & \Big| & \mathcal{C}(\boldsymbol{\Lambda}) \\
 & P_{Dis(\mathfrak{X})} & \supset & P_{i_0} & \supset & & P_{k_0\xi_0} & \subset & P_{\boldsymbol{\Lambda}} & \\
 & & & & & P_0 & & & &
\end{array}$$

with $P_{i_0} \supset P_0 \subset P_{k_0\xi_0}$.

Here, P_0 is a Γ-percuspidal (modify the notation if necessary).

Moreover, the parabolic containments are actually dominations.

It should also be kept in mind that

$$\check{a}_{Dis(\mathfrak{X})} = \mathrm{Re}(\mathfrak{X}(P_{i_0}, A_{i_0})^{\sim}).$$

Because of our hypothesis on Λ_{i_0}, by subtraction we have

$$w(\mathfrak{X}(P_{i_0}, A_{i_0})^{\sim}) \subset \check{a}_{\boldsymbol{\Lambda}} + \sqrt{-1}\check{a}_{\boldsymbol{\Lambda}},$$

from which it follows that w induces a map

$$A_{Dis(\mathfrak{X})} \rightarrow A_{\boldsymbol{\Lambda}}.$$

These points made, a parabolic generality will complete the proof.

<u>Lemma 8.5</u> <u>Let</u> P_0 <u>be a</u> Γ-<u>percuspidal split parabolic subgroup of</u> G; <u>let</u> P_1, P_2 <u>be associate</u> Γ-<u>cuspidal split parabolic subgroups of</u> G <u>such that</u>

$$\begin{cases} P_1 \succeq P_0 \\ P_2 \succeq P_0. \end{cases}$$

<u>Suppose that</u> P',P'' <u>are</u> Γ-<u>cuspidal split parabolic subgroups of</u> G <u>with</u>

$$\begin{cases} P' \succeq P_1 \\ P'' \succeq P_2 \end{cases}$$

<u>and suppose that</u> w' <u>is an element of</u> $W(A_1, A_2)$ <u>with</u>

$$w'(A') \subset A''.$$

<u>Then there is an associate</u> Q'' <u>of</u> P'' <u>such that</u>

$$P' \succeq Q''.$$

<u>Proof</u> We can and will assume that w' is the restriction of some element of $W(A_0)$. This done, fix a point

$$H'' \in w'(\mathcal{C}_{P'}(\mathfrak{a}'))$$

and a line segment

$$\ell : [0,1] \to \mathfrak{a}''$$

for which

$$\begin{cases} \ell(0) = H'' \\ \ell(]0,1]) \subset \mathfrak{a}''_r. \end{cases}$$

The second condition means that $\ell(]0,1])$ lies in a certain chamber of $\mathfrak{a}''$. Consequently (see Lemma 2 on p. 67 of TES), there is a $w''\in W(A_0)$ having the property that it sends this chamber to the positive chamber of some associate Q'' of P''. But then $w''w'(\mathfrak{C}_{P'}(\mathfrak{a}'))$ is the positive chamber of a dominant of Q'', which we claim must be P' itself. In fact, $w''w'$ sends $\mathfrak{C}_{P'}(\mathfrak{a}')$ to $w''w'(\mathfrak{C}_{P'}(\mathfrak{a}'))$, hence, upon taking spans, one forces the association of P' and the dominant and, finally, by uniqueness, their equality. //

To say more, we shall have to assume more. So let $\mathfrak{C}_{j_0}$ be another G-conjugacy class in $\mathfrak{C}_0$, $\mathfrak{Y}$ an equivariant system of admissible affine subspaces attached to $\mathfrak{C}_{j_0}$ which, like $\mathfrak{X}$, we take to be principal. Again, we suppose given an Eisenstein system $\{E,\nabla\}$ belonging to $\mathfrak{Y}$ of "residue type", specializing now the g above to

$$g|_{\Lambda_{j_0}} = E(\mathfrak{Y}:G|\{1\}:P_{j_0}|A_{j_0}:T_{j_0}:\Lambda_{j_0}:?).$$

The problem then will be to study

$$\underset{\sim}{E}(f|_{\Lambda_{i_0}},g|_{\Lambda_{j_0}})$$

under suitable restrictions on Λ_{i_0} and Λ_{j_0}.

The parameters

$$\begin{cases} \Lambda_{i_0} & \text{per } f|_{\Lambda_{i_0}} \\ \Lambda_{j_0} & \text{per } g|_{\Lambda_{j_0}} \end{cases}$$

are said to be in general position with respect to $(\mathfrak{X},\mathfrak{Y})$ provided the following condition is met :

$$\begin{cases} \forall k_0, \forall \xi_0, \forall w_{k_0\xi_0:i_0}, \forall w_{k_0\xi_0:j_0} \\ \forall \lambda_{k_0\xi_0} \in \Sigma^{0}_{P_{k_0\xi_0}}(\mathfrak{g}, \mathfrak{a}_{k_0\xi_0}) \text{ st} \end{cases}$$

$$(w_{k_0\xi_0:i_0}(\mathfrak{X}(P_{i_0},A_{i_0})) + w_{k_0\xi_0:j_0}(\mathfrak{Y}(P_{j_0},A_{j_0})),\lambda_{k_0\xi_0}) \not\equiv 0$$

$$\Longrightarrow (w_{k_0\xi_0:i_0}\Lambda_{i_0} + w_{k_0\xi_0:j_0}\bar{\Lambda}_{j_0},\lambda_{k_0\xi_0}) \neq 0.$$

Observe that if the pair $(\Lambda_{i_0},\Lambda_{j_0})$ is in general position per $(\mathfrak{X},\mathfrak{Y})$, then

$$\begin{cases} \Lambda_{i_0} & \text{is in general position per } g|_{\Lambda_{j_0}} \\ \Lambda_{j_0} & \text{is in general position per } f|_{\Lambda_{i_0}}, \end{cases}$$

the converse being false in general.

Theorem 8.6 (Second Theorem of Control) *Suppose that the pair* $(\Lambda_{i_0},\Lambda_{j_0})$ *is* $(\mathfrak{X},\mathfrak{Y})$-*general with*

$$\begin{cases} \Lambda_{i_0} \in X(P_{i_0},A_{i_0}) + \sqrt{-1}\check{\mathfrak{a}}_{\mathrm{Dis}(\mathfrak{X})} \\ \Lambda_{j_0} \in Y(P_{j_0},A_{j_0}) + \sqrt{-1}\check{\mathfrak{a}}_{\mathrm{Dis}(\mathfrak{Y})} \end{cases}.$$

Let

$\underset{\sim}{\Lambda} \in \underset{\sim}{E}(f|_{\Lambda_{i_0}}, g|_{\Lambda_{j_0}})$.

<u>Then</u> $\underset{\sim}{\Lambda}$ <u>is pure imaginary iff</u>

$$\mathcal{C}(\underset{\sim}{\Lambda}) = \begin{cases} \mathcal{C}(\mathrm{Dis}(\mathfrak{X})) \\ \quad \| \\ \mathcal{C}(\mathrm{Dis}(\mathfrak{Y})). \end{cases}$$

We shall spell out the proof in the lines below.

Let us start off by assuming that

$$\mathcal{C}(\underset{\sim}{\Lambda}) = \begin{cases} \mathcal{C}(\mathrm{Dis}(\mathfrak{X})) \\ \quad \| \\ \mathcal{C}(\mathrm{Dis}(\mathfrak{Y})) \end{cases} .$$

Fix a $\underset{\sim}{\Lambda} \in \underset{\sim}{E}(f|_{\Lambda_{i_0}}, g|_{\Lambda_{j_0}})$, say (cf. Proposition 7.3)

$$\underset{\sim}{\Lambda} = \underset{\sim\sim\sim\sim\sim\sim\sim\sim\sim\sim\sim\sim\sim\sim\sim}{(w_{k_0\xi_0:i_0}\Lambda_{i_0} + w_{k_0\xi_0:j_0}\bar{\Lambda}_{j_0})} P_{k_0\xi_0} .$$

Then, from

$$P_{\underset{\sim}{\Lambda}} \succeq P_{k_0\xi_0} \qquad\qquad (P_{\underset{\sim}{\Lambda}} \in \mathcal{C}_i(\underset{\sim}{\Lambda})),$$

the general position hypothesis entails

$$w_{k_0\xi_0:i_0}(\mathfrak{X}(P_{i_0}, A_{i_0})) + w_{k_0\xi_0:j_0}(\mathfrak{Y}(P_{j_0}, A_{j_0})) \subset \check{a}_{\underset{\sim}{\Lambda}} + \sqrt{-1}\check{a}_{\underset{\sim}{\Lambda}} ,$$

so, by subtraction,

$$w_{k_0\xi_0:i_0}(\mathfrak{X}(P_{i_0}, A_{i_0})^{\sim}) + w_{k_0\xi_0:j_0}(\mathfrak{Y}(P_{j_0}, A_{j_0})^{\sim}) \subset \check{a}_{\underset{\sim}{\Lambda}} + \sqrt{-1}\check{a}_{\underset{\sim}{\Lambda}} ,$$

meaning that actually, for reasons of dimension,

$$w_{k_0\xi_0:i_0}(\mathfrak{X}(P_{i_0}, A_{i_0})^{\sim}) = w_{k_0\xi_0:j_0}(\mathfrak{Y}(P_{j_0}, A_{j_0})^{\sim}).$$

On the other hand,

$$\begin{cases} w_{k_0\xi_0:i_0}(X(P_{i_0},A_{i_0})) \perp w_{k_0\xi_0:i_0}(\mathfrak{X}(P_{i_0},A_{i_0})^{\sim}) \\ w_{k_0\xi_0:j_0}(Y(P_{j_0},A_{j_0})) \perp w_{k_0\xi_0:j_0}(\mathfrak{Y}(P_{j_0},A_{j_0})^{\sim}), \end{cases}$$

while

$$w_{k_0\xi_0:i_0}(X(P_{i_0},A_{i_0})) + w_{k_0\xi_0:j_0}(Y(P_{j_0},A_{j_0})) \in \check{a}_{\underset{\sim}{\Lambda}} + \sqrt{-1}\check{a}_{\underset{\sim}{\Lambda}} .$$

Conclusion:

$$w_{k_0\xi_0:i_0}(X(P_{i_0},A_{i_0})) + w_{k_0\xi_0:j_0}(Y(P_{j_0},A_{j_0})) = 0.$$

Therefore $\underset{\sim}{\Lambda}$ is pure imaginary.

To go the other way, let us deny our contention. Thus, on the basis of the first theorem of control, we have, e.g.,

$$\text{rank}(\mathfrak{C}(\underset{\sim}{\Lambda})) > \text{rank}(\mathfrak{C}(\text{Dis}(\mathfrak{X}))).$$

Put $P = P_{\underset{\sim}{\Lambda}}$ and choose a $\Lambda \in \check{a} + \sqrt{-1}\check{a}$ such that $\underset{\sim}{\Lambda} = \underset{\sim}{\Lambda}_P$ -- then (cf. Theorem 6.2)

$$\Lambda = \Lambda_f + \tilde{\Lambda}_g,$$

where

$$\begin{cases} \Lambda_f \in E_P(f) \\ \Lambda_g \in E_P(g) . \end{cases}$$

Furthermore, thanks to Proposition 6.3 (and, of course, the hypotheses on the E-functions),

$$\begin{cases} \text{Re}(\Lambda_f) \in \mathfrak{D}_P(\check{a})^- \\ \text{Re}(\Lambda_g) \in \mathfrak{D}_P(\check{a})^- . \end{cases}$$

Accordingly, it suffices to show that $\mathrm{Re}(\Lambda_f) \neq 0$. By definition, $\mathrm{Re}(\Lambda_f)$ is the image under the canonical projection

$$\check{a}_{k_0\xi_0} \to \check{a}$$

of

$$\mathrm{Re}(w_{k_0\xi_0:i_0}\Lambda_{i_0}),$$

i.e., of

$$w_{k_0\xi_0:i_0}(X(P_{i_0},A_{i_0})).$$

In what follows, write w in place of $w_{k_0\xi_0:i_0}$ and retain the k_0 but suppress the ξ_0 from the notation. Taking into account conditions E-S:I and Geom:III in TES, we then have

$$wX(P_{i_0},A_{i_0}) \in \mathfrak{D}_{P_{\mathrm{Dis}(\mathfrak{X}_w)}^{k_0}}(\check{a}_{\mathrm{Dis}(\mathfrak{X}_w)}^{k_0}).$$

In this connection, recall that

$$(P_{\mathrm{Dis}(\mathfrak{X}_w)},S_{\mathrm{Dis}(\mathfrak{X}_w)};A_{\mathrm{Dis}(\mathfrak{X}_w)}) \succeq (P_{k_0},S_{k_0};A_{k_0}),$$

the daggering procedure determining parabolic data

$$(P_{\mathrm{Dis}(\mathfrak{X}_w)}^{k_0},S_{\mathrm{Dis}(\mathfrak{X}_w)}^{k_0};A_{\mathrm{Dis}(\mathfrak{X}_w)}^{k_0})$$

in $M_{\mathrm{Dis}(\mathfrak{X}_w)}$. Generically, let λ^{k_0} be the root dual to λ_{k_0}, $\lambda_{\mathrm{Dis}}^{k_0}$ its image under the canonical projection

$$\check{a}_{k_0} \to \check{a}_{\mathrm{Dis}(\mathfrak{X}_w)}^{k_0}.$$

Because

$$\mathrm{Dis}(\mathfrak{X}_w)(P_{k_0},A_{k_0})$$

is spanned by the

$$\lambda^{k_0} \in w(\mathfrak{X}(P_{i_0},A_{i_0})^{\sim}),$$

for all

$$\lambda^{k_0} \notin w(\mathfrak{X}(P_{i_0},A_{i_0})^{\sim}),$$

$$(\lambda^{k_0},wX(P_{i_0},A_{i_0}))$$
$$= (\lambda^{k_0}_{Dis},wX(P_{i_0},A_{i_0})) > 0.$$

On the other hand, $\check{a}$ is spanned by the

$$\lambda^{k_0} \in \check{a}.$$

To get a contradiction, assume now that $Re(\Lambda_f) = 0$ -- then the projection of $wX(P_{i_0},A_{i_0})$ onto $\check{a}$ is null, hence

$$\Rightarrow \begin{matrix} \lambda^{k_0} \notin w(\mathfrak{X}(P_{i_0},A_{i_0})^{\sim}) \\ \lambda^{k_0} \notin \check{a}. \end{matrix}$$

So, the set of dual roots which project nontrivially onto $\check{a}^{k_0}_{Dis(\mathfrak{X}_w)}$ is disjoint from the set of dual roots which lie in $\check{a}$. The cardinality of the first set is

$$\dim(a^{k_0}_{Dis(\mathfrak{X}_w)});$$

the cardinality of the second set is

$$\dim(a).$$

Consequently,

$$\ell_0 \geq \dim(a^{k_0}_{Dis(\mathfrak{X}_w)}) + \dim(a)$$
$$= \ell_0 - \mathrm{rank}(\mathbf{C}(Dis(\mathfrak{X}_w))) + \mathrm{rank}(\mathbf{C}(\underset{\sim}{\Lambda})),$$

that is,

$$\operatorname{rank}(\mathfrak{C}(\mathrm{Dis}(\mathfrak{X}))) \geq \operatorname{rank}(\mathfrak{C}(\mathrm{Dis}(\mathfrak{X}_w))) \geq \operatorname{rank}(\mathfrak{C}(\underset{\sim}{\Lambda})),$$

an impossibility.

The proof of the second theorem of control is therefore complete.

Remarks (1) It is not required that the dimensions of $\mathfrak{X}$ and $\mathfrak{Y}$ be equal. Of course, in such a situation, there can be no pure imaginary elements in

$$\underset{\sim}{E}(f|_{\Lambda_{i_0}}, g|_{\Lambda_{j_0}}).$$

(2) If $\underset{\sim}{\Lambda}$ is pure imaginary, say $\underset{\sim}{\Lambda} = \underset{\sim}{\Lambda}_P$ with

$$\Lambda = \Lambda_f + \bar{\Lambda}_g,$$

where

$$\begin{cases} \Lambda_f \in E_P(f), & \Lambda_f = \mathrm{pro}(w\Lambda_{i_0}) \qquad (w = w_{k_0\xi_0:i_0}) \\ \Lambda_g \in E_P(g), & \Lambda_g = \mathrm{pro}(w\Lambda_{j_0}) \qquad (w = w_{k_0\xi_0:j_0}), \end{cases}$$

then both

$$\begin{cases} \mathfrak{X}_w \\ \mathfrak{Y}_w \end{cases}$$

are still principal.

Our theorems of control can be applied to the inner product problem for residual Eisenstein series. Referring to §5 of

[2-(c)] for a complete elaboration of the necessary background, let (P_1,S_1), (P_2,S_2) be, as there, Γ-cuspidal split parabolic subgroups of G with special split components A_1, A_2 -- then the issue is the explicit computation of the inner product

$$(\underset{\sim}{Q}^H E(P_1|A_1:\Phi_1:\Lambda_1:?),\ \underset{\sim}{Q}^H E(P_2|A_2:\Phi_2:\Lambda_2:?)),$$

where

$$\begin{cases} \Phi_1 \in E(\delta, \mathcal{O}_1 : \mathfrak{C}_0) \cap E(\delta, \underset{\sim}{\mathcal{O}_0}; \underset{\sim}{\mathfrak{X}}: P_1|A_1) \\ \Phi_2 \in E(\delta, \mathcal{O}_2 : \mathfrak{C}_0) \cap E(\delta, \underset{\sim}{\mathcal{O}_0}; \underset{\sim}{\mathfrak{Y}}: P_2|A_2), \end{cases}$$

the association class $\mathfrak{C}_0$ being subject to the condition

$$\begin{cases} \mathfrak{C}_{k_1} \succeq \mathfrak{C}_0 \\ \mathfrak{C}_{k_2} \succeq \mathfrak{C}_0 \end{cases} .$$

Here, the levels

$$\begin{cases} \text{rank}(\mathfrak{C}_0) - \text{rank}(\mathfrak{C}_1) \\ \text{rank}(\mathfrak{C}_0) - \text{rank}(\mathfrak{C}_2) \end{cases}$$

are positive. There is no loss of generality in supposing that for some principal $\mathfrak{X}$

$$E(P_1|A_1:\Phi_1:\Lambda_1:?) \quad (\equiv f)$$

$$= E(\mathfrak{X}:G|\{1\}:P_{i_0}|A_{i_0}:T_{i_0}:X(P_{i_0},A_{i_0}) + \Lambda_1:?)$$

and that for some principal $\mathfrak{Y}$

$$E(P_2|A_2:\Phi_2:\Lambda_2:?:) \quad (\equiv g)$$

$$= E(\mathfrak{Y}:G|\{1\}:P_{j_0}|A_{j_0}:T_{j_0}:Y(P_{j_0},A_{j_0}) + \Lambda_2:?).$$

Lest there be any confusion, it should be kept in mind that

$$\begin{cases} P_{\mathrm{Dis}(\mathfrak{X})} = P_1 \\ P_{\mathrm{Dis}(\mathfrak{Y})} = P_2, \end{cases}$$

while

$$\begin{cases} \mathfrak{X} \in \underset{\sim}{\mathfrak{X}}_{i_0}(\mathrm{rank}(\mathfrak{C}_1)) \\ \mathfrak{Y} \in \underset{\sim}{\mathfrak{Y}}_{j_0}(\mathrm{rank}(\mathfrak{C}_2)). \end{cases}$$

In addition, we assume, as we may, that Λ_1 and Λ_2 are pure imaginary, tacitly taking the corresponding pair $(\Lambda_{i_0}, \Lambda_{j_0})$, i.e.,

$$(X(P_{i_0}, A_{i_0}) + \Lambda_1, Y(P_{j_0}, A_{j_0}) + \Lambda_2)$$

in general position relative to $(\mathfrak{X}, \mathfrak{Y})$.

Having set the stage, denote by

$$\underset{\sim}{E}_{PI}(f|_{\Lambda_{i_0}}, g|_{\Lambda_{j_0}})$$

the set of pure imaginary elements in

$$\underset{\sim}{E}(f|_{\Lambda_{i_0}}, g|_{\Lambda_{j_0}}).$$

If the levels

$$\begin{cases} \mathrm{rank}(\mathfrak{C}_0) - \mathrm{rank}(\mathfrak{C}_1) \\ \mathrm{rank}(\mathfrak{C}_0) - \mathrm{rank}(\mathfrak{C}_2) \end{cases}$$

are different, or the same but $\mathfrak{C}_1 \neq \mathfrak{C}_2$, then

$$\underset{\sim}{E}_{PI}(f|_{\Lambda_{i_0}}, g|_{\Lambda_{j_0}})$$

is empty and the inner product

$$(Q^{\underset{\sim}{H}}E(P_1|A_1:\Phi_1:\Lambda_1?),\ Q^{\underset{\sim}{H}}E(P_2|A_2:\Phi_2:\Lambda_2?))$$

tends to 0 as $\underset{\sim}{H} \to -\infty$ in $\underset{\sim}{a}_Q$ (see §10). That being, consider now the remaining case, namely when $\mathcal{C}_1 = \mathcal{C}_2$, say $\mathcal{C}$. It will not be restrictive, of course, to suppose that O_1 and O_2 are associate.

Proposition 8.7 Under the above conditions, for all $\underset{\sim}{H}\in\underset{\sim}{a}_Q$,

$$\sum_{\underset{\sim}{\Lambda}\in\underset{\sim}{E}_{PI}}(f|_{\Lambda_{i_0}}, g|_{\Lambda_{j_0}})p_{\underset{\sim}{\Lambda}}(\underset{\sim}{H})\cdot e^{\langle \underset{\sim}{H},\underset{\sim}{\Lambda}\rangle}$$

is equal to the sum

$$(-1)^{\ell}\cdot \mathrm{vol}(\mathcal{C})\cdot \sum_{i=1}^{r}\ \sum_{\mu=1}^{r_i}\ \sum_{w_{i\mu:2}\in W(A_{i\mu},A_2)}\ \sum_{w_{i\mu:1}\in W(A_{i\mu},A_1)}$$

of the product of

$$\exp(\langle \underset{\sim}{H}(P_{i\mu}), w_{i\mu:2}\bar{\Lambda}_2 + w_{i\mu:1}\Lambda_1\rangle)$$

$$\times\ (1/\prod\nolimits_{\lambda_{i\mu}}(w_{i\mu:2}\bar{\Lambda}_2 + w_{i\mu:1}\Lambda_1, \lambda_{i\mu}))$$

with

$$(c_{res}(P_{i\mu}|A_{i\mu}:P_1|A_1:w_{i\mu:1}:\Lambda_1)\Phi_1, c_{res}(P_{i\mu}|A_{i\mu}:P_2|A_2:w_{i\mu:2}:\Lambda_2)\Phi_2).$$

Proof Taking into account the Appendix to Chap. 6 in TES (and Theorem 8.3), the right hand side is comprised of (some) of those terms for which (potentially) $\mathcal{C}_i(\underset{\sim}{\Lambda}) \subset \mathcal{C}$. With respect to a generic domination $P_{i\mu} \succeq P_{k_0\xi_0}$, the differences

$$\begin{cases} f_{P_{i\mu}}(x) - \sum_{w_{i\mu:1}\in W(A_{i\mu},A_1)} c_{res}(\ldots)\Phi_1(x) \\ g_{P_{i\mu}}(x) - \sum_{w_{i\mu:2}\in W(A_{i\mu},A_2)} c_{res}(\ldots)\Phi_2(x), \end{cases}$$

qua functions on $G/\Gamma\cap P_{i\mu}$, have a constant term along $P^{\dagger}_{k_0\xi_0}$ given solely by the

$$\begin{cases} w\Lambda_{i_0} & (w = w_{k_0\xi_0:i_0}) \\ w\Lambda_{j_0} & (w = w_{k_0\xi_0:j_0}) \end{cases}$$

such that

$$\begin{cases} \mathfrak{X}_w \\ \mathfrak{Y}_w \end{cases}$$

are not principal. Thus the exponential contributions to the constant term per $P_{i\mu}$ <u>not</u> involving a c_{res} necessarily have a nonzero real part and so, being in $\mathfrak{D}_{P_{i\mu}}(\check{\mathfrak{a}}_{i\mu})^{-}$, the associated truncation exponents $\underset{\sim}{\Lambda}$ are likewise nonzero. But then, in view of what has been said above, the corresponding $p_{\underset{\sim}{\Lambda}}$ must vanish identically. //

It is worth mentioning that the preceding assertion extends at once to when only

$$\begin{cases} \Phi_1\in E(\delta,\mathcal{O}_1:\mathbb{C}_0) \\ \Phi_2\in E(\delta,\mathcal{O}_2:\mathbb{C}_0) \end{cases} \quad (\mathcal{O}_1 \text{ and } \mathcal{O}_2 \text{ associate}).$$

Furthermore, in the $\mathfrak{X}$-notation of [2-(c)] (and the language of the weak constant term), our conclusion can be compactly summarized by the statement that for all $H \in \mathfrak{a}_Q$,

$$\sum_{\Lambda \in E_{PI}(f|_{\Lambda_{i_0}}, g|_{\Lambda_{j_0}})} p_\Lambda(H) \cdot e^{\langle H, \Lambda \rangle}$$

is equal to

$$(-1)^{\ell} \cdot \mathrm{vol}(\mathfrak{C})$$

times

$$\mathbb{I}(\mathfrak{C} : P_{i\mu}|A_{i\mu} : E^W_{P_{i\mu}}(P_1|A_1 : \Phi_1 : \Lambda_1 : ?), E^W_{P_{i\mu}}(P_2|A_2 : \Phi_2 : \Lambda_2 : ?) : H),$$

summed over i and μ.

§9. Holomorphic Extension

The purpose of this § is to consider a certain class of Detroit families which arise in the theory of c-functions, the point of the procedure being that the powerful machinery developed in §4 can then be applied to discuss the holomorphicity on the imaginary axis of the leading terms in the inner product formula.

Let us place ourselves in the circumstances surrounding Proposition 8.7. Thus suppose that

$$\begin{cases} \Phi_1 \in \mathcal{E}(\delta, \mathcal{O}_1 : \mathbf{C}_0) \\ \Phi_2 \in \mathcal{E}(\delta, \mathcal{O}_2 : \mathbf{C}_0) \end{cases} \quad (\mathcal{O}_1 \text{ and } \mathcal{O}_2 \text{ associate}).$$

Then, as we have seen, the "pure imaginary" contribution to the inner product

$$(Q^{\underset{\sim}{H}} E(P_1|A_1:\Phi_1:\Lambda_1:?),\ Q^{\underset{\sim}{H}} E(P_2|A_2:\Phi_2:\Lambda_2:?))$$

is equal to the sum

$$(-1)^{\ell} \cdot \text{vol}(\mathbf{C}) \cdot \sum_{i=1}^{r} \sum_{\mu=1}^{r_i} \sum_{w_{i\mu:2} \in W(A_{i\mu}, A_2)} \sum_{w_{i\mu:1} \in W(A_{i\mu}, A_1)}$$

of the product of

$$\exp(\langle \underset{\sim}{H}(P_{i\mu}), w_{i\mu:2}\bar{\Lambda}_2 + w_{i\mu:1}\Lambda_1 \rangle)$$

$$\times\ (1/\textstyle\prod_{\lambda_{i\mu}} (w_{i\mu:2}\bar{\Lambda}_2 + w_{i\mu:1}\Lambda_1, \lambda_{i\mu}))$$

with

$$(c_{res}(P_{i\mu}|A_{i\mu}:P_1|A_1:w_{i\mu:1}:\Lambda_1)\Phi_1, c_{res}(P_{i\mu}|A_{i\mu}:P_2|A_2:w_{i\mu:2}:\Lambda_2)\Phi_2).$$

Here, Λ_1 and Λ_2, while pure imaginary, are not absolutely arbitrary; in particular

$$\prod_{\lambda_{i\mu}}(w_{i\mu:2}\bar{\Lambda}_2 + w_{i\mu:1}\Lambda_1,\lambda_{i\mu}) \neq 0.$$

We therefore ask: Is

$$\sum_i\sum_\mu\sum_{w_{i\mu:2}}\sum_{w_{i\mu:1}}\cdots$$

a holomorphic function of $(\Lambda_1,\bar{\Lambda}_2)$ on $\sqrt{-1}\check{a}_1 \times \sqrt{-1}\check{a}_2$? Naturally, since ostensible singularities exist, to answer the question we must devise a way to remove them. The reader who is familiar with the first few pages of §6 in [2-(c)] will realize that the method employed there to treat a slightly less general problem can be readily adopted to the present situation, the upshot being an affirmative response. This important conclusion can also be arrived at in a less ad hoc fashion via the theory of Detroit families. Because the issue at hand is primarily methodological in character, we shall feel free to omit some of the (routine) details.

Let i be any index between 1 and r. Given

$$\begin{cases} w_{i:1}\in W(A_i,A_1) \\ w_{i:2}\in W(A_i,A_2), \end{cases}$$

write

$$\text{Ю}_i(\Lambda_1,\Lambda_2;w_{i:1},w_{i:2};\Phi_1,\Phi_2:\underset{\sim}{H})$$

for

$$\sum_{\mu=1}^{r_i} \exp(\langle \underset{\sim}{H}(P_{i\mu}), w_{i\mu:2}\bar{\Lambda}_2 + w_{i\mu:1}\Lambda_1\rangle)$$

$$\times\ (c_{res}(P_{i\mu}|A_{i\mu}:P_1|A_1:w_{i\mu:1}:\Lambda_1)\Phi_1, c_{res}(P_{i\mu}|A_{i\mu}:P_2|A_2:w_{i\mu:2}:\Lambda_2)\Phi_2).$$

Then the $\mathbb{D}_i$ are meromorphic functions of $(\Lambda_1, \bar{\Lambda}_2)$ which are, moreover, holomorphic on $\sqrt{-1}\check{\mathfrak{a}}_1 \times \sqrt{-1}\check{\mathfrak{a}}_2$. Our claim is that the same is true of

$$\sum_{i=1}^{r} \sum\nolimits_{w_{i:1}\in W(A_i,A_1)} \sum\nolimits_{w_{i:2}\in W(A_i,A_2)} \frac{\mathbb{D}_i(\Lambda_1,\Lambda_2;w_{i:1},w_{i:2};\Phi_1,\Phi_2:\underset{\sim}{H})}{\prod_{\lambda_i}(w_{i:2}\bar{\Lambda}_2 + w_{i:1}\Lambda_1,\lambda_i)} .$$

Omitting henceforth Φ_1, Φ_2, and $\underset{\sim}{H}$ from the notation, we shall also set $A = A_1$ and thus agree to write w_i in place of $w_{i:1}$. Given $w_{12}\in W(A_1,A_2)$, put

$$T(\Lambda_1,\Lambda_2;w_{12}) = \sum_{i=1}^{r} \sum\nolimits_{w_i\in W(A_i,A)} \frac{\mathbb{D}_i(\Lambda_1,\Lambda_2;w_i,w_iw_{12})}{\prod_{\lambda_i}(w_iw_{12}\bar{\Lambda}_2 + w_i\Lambda_1,\lambda_i)} .$$

Reformulated, the claim supra then amounts to the assertion that

$$\sum\nolimits_{w_{12}\in W(A_1,A_2)} T(\Lambda_1,\Lambda_2;w_{12})$$

is a meromorphic function of $(\Lambda_1,\bar{\Lambda}_2)$, holomorphic on $\sqrt{-1}\check{\mathfrak{a}}_1 \times \sqrt{-1}\check{\mathfrak{a}}_2$. To establish that this is in fact the case, it suffices to analyze each constituent of the sum separately.

So fix a w_{12}. Freeze Λ_2 and put

$$\Lambda = \Lambda_1 + w_{12}\bar{\Lambda}_2.$$

The space $\check{\mathfrak{a}}(\equiv V)$ comes supplied with the structure of a g.r.s., the chambers being in a natural one-to-one correspondence with the elements of the set

$$\coprod_{i=1}^{r} \{i\} \times W(A_i, A),$$

as follows from Lemma 2 on p. 67 of TES. Supposing that $\mathfrak{C}$ is such a chamber, say

$$\mathfrak{C} \longleftrightarrow \{i\} \times \{w_i\},$$

let

$$\begin{aligned} \phi_{\mathfrak{C}}(\Lambda)(= \phi_{\mathfrak{C}}(\Lambda:\Lambda_2)) \\ = \mathbb{E}_i(\Lambda - w_{12}\bar{\Lambda}_2, \Lambda_2; w_i, w_i w_{12}). \end{aligned}$$

The $\phi_{\mathfrak{C}}$ are, of course, meromorphic and, by fundamentals, are holomorphic on $\sqrt{-1}\check{\mathfrak{a}}$.

Lemma 9.1 The $\phi_{\mathfrak{C}}$ constitute a Detroit family $\underset{\sim}{\phi}$.

[We shall not stop to provide the verification. It is based on the functional equations and runs along essentially familiar lines.]

Form now $\mathrm{III}_{\underset{\sim}{\phi}}$ from $\underset{\sim}{\phi}$ (cf. §4) -- then Proposition 4.5 implies that $\mathrm{III}_{\underset{\sim}{\phi}}$ is a holomorphic function of Λ in $\sqrt{-1}\check{\mathfrak{a}}$. On the other hand, an easy examination of the definitions reveals that modulo a constant (= canonical volume factor),

$$T(\Lambda_1,\Lambda_2;w_{12})$$

is the same as

$$Ш_{\phi}(\Lambda).$$

Accordingly, $T(?,?;w_{12})$ is holomorphic in Λ_1 for each Λ_2; reversing the roles, $T(?,?;w_{12})$ is holomorphic in Λ_2 for each Λ_1. By Hartog's theorem, therefore, $T(?,?;w_{12})$ is holomorphic at each point

$$(\Lambda_1,\Lambda_2)\in\sqrt{-1}\check{a}_1\times\sqrt{-1}\check{a}_2.$$

[Note: We have deliberately glossed over the details centering on domains.]

Remark To make the preceding argument go, one need not work exclusively on the imaginary axis. Instead, it is enough to look at points (Λ_1,Λ_2) with the property that

$$Ю_i(?,?;w_i,w_iw_{12})\Big|_{(\Lambda_1,\Lambda_2)}$$

is holomorphic for all i and w_i. In the notation of §6 in [2-(c)], the (continued) value

$$T(\Lambda_1,\Lambda_2;w_{12})$$

is the sum

$$\sum_{0\leq k\leq \ell_{w_{12}}(\Lambda_1,\Lambda_2)}\sum_i\sum_{w_i:\iota(\Lambda_1,\Lambda_2;w_i,w_{12})=k}$$

of

$$Я_i^{(k)}(\Lambda_1,\Lambda_2;w_i,w_iw_{12}).$$

§10. Asymptotics

The purpose of this § is to discuss the difference between

$$(Q_{\underset{\sim}{H}}E(P_1|A_1:\Phi_1:\Lambda_1:?),\ Q_{\underset{\sim}{H}}E(P_2|A_2:\Phi_2:\Lambda_2:?))$$

and

$$(-1)^{\ell}\cdot\mathrm{vol}(\mathfrak{C})$$

times

$$\sum_{w_{12}\in W(A_1,A_2)} T(\Lambda_1,\Lambda_2;w_{12})$$

as $\underset{\sim}{H} \to -\infty$. Here, as before, the setting is residual and, anticipating the eventual applications, Λ_1 and Λ_2 will be taken pure imaginary. Our main conclusion then turns out to be that the difference between the two tends to 0 as $\underset{\sim}{H} \to -\infty$ (in a suitable sense), there even being certain uniformities per Λ_1 and Λ_2. It is the latter point at which a definite difficulty appears: Proposition 8.7 imposes a general position hypothesis and this must be eliminated before anything sensible can be said. It is for this reason that some rather technical preliminaries must be dispensed with at the beginning.

Suppose that, in the notation of the second theorem of control,

$$\begin{matrix} \mathfrak{C}(\mathrm{Dis}(\mathfrak{X})) & = & \mathfrak{C}(\mathrm{Dis}(\mathfrak{Y})) \\ \| & & \| \\ & \mathfrak{C} & \end{matrix}.$$

If, as there, the pair $(\Lambda_{i_0},\Lambda_{j_0})$ is $(\mathfrak{X},\mathfrak{Y})$-general with

$$\begin{cases} \Lambda_{i_0} \in X(P_{i_0},A_{i_0}) + \sqrt{-1}\check{\underset{\sim}{a}}_{\mathrm{Dis}(\mathfrak{X})} \\ \Lambda_{j_0} \in Y(P_{j_0},A_{j_0}) + \sqrt{-1}\check{\underset{\sim}{a}}_{\mathrm{Dis}(\mathfrak{Y})}, \end{cases}$$

then the inner product

$$(Q^{\underset{\sim}{H}}f|_{\Lambda_{i_0}}, g|_{\Lambda_{j_0}})$$

is an exponential polynomial on $\underset{\sim}{a}_Q$ and, by the first theorem of control,

$$\forall \underset{\sim}{\Lambda} \in \underset{\sim}{E}(f|_{\Lambda_{i_0}}, g|_{\Lambda_{j_0}}),$$

$$\mathcal{C} \succeq \mathcal{C}(\underset{\sim}{\Lambda}) \succeq \mathcal{C}_0.$$

Furthermore, if we measure the purity of $\underset{\sim}{\Lambda}$ at the $\mathcal{C}(\underset{\sim}{\Lambda})$ level, then the possibilities for the corresponding real parts are finite in number, <u>independently</u> of $(\Lambda_{i_0},\Lambda_{j_0})$ (see the Scholium in §6). That is, one can write down a (minimal) list

$$\underset{\sim}{L}: \underset{\sim}{\Lambda}_0 = \underset{\sim}{0}, \underset{\sim}{\Lambda}_1, \ldots, \underset{\sim}{\Lambda}_N$$

of elements in $\check{\underset{\sim}{a}}$ (the real affine dual of $\underset{\sim}{a}$) characterized by the property that

$$\forall \underset{\sim}{\Lambda}, \quad \mathrm{Re}(\underset{\sim}{\Lambda}) \in \underset{\sim}{L}.$$

Which of these phenomena persist if the assumption of $(\mathfrak{X},\mathfrak{Y})$-general is dropped? Of course,

$$(Q^{\underset{\sim}{H}}f|_{\Lambda_{i_0}}, g|_{\Lambda_{j_0}})$$

will still be an exponential polynomial in $\underset{\sim}{H}$ but, unfortunately, neither the first nor the second theorem of control need be true. However, what is true (and ultimately decisive) is that the list $\underset{\sim}{L}$ does not increase when the parameters are moved from general position. As might be expected, the justification is not immediate.

Thanks to what has been said in §7 of [2-(c)],

$$(\underset{\sim}{Q}^H f|_{\Lambda_{i_0}}, g|_{\Lambda_{j_0}})$$

is a meromorphic function of $(\Lambda_{i_0}, \bar{\Lambda}_{j_0})$, holomorphic on a certain open domain containing

$$(X(P_{i_0}, A_{i_0}) + \sqrt{-1}\check{a}_{Dis(\mathfrak{X})}) \times (Y(P_{j_0}, A_{j_0}) + \sqrt{-1}\check{a}_{Dis(\mathfrak{Y})}).$$

Using Proposition 7.3, we can actually express

$$(\underset{\sim}{Q}^H f|_{\Lambda_{i_0}}, g|_{\Lambda_{j_0}})$$

by means of a sum

$$(-1)^{\ell_0} \cdot \mathrm{vol}(\mathfrak{C}_0) \cdot \sum_{k_0=1}^{r_0} \sum_{\xi_0=1}^{r_{k_0}} \sum_{w_{k_0\xi_0:i_0} \in W(A_{k_0\xi_0}, A_{i_0})} \sum_h$$

of the

$$Q_h(\mathfrak{X}:\underset{\sim}{H}(P_{k_0\xi_0}):w_{k_0\xi_0:i_0}:\Lambda_{i_0}) \cdot \exp(\langle \underset{\sim}{H}(P_{k_0\xi_0}), w_{k_0\xi_0:i_0}\Lambda_{i_0} + \bar{\Lambda}_h \rangle).$$

In this connection, recall that

$$Q_h(\mathfrak{X}:H:w_{k_0\xi_0:i_0}:\Lambda_{i_0}) \in \mathfrak{m}_{i_0,\mathfrak{X}}[H],$$

while

$$\sum_h = \sum_{w_{k_0 \xi_0 : j_0}},$$

where

$$\Lambda_h \leftrightarrow w_{k_0 \xi_0 : j_0} \Lambda_{j_0}.$$

There will be no need to stress Λ_{j_0} in the notation since it plays only a secondary role in what follows. Put

$$w = w_{k_0 \xi_0 : i_0}$$

and choose, as is clearly possible, a finite set $\{p_n\}$ of polynomials such that

$$(Q^{\underset{\sim}{H}} f|_{\Lambda_{i_0}}, g|_{\Lambda_{j_0}})$$

is the sum

$$\sum_{k_0} \sum_{\xi_0} \sum_w \sum_h \sum_n$$

of the

$$\phi_n(\mathfrak{X}:w:h:\Lambda_{i_0}) \cdot p_n(\underset{\sim}{H}(P_{k_0 \xi_0})) \cdot \exp(\langle \underset{\sim}{H}(P_{k_0 \xi_0}), w\Lambda_{i_0} + \bar{\Lambda}_h \rangle).$$

Here,

$$\phi_n(\mathfrak{X}:w:h:\Lambda_{i_0}) \in \mathfrak{M}_{i_0, \mathfrak{X}}.$$

If now the pair $(\Lambda_{i_0}, \Lambda_{j_0})$ is $(\mathfrak{X}, \mathfrak{Y})$-general with

$$\begin{cases} \Lambda_{i_0} \in X(P_{i_0}, A_{i_0}) + \sqrt{-1}\check{\mathfrak{a}}_{\mathrm{Dis}(\mathfrak{X})} \\ \Lambda_{j_0} \in Y(P_{j_0}, A_{j_0}) + \sqrt{-1}\check{\mathfrak{a}}_{\mathrm{Dis}(\mathfrak{Y})}, \end{cases}$$

then the foregoing sum can be cut down considerably. Indeed, all pairs (w,h) for which

$$\mathrm{Re}((\underset{\sim}{w\Lambda_{i_0} + \bar{\Lambda}_h})P_{k_0\xi_0}) \notin \underset{\sim}{L}$$

can be thrown out. Denote by

$$\sum^{[w,h]}$$

the subsum arising thereby -- then, with the understanding that any other automatic cancellations are to be suppressed,

$$\sum_{k_0,\xi_0} \sum^{[w,h]} \sum_{h} \ . \ . \ .$$

gives

$$(Q_{\sim}^{H} f|_{\Lambda_{i_0}}, g|_{\Lambda_{j_0}}) \ .$$

To exploit this point, let us assume in addition that Λ_{i_0} is "regular" in the sense that it also lies on no singular hyperplane of the ϕ_n. Pick a real unit vector λ_0 such that $\forall n$, ϕ_n is holomorphic at $\Lambda_{i_0} + \zeta\lambda_0(|\zeta|\leq\delta)$ -- then, for δ sufficiently small and positive,

$$(Q_{\sim}^{H} f|_{\Lambda_{i_0}}, g|_{\Lambda_{j_0}})$$

is equal to

$$\frac{1}{2\pi\sqrt{-1}} \int_{|\zeta|=\delta} \frac{1}{\zeta} \{\ldots\} d\zeta,$$

where $\{\ldots\}$ is the sum

$$\sum_{k_0,\xi_0} \sum^{[w,h]} \sum_{n}$$

of the

$$\phi_n(\mathfrak{X}:w:h:\Lambda_{i_0} + \zeta\lambda_0)\cdot p_n(\underset{\sim}{H}(P_{k_0\xi_0}))\cdot \exp(\langle \underset{\sim}{H}(P_{k_0\xi_0}), w\Lambda_{i_0} + \bar{\Lambda}_h + \zeta w\lambda_0\rangle)$$

or still, upon pulling the sum to the outside, is equal to

$$\sum_{k_0,\xi_0} \sum^{[w,h]} \sum_n p_n(\underset{\sim}{H}(P_{k_0\xi_0}))$$

$$\times \frac{1}{2\pi\sqrt{-1}} \int_{|\zeta|=\delta} \left[\frac{1}{\zeta}\,\phi_n(\mathfrak{X}:w:h:\Lambda_{i_0}+\zeta\lambda_0)\right]\cdot \exp(\langle \underset{\sim}{H}(P_{k_0\xi_0}), w\Lambda_{i_0}+\bar{\Lambda}_h+\zeta w\lambda_0\rangle)d\zeta.$$

Each of these integrals can be calculated by the residue theorem. Thus fix w, h, and n; let $\{\zeta_\nu\}$ be the associated set of poles apart from the one at 0 -- then, in a suggestive notation, our evaluation takes the form

$$\phi_n(\mathfrak{X}:w:h:\Lambda_{i_0})\cdot \exp(\langle \underset{\sim}{H}(P_{k_0\xi_0}), w\Lambda_{i_0} + \bar{\Lambda}_h\rangle)$$

$$+ \sum_\nu \psi_{n\nu}(\mathfrak{X}:w:h:\Lambda_{i_0})\cdot q_{n\nu}(\underset{\sim}{H}(P_{k_0\xi_0}))\cdot \exp(\langle \underset{\sim}{H}(P_{k_0\xi_0}), w\Lambda_{i_0}+\bar{\Lambda}_h+\zeta_\nu w\lambda_0\rangle).$$

For the sake of simplicity, let

$$E_h(\mathfrak{X}:H:w:\Lambda_{i_0}) = \sum_n p_n(H)\sum_\nu \psi_{n\nu}\cdot q_{n\nu}\cdot \exp_{n\nu}\ .$$

We remark that

$$E_h(\mathfrak{X}:?:w:\Lambda_{i_0})$$

is an exponential polynomial on $\underset{\sim}{a}_Q$, its exponents being in the set

$$\{(\underbrace{w\Lambda_{i_0} + \bar{\Lambda}_h + \zeta_\nu w\lambda_0}) P_{k_0\xi_0}\}\ .$$

To summarize:

Lemma 10.1 *Suppose that the pair* $(\Lambda_{i_0}, \Lambda_{j_0})$ *is* $(\mathfrak{X},\mathfrak{Y})$-*general with*

$$\begin{cases} \Lambda_{i_0} \in X(P_{i_0}, A_{i_0}) + \sqrt{-1}\check{\mathfrak{a}}_{\mathrm{Dis}(\mathfrak{X})} \\ \Lambda_{j_0} \in Y(P_{j_0}, A_{j_0}) + \sqrt{-1}\check{\mathfrak{a}}_{\mathrm{Dis}(\mathfrak{Y})} \end{cases}$$

and Λ_{i_0} *regular* -- *then*

$$(Q^{H}f|_{\Lambda_{i_0}}, g|_{\Lambda_{j_0}})$$

is the sum

$$\sum_{k_0,\xi_0} \sum^{[w,h]}$$

of the

$$Q_h(\mathfrak{X}:\underset{\sim}{H}(P_{k_0\xi_0}):w:\Lambda_{i_0})\cdot\exp(\langle \underset{\sim}{H}(P_{k_0\xi_0}), w\Lambda_{i_0} + \bar{\Lambda}_h\rangle)$$
$$+ E_h(\mathfrak{X}:\underset{\sim}{H}(P_{k_0\xi_0}):w:\Lambda_{i_0}).$$

[Note: The lemma is not a tautology: The E_h, when added up, must vanish.]

Freeze a

$$\Lambda_{j_0} \in Y(P_{j_0}, A_{j_0}) + \sqrt{-1}\check{\mathfrak{a}}_{\mathrm{Dis}(\mathfrak{Y})}$$

with the property that there exists a

$$\Lambda_{i_0} \in X(P_{i_0}, A_{i_0}) + \sqrt{-1}\check{\mathfrak{a}}_{\mathrm{Dis}(\mathfrak{X})}$$

for which $(\Lambda_{i_0}, \Lambda_{j_0})$ is $(\mathfrak{X},\mathfrak{Y})$-general. Note that $(\Lambda_{i_0}, \Lambda_{j_0})$ is then $(\mathfrak{X},\mathfrak{Y})$-general for a.e.

$$\Lambda_{i_0} \in X(P_{i_0}, A_{i_0}) + \sqrt{-1}\check{\mathfrak{a}}_{\mathrm{Dis}(\mathfrak{X})}.$$

Bear in mind too that

$$(Q^{\underset{\sim}{H}} f|_{?}, g|_{\Lambda_{j_0}})$$

depends holomorphically on ?. Consider now an arbitrary

$$\Lambda_{i_0}^0 \in X(P_{i_0}, A_{i_0}) + \sqrt{-1}\check{\underset{\sim}{a}}_{Dis(\mathfrak{X})} .$$

So: The pair $(\Lambda_{i_0}^0, \Lambda_{j_0})$ need not be $(\mathfrak{X}, \mathfrak{Y})$-general. Suppose, however, we move to a nearby Λ_{i_0}, maintaining generality and regularity -- then the preceding lemma is applicable and

$$Q_h \cdot \exp + E_h,$$

being a contour integral, must extend holomorphically past $\Lambda_{i_0}^0$.

In other words:

$$\sum_{k_0, \xi_0} \sum^{[w,h]} (Q_h \cdot \exp + E_h),$$

when continued to $\Lambda_{i_0}^0$, gives

$$(Q^{\underset{\sim}{H}} f|_{\Lambda_{i_0}^0}, g|_{\Lambda_{j_0}}).$$

The objective will be to utilize this fact in order to confine the exponents.

Given an open subset $\underset{\sim}{U}$ of $\check{\underset{\sim}{a}} + \sqrt{-1}\check{\underset{\sim}{a}}$ (the affine dual of $\underset{\sim}{a}$), write

$$\sum_{\underset{\sim}{U}}^{[w,h]}$$

for the subsum of $\sum^{[w,h]}$ consisting of those exponents $w\Lambda_{i_0} + \bar{\Lambda}_h$ such that

$$\underset{\sim}{(w\Lambda_{i_0} + \bar{\Lambda}_h)}_{P_{k_0 \xi_0}} \in \underset{\sim}{U}.$$

Matters can always be arranged so as to ensure that their δ-neighborhoods are contained in $\underset{\sim}{U}$, hence, in the notation supra,

$$\underset{\sim}{(w\Lambda_{i_0} + \bar{\Lambda}_h + \zeta_\nu w\lambda_0)} P_{k_0\xi_0} \in \underset{\sim}{U}$$

too.

Let us assume that

$$\sum^{[w,h]} = \sum_u \sum_{\underset{\sim}{U}_u}^{[w,h]},$$

the $\underset{\sim}{U}_u$ being disjoint with

$$\underset{\sim}{(w\Lambda_{i_0} + \bar{\Lambda}_h)} P_{k_0\xi_0} \in \underset{\sim}{U}_u \qquad (\exists u).$$

Then

$$\begin{aligned} 0 &= \sum_{k_0,\xi_0} \sum^{[w,h]} E_h \\ &= \sum_{k_0,\xi_0} \sum_u \sum_{\underset{\sim}{U}_u}^{[w,h]} E_h \\ \Rightarrow \quad 0 &= \sum_{k_0,\xi_0} \sum_{\underset{\sim}{U}_u}^{[w,h]} E_h \qquad (\forall u). \end{aligned}$$

Now specialize the $\underset{\sim}{U}_u$ by first taking disjoint open neighborhoods $\underset{\sim}{U}_u^0$ in $\underset{\sim}{\check{a}}$ of the

$$\underset{\sim}{\Lambda}_u \in \underset{\sim}{L} \ (u = 0,1,\ldots,N)$$

and then setting

$$\underset{\sim}{U}_u = \underset{\sim}{U}_u^0 + \sqrt{-1}\underset{\sim}{\check{a}}.$$

To calculate the real parts of the

$$\underset{\sim}{\Lambda} \in \underset{\sim}{E}(f|_{\Lambda_{i_0}^0}, g|_{\Lambda_{j_0}}),$$

write

$$(Q^H f|_{\Lambda_{i_0}^0}, g|_{\Lambda_{j_0}})$$

as the extension to $\Lambda_{i_0}^0$ of

$$\sum_{k_0,\xi_0} \sum^{[w,h]} (Q_h \cdot \exp + E_h)$$

or still, as the extension to $\Lambda_{i_0}^0$ of

$$\sum_{k_0,\xi_0} \sum_{u} \sum_{U_u}^{[w,h]} (Q_h \cdot \exp + E_h).$$

Fix a u -- then

$$\sum_{k_0,\xi_0} \sum_{U_u}^{[w,h]} (Q_h \cdot \exp + E_h)$$

$$= \sum_{k_0,\xi_0} \sum_{U_u}^{[w,h]} Q_h \cdot \exp.$$

The latter, when extended to $\Lambda_{i_0}^0$, is an exponentional polynomial on a_Q; all its exponents are in U_u and, since U_u can be shrunk at will, each has real part equal to Λ_u.

So, dropping the $\Lambda_{i_0}^0$ from the notation, the real parts of the

$$\Lambda \in E(f|_{\Lambda_{i_0}}, g|_{\Lambda_{j_0}})$$

have been completely accounted for modulo the initial supposition on Λ_{j_0}. A reversal of roles easily eliminates this asymmetry.

We have therefore proved:

Theorem 10.2 Let

$$\begin{cases} \Lambda_{i_0} \in X(P_{i_0}, A_{i_0}) + \sqrt{-1}\check{\mathfrak{a}}_{Dis(\mathfrak{X})} \\ \Lambda_{j_0} \in Y(P_{j_0}, A_{j_0}) + \sqrt{-1}\check{\mathfrak{a}}_{Dis(\mathfrak{Y})}. \end{cases}$$

Then

$$\forall \underset{\sim}{\Lambda} \in \underset{\sim}{E}(f|_{\Lambda_{i_0}}, g|_{\Lambda_{j_0}}),$$
$$\operatorname{Re}(\underset{\sim}{\Lambda}) \in \underset{\sim}{L}.$$

There is an incidental bonus to all of this, implicit in what has been said above. Denote by

$$\underset{\sim}{E}_u(f|_{\Lambda_{i_0}}, g|_{\Lambda_{j_0}})$$

the set of elements $\underset{\sim}{\Lambda}$ in

$$\underset{\sim}{E}(f|_{\Lambda_{i_0}}, g|_{\Lambda_{j_0}})$$

for which

$$\operatorname{Re}(\underset{\sim}{\Lambda}) = \underset{\sim}{\Lambda}_u \qquad (u = 0, 1, \ldots, N).$$

In particular (cf. §8):

$$\underset{\sim}{E}_0(f|_{\Lambda_{i_0}}, g|_{\Lambda_{j_0}}) = \underset{\sim}{E}_{PI}(f|_{\Lambda_{i_0}}, g|_{\Lambda_{j_0}}).$$

Thus, for all $\underset{\sim}{H} \in \underset{\sim}{a}_Q$, we have

$$(Q^{\underset{\sim}{H}} f|_{\Lambda_{i_0}}, g|_{\Lambda_{j_0}})$$
$$= \sum_{\underset{\sim}{\Lambda} \in \underset{\sim}{E}(f|_{\Lambda_{i_0}}, g|_{\Lambda_{j_0}})} p_{\underset{\sim}{\Lambda}}(\underset{\sim}{H}) \cdot e^{\langle \underset{\sim}{H}, \underset{\sim}{\Lambda} \rangle}$$
$$= \sum_u \sum_{\underset{\sim}{\Lambda} \in \underset{\sim}{E}_u(f|_{\Lambda_{i_0}}, g|_{\Lambda_{j_0}})} p_{\underset{\sim}{\Lambda}}(\underset{\sim}{H}) \cdot e^{\langle \underset{\sim}{H}, \underset{\sim}{\Lambda} \rangle} \quad .$$

Furthermore,

$$\sum_{\underset{\sim}{\Lambda}\in \underset{\sim}{E}_u} (f|_{\Lambda_{i_0}}, g|_{\Lambda_{j_0}}) p_{\underset{\sim}{\Lambda}}(\underset{\sim}{H}) \cdot e^{\langle \underset{\sim}{H}, \underset{\sim}{\Lambda}\rangle}$$

is holomorphic in $(\Lambda_{i_0}, \bar{\Lambda}_{j_0})$ on

$$(X(P_{i_0}, A_{i_0}) + \sqrt{-1}\check{a}_{\text{Dis}(\underset{\sim}{X})}) \times (Y(P_{j_0}, A_{j_0}) + \sqrt{-1}\check{a}_{\text{Dis}(\underset{\sim}{Y})}).$$

[Note: Take $u = 0$ -- then we get still another way of looking at the results in §9.]

The preparation behind us, we are finally in a position to take up the asymptotics.

Let $\underset{\sim}{[}_0$ be the subset of $\underset{\sim}{a}_0$ comprised of those elements $\underset{\sim}{H}_0$ such that

$$\forall \underset{\sim}{H} \in \underset{\sim}{a},$$

$$\underset{\sim}{H} < \underset{\sim}{H} + \underset{\sim}{H}_0 \Leftrightarrow \underset{\sim}{H}_0 \in \underset{\sim}{[}_0.$$

Let $\underset{\sim}{F}_0$ be a basis for $\underset{\sim}{a}_0$ made up of elements from $\underset{\sim}{[}_0$. Given $\underset{\sim}{H}_1, \underset{\sim}{H}_2 \in \underset{\sim}{a}$, write

$$\underset{\sim}{H}_1 <_0 \underset{\sim}{H}_2$$

iff

$$\underset{\sim}{H}_2 - \underset{\sim}{H}_1$$

$$= \sum_{\underset{\sim}{H}_0 \in \underset{\sim}{F}_0} c_{\underset{\sim}{H}_0} \cdot \underset{\sim}{H}_0 \quad (c_{\underset{\sim}{H}_0} \geq 0 \ (\forall \underset{\sim}{H}_0),\ c_{\underset{\sim}{H}_0} > 0 \ (\exists \underset{\sim}{H}_0)).$$

Obviously,

$$\underset{\sim}{H}_1 <_0 \underset{\sim}{H}_2 \Rightarrow \underset{\sim}{H}_1 < \underset{\sim}{H}_2.$$

By $\lim_0$, we shall understand limit in the $<_0$ ordering (as compared to limit taken with respect to the $<$ ordering).

[Our parameter space $\underset{\sim}{a}$ is directed by either of these orderings (although not necessarily by $\prec$). In all the earlier work ([2-(a)], [2-(b)], [2-(c)]), $<$, while weak, was sufficient. Regrettably, more is now needed, primarily because of the presence of polynomials at the higher levels.]

Proposition 10.3 Fix $\Lambda_u \in L$, $\Lambda_u \neq 0$ -- then

$$\lim_{H \to -\infty}{}_0 \sum_{\Lambda \in E_u} (f|_{\Lambda_{i_0}}, g|_{\Lambda_{j_0}}) p_\Lambda(H) \cdot e^{\langle H, \Lambda \rangle} = 0$$

uniformly on compacta in

$$(X(P_{i_0}, A_{i_0}) + \sqrt{-1}\check{a}_{Dis(\mathfrak{X})}) \times (Y(P_{j_0}, A_{j_0}) + \sqrt{-1}\check{a}_{Dis(\mathfrak{Y})}).$$

[Note: For this statement, H need not lie in a_Q.]

The idea behind the proof is simple enough. Start off by fixing

$$\begin{cases} \Lambda_{i_0} \in X(P_{i_0}, A_{i_0}) + \sqrt{-1}\check{a}_{Dis(\mathfrak{X})} \\ \Lambda_{j_0} \in Y(P_{j_0}, A_{j_0}) + \sqrt{-1}\check{a}_{Dis(\mathfrak{Y})} \end{cases}$$

and then restrict the discussion to a small compact neighborhood of $(\Lambda_{i_0}, \Lambda_{j_0})$. Of course, no hypothesis of general position is in force now. Next, decompose

$$\sum_{\Lambda \in E_u} (f|_{\Lambda_{i_0}}, g|_{\Lambda_{j_0}}) p_\Lambda(H) \cdot e^{\langle H, \Lambda \rangle}$$

as a sum

$$\sum_{k_0, \xi_0} \sum_{\underset{\sim}{U}u}^{[w,h]} \sum_{n} ,$$

where, in an abbreviated notation, a typical summand is a polynomial times a contour integral

$$\int_{|\zeta|=\delta} \frac{[\ldots]}{\zeta} \cdot \exp \cdot d\zeta .$$

Here, thanks to joint holomorphicity in $(\Lambda_{i_0}, \bar{\Lambda}_{j_0})$, the argument of $[\ldots]$ has the form

$$\begin{cases} \Lambda_{i_0} + \zeta \cdot ?_{i_0} \\ \Lambda_{j_0} + \bar{\zeta} \cdot ?_{j_0}, \end{cases}$$

the pair $(?_{i_0}, ?_{j_0})$ being a real unit vector in $\check{a}_{Dis(X)} \times \check{a}_{Dis(Y)}$ selected in such a way that all the relevant singularities are avoided. Moreover, the argument of the exponential is confined to a compact set. Because $\frac{[\ldots]}{\zeta}$ is evidently bounded, if we take coordinates and transfer the question to $\underset{\sim}{R}^n$, everything is seen to come down to the following elementary lemma, the proof of which may be safely omitted.

Lemma 10.4 Let

$$f_z(x_1,\ldots,x_n) = p(x_1,\ldots,x_n)e^{x_1 z_1 + \ldots + x_n z_n}$$

be an exponential polynomial. Suppose that $\mathrm{Re}(z_i) > 0$ $\forall i$ and the z_i stay bounded -- then

$$\forall \varepsilon > 0 \quad \exists x_\varepsilon \quad \text{st} \quad x <_0 x_\varepsilon$$

$$\Longrightarrow |f_z(x)| < \varepsilon.$$

The above theory admits of an immediate application to Eisenstein series. Let, as usual

$$\begin{cases} \Phi_1 \in E(\delta, \mathcal{O}_1 : \mathbf{C}_0) \\ \\ \Phi_2 \in E(\delta, \mathcal{O}_2 : \mathbf{C}_0) \end{cases} \qquad (\mathcal{O}_1 \text{ and } \mathcal{O}_2 \text{ associate}).$$

Theorem 10.5 <u>For all</u> $H \in \mathfrak{a}_Q$, <u>the inner product</u>

$$(Q^{H} E(P_1 | A_1 : \Phi_1 : \Lambda_1 : ?),\ Q^{H} E(P_2 | A_2 : \Phi_2 : \Lambda_2 : ?))$$

<u>is equal to the sum</u>

$$(-1)^{\ell} \cdot \mathrm{vol}(\mathbf{C}) \cdot \sum_{w_{12} \in W(A_1, A_2)} \sum_{0 \le k \le \ell_{w_{12}}(\Lambda_1, \Lambda_2)} \sum_{i} \sum_{w_i : \iota(\Lambda_1, \Lambda_2; w_i, w_{12}) = k}$$

<u>of</u>

$$\mathcal{R}_i^{(k)}(\Lambda_1, \Lambda_2; w_i, w_i w_{12}; \Phi_1, \Phi_2 : H)$$

<u>PLUS</u>

$$o(\Lambda_1, \Lambda_2 : H),$$

<u>where</u>

$$\lim_{H \to -\infty} {}_0 \quad o(\Lambda_1, \Lambda_2 : H) = 0$$

<u>uniformly on compacta in</u>

$$\sqrt{-1}\check{\mathfrak{a}}_1 \times \sqrt{-1}\check{\mathfrak{a}}_2 .$$

[The reader may find it instructive to compare this result with Theorem 6.2 in [2-(c)].]

References

Arthur, J.: [1-(a)] The trace formula in invariant form, Ann. of Math., vol. 114(1981), pp. 1-74.

[1-(b)] On the inner product of truncated Eisenstein series, Duke Math. J., vol. 49(1982), pp. 35-70.

Osborne, M.S. and Warner, G.: [2-(a)] The Selberg trace formula I: Γ-rank one lattices, Crelle's J., vol. 324(1981), pp. 1-113.

[2-(b)] The Selberg trace formula II: Partition, reduction, truncation, Pacific J., Vol. 106 (1983), pp. 307-496.

[2-(c)] The Selberg trace formula III: Inner product formulae (initial considerations), Memoirs A.M.S., To Appear (1983).

UNDERSTANDING THE UNITARY DUAL

David A. Vogan, Jr.
Department of Mathematics
Massachusetts Institute of Technology
Cambridge, Massachusetts 02139

1. Introduction

Let G be a semisimple Lie group. One of the basic unsolved problems in non-commutative harmonic analysis is to describe the set $\hat{G}_u$ of equivalence classes of irreducible unitary representations of G. In those notes I want to advertise some conjectures about the general form of this solution: one old (Conjecture 4.4), one new (Conjecture 6.4), one borrowed (Conjecture 6.2), and one at least that is likely to be true (Conjecture 4.5). Before embarking on a technical discussion of them, I will try to explain why such conjectures are worth considering, since their resolution may not contribute much to the determination of $\hat{G}_u$.

An analogy with the theory of highest weight modules may be helpful. One of the fundamental problems in that theory is the character problem: determination of the formal characters of the irreducible highest weight modules. It is not very hard to show that, for a fixed semisimple Lie algebra $\mathfrak{g}$, this problem can be solved by a finite algebraic computation: one has to determine the ranks of a finite number of matrices, which depend on the structure constants of $\mathfrak{g}$ and on the highest weights in question. As a practical project, this is unpleasant even for $\mathfrak{sl}(3)$; but it is a kind of answer. Using more sophisticated ideas, various people have tabulated the formal characters for $\mathfrak{g}$ of dimension at most 51. (The ideas are in [11].) Such tables are another kind of solution of the character problem, for those $\mathfrak{g}$ for which they exist. For most mathematicians, the best solution is the Kazhdan-Lusztig conjecture, proved by Kazhdan-Lusztig, Beilinson-Bernstein, and Brylinski-Kashiwara ([12], [13], [5], [7]). This is an algorithm for computing formal characters from the "structure constants" of Weyl groups. As a tool for hand computation in small Lie algebras, it is not as good as Jantzen's techniques in [11]. It is easier to use than the method of computing ranks; but that is not a very important point. What sets it apart is that it makes sense; it connects the formal characters with other mathematical objects, and reveals structure in them that was not obvious before.

*Supported in part by NSF grant MCS-8202127

All of that is important, but hardly new. The point is that the problem of finding $\hat{G}_u$ presents similar difficulties. It can be reduced (for a fixed G) to the solution of a countable set of polynomial inequalities on vector spaces. Experience indicates that when these inequalities are written explicitly, they can be solved by a finite process. If this were proved (and it should not be difficult), one would have a computable description of $\hat{G}_u$. Partly because of interest from physics, such calculations have become a fairly popular indoor sport. They lead to lists of unitary representations, and complete algebraic information about them. These lists -- notably Bargmann's [4], but also the later ones -- have been critical to the development of the subject, but they have obvious drawbacks. One doesn't usually want complete information, and having it can be a hindrance.

This problem was alleviated by Langlands' classification of $\hat{G}$ (the possibly non-unitary admissible representations) in [16]. One can now try to describe $\hat{G}_u$ as a subset of $\hat{G}$, and the answer is often more comprehensible than a description of $\hat{G}_u$ by itself. (The unitary spherical series for $SL(2,\mathbb{R})$ is represented by the "unitary cross" in this way, rather than as a certain allowed range of eigenvalues for the Casimir operator.) This technology, which is described in [14], is analogous to Jantzen's work on highest weight modules. As it stands, however, it does not determine $\hat{G}_u$ in general; and it would not be completely satisfactory if it did. (The reason is that there are nice families of unitary representations, such as those found by Flensted-Jensen in [9], which do not look like families in the Langlands parameters.)

What is needed is some major insight analogous to the Kazhdan-Lusztig conjecture for the character problem. Perhaps the best existing candidate is the Kirillov-Kostant "orbit method", which suggests that $\hat{G}_u$ is related to the set of integral orbits in the real dual $\mathfrak{g}_0^*$ of the Lie algebra $\mathfrak{g}_0$ of G. One problem with this method is that complementary series representations are often not attached to any orbit; but since complementary series are a thorn in the side of most known approaches, this objection is not too important. More serious is the fact that some integral orbits seem to have no attached representation (see [21]); and the orbit-to-representation correspondence is difficult to make precise in any case.

The conclusion I draw from this is that no one knows what $\hat{G}_u$ should look like. If this is granted, one should try to guess as much as possible about $\hat{G}_u$, to narrow and direct the search for its complete structure. The idea behind all of the conjectures here is Langlands' functoriality principle. This can be interpreted (optimistically) as suggesting that under certain conditions one can establish a correspondence between representations of another reductive group L and those of G. Still more optimistically, one can hope that, under more restrictive conditions, unitary representations will correspond to unitary representations.

The greatest success of this idea is in Langlands' classification of $\hat{G}$. He shows that every irreducible representation of G corresponds to a Cartan subgroup H of G, and a character $\Gamma \in \hat{H}$. (This is made precise in Section 2.) If H is fixed, this defines a correspondence between $\hat{H}$ and a subset of $\hat{G}$. In this case unitary characters of H do correspond to unitary (in fact to tempered) representations of G. The difficulty is that non-tempered unitary representations of G correspond to non-unitary characters Γ of H; the most obvious example is the trivial representation of G. If there were no other examples, we would be happy. What happens, however, is that there are other unitary representations of G, neither tempered nor trivial. What we want to do is make these intermediate cases look like they are attached to unitary representations of groups intermediate between H and G. Stated a little less ambitiously, the problem is this: which unitary representations of G are built from unitary representations of smaller groups? (The interpretation of the word "built" is to be regarded as part of the problem.) My claim is that a good description of $\hat{G}_u$ must be compatible with this problem; and, therefore, that an understanding of this problem may shed some light on the structure of $\hat{G}_u$.

The kind of correspondence we want to consider is this. Fix a Cartan subgroup H of G. Let L be another reductive group containing a Cartan subgroup isomorphic to (and henceforth identified with) H. The group L should be smaller than (though not necessarily a subgroup of) G. To each character $\Gamma \in \hat{H}$, there correspond irreducible representations X^L and X^G of L and of G. (Here we are being vague, since the Langlands classification is a little more subtle than we have indicated so far.) There are at least two things we can ask of (G,H,L,Γ):

(1.1) the algebraic structure of X^L determines that of X^G

(1.2) X^L unitary $\Longrightarrow$ X^G unitary

(For "algebraic structure", one may read "distribution character".) The conjectures describe various hypotheses on (G,H,L,Γ), which should make one or both of these assertions true. The idea that such hypotheses exist is what I mean by "Langlands functoriality" for unitary representations.

The first example of such a phenomenon is when L is a Levi factor of a real parabolic subgroup P of G. In that case

$$X^G \subseteq \mathrm{Ind}_P^G X^L,$$

so (1.2) is automatic. (1.1) means that the induced representation should be irreducible; many conditions are known which will guarantee that. This situation is discussed in Section 3.

The second example is when L is the centralizer of a compact torus in G. In that case, [17] provides good conditions for (1.1). Essentially the same conditions should give (1.2); this conjecture, and a little of the evidence for it, are described in Section 4.

Section 5 examines what these ideas imply about $\hat{G}_u$. The case of Sp(n,1) is considered (following [1]); and we find several unitary representations still unaccounted for. Section 6 is a collection of incomplete ideas which deal with these mysterious representations and related problems. There is one precise conjecture about Langlands functoriality for groups L not contained in G (Conjecture 6.4), but it is intended only as a hint about this intriguing phenomenon.

Since the main point is to formulate conjectures, almost all proofs are omitted. The ideas owe a great deal to conversations with, and the work of, almost everyone in the bibliography. (I hope they will be satisfied with my decision to attribute very little to anyone, rather than risking errors and omissions.) It is a pleasure particularly to thank Tony Knapp and Jeff Adams, with whom I have had extensive discussions as this paper was being written.

2. Notation and the Langlands classification

Let G be a reductive group in Harish-Chandra's class (see [10]; in particular, G is allowed to be non-linear or disconnected). Fix a Cartan involution θ of G, with fixed point set K (a maximal compact subgroup). Put

$$\begin{aligned} G_0 &= \text{identity component of } G \\ \mathfrak{g}_0 &= \mathrm{Lie}(G) \\ \mathfrak{g} &= \mathfrak{g}_0 \otimes_{\mathbb{R}} \mathbb{C} \\ U(\mathfrak{g}) &= \text{universal enveloping algebra of } \mathfrak{g} \\ Z(\mathfrak{g}) &= \text{center of } U(\mathfrak{g}). \end{aligned} \tag{2.1}$$

We consider only $(\mathfrak{g},K)$ modules (instead of the corresponding group representations). Put

$$F(\mathfrak{g},K) = \text{category of } (\mathfrak{g},K) \text{ modules of finite length} \tag{2.2}$$

(see [22]). Then $\hat{G}$ is the set of equivalence classes of irreducibles in $\mathrm{Ob}(F(\mathfrak{g},K))$. Fix a Cartan subalgebra $\mathfrak{h} \subseteq \mathfrak{g}$, and define

$$\begin{aligned} \Delta(\mathfrak{g},\mathfrak{h}) &= \text{root system of } \mathfrak{h} \text{ in } \mathfrak{g} \\ W(\mathfrak{g},\mathfrak{h}) &= \text{Weyl group of } \mathfrak{h} \text{ in } \mathfrak{g} \\ \xi: \quad Z(\mathfrak{g}) &\to S(\mathfrak{h})^{W(\mathfrak{g},\mathfrak{h})} \quad \text{(Harish-Chandra map).} \end{aligned} \tag{2.3}$$

Suppose $\lambda \in \mathfrak{h}^*$. We say that a $\mathfrak{g}$ module has <u>infinitesimal character λ</u> if $Z(\mathfrak{g})$ acts on it by ξ composed with evaluation at λ.

<u>Definition 2.4</u> Suppose $\mathfrak{h}_0 \subseteq \mathfrak{g}_0$ is a θ-stable Cartan subalgebra, and H is its centralizer in G. Write

$$\mathfrak{h}_0 = \mathfrak{t}_0 + \mathfrak{a}_0$$

for the +1 and -1 eigenvalues of θ, and

$$T = H \cap K$$
$$A = \exp(\mathfrak{a}_0).$$

Then A is a vector group, and

$$H = T \times A,$$

a direct product. A <u>regular pseudocharacter</u> of H (briefly, a <u>regular</u> <u>character</u>) is a pair

$$\gamma = (\Gamma, \bar{\gamma})$$

subject to the following conditions.

R-1) $\Gamma \in \hat{H}$ (that is, Γ is an irreducible representation of H; recall that it may be non-abelian) and $\bar{\gamma} \in \mathfrak{h}^*$.

R-2) Suppose α is an imaginary root of $\mathfrak{h}$ in $\mathfrak{g}$. Then $\langle\alpha,\bar{\gamma}\rangle$ is real and non-zero. Write Ψ for the unique system of positive imaginary roots making $\bar{\gamma}$ dominant.

R-3) Write $\rho(\Psi)$ for half the sum of all the roots in Ψ, and $\rho_c(\Psi)$ for half the sum of the compact ones. Then

$$d\Gamma = \bar{\gamma} + \rho(\Psi) - 2\rho_c(\Psi).$$

Write <u>$P(H)$</u> (or <u>$P^G(H)$</u>) for the set of regular characters of H.

This definition is discussed in [17] or [22]. ($P(H)$ was called $\hat{H}'$ there, a notation which has proved disastrous.) Write L for the centralizer of A in G. If $\gamma \in \mathcal{P}(H)$, then $\bar{\gamma}$ is the Harish-Chandra parameter of a discrete series representation X^L of L. We can choose X^L so that Γ is the highest weight of the lowest $L \cap K$-type of X^L; this determines X^L uniquely. Choose a real parabolic subgroup $P = LN$ of G, with Levi factor L, in such a way that $\mathrm{Re}(\bar{\gamma}|_{\mathfrak{a}_0})$ is negative on the roots of $\mathfrak{a}$ in $\mathfrak{n}$. Then we define

$$X(\gamma) = X^G(\gamma) = \mathrm{Ind}_{LN}^G(X^L \otimes 1), \tag{2.5}$$

the <u>standard representation</u> with parameter γ ([17]). $X(\gamma)$ has finite length, and infinitesimal character $\bar{\gamma}$. It is convenient to have a few more standard representations.

<u>Definition 2.6</u> Suppose $H = TA$ is a θ-stable Cartan subgroup of G. A <u>limit pseudocharacter</u> (or <u>limit character</u>) of H is a triple

$$\gamma = (\Psi, \Gamma, \bar{\gamma})$$

with the following properties.

L-1) Ψ is a positive system for the imaginary roots of $\mathfrak{h}$ in $\mathfrak{g}$, $\Gamma \in \hat{H}$, and $\bar{\gamma} \in \mathfrak{h}^*$.

L-2) If $\alpha \in \Psi$, then $\langle \alpha, \bar{\gamma} \rangle \geq 0$.

L-3) $d\Gamma = \bar{\gamma} + \rho(\Psi) - 2\rho_c(\Psi)$.

Write <u>$\mathcal{P}_{\mathrm{lim}}(H)$</u> (or $\mathcal{P}^G_{\mathrm{lim}}(H)$) for the set of limit characters on H. A limit character γ is called <u>final</u> if it also satisfies

F-1) Suppose α is a simple root of Ψ, and $\langle \alpha, \bar{\gamma} \rangle = 0$. Then α is noncompact.

F-2) Suppose α is a real root of $\mathfrak{h}$ in $\mathfrak{g}$, and $\langle \alpha, \bar{\gamma} \rangle = 0$. Then α does not satisfy the parity condition ([22], Definition 8.3.11).

Write <u>$\mathcal{P}_f(H)$</u> (or $\mathcal{P}^G_f(H)$) for the set of final limit characters on H.

Attached to every limit character γ is a <u>standard</u> (<u>limit</u>) <u>representation</u>

$$X(\gamma) = X^G(\gamma) \tag{2.7}$$

defined in analogy with (2.5). It is induced from a limit of discrete series on G^A; a discussion may be found in [17] or [15].

Proposition 2.8 Suppose $\gamma \in P_{lim}(H)$ (Definition 2.6).

a) γ satisfies F-1) $\Longleftrightarrow$ $X(\gamma) \neq 0$.

b) If γ does not satisfy F-2), then $X(\gamma)$ is a direct sum of two standard limit representations attached to a more compact Cartan subgroup.

c) If γ is final, then $X(\gamma)$ has a unique irreducible submodule.

For connected linear G, this is proved in [15]. We omit the general proof.

Whenever γ satisfies F-1), write

$$\begin{aligned} \overline{X}^G(\gamma) &= \overline{X}(\gamma) = \mathrm{soc}(X(\gamma)) \\ &= \text{largest completely reducible submodule of } X(\gamma). \end{aligned} \tag{2.9)(a}$$

Proposition 2.8 says that $\overline{X}(\gamma)$ is irreducible if γ is final. If γ does not satisfy F-1), write

$$\overline{X}(\gamma) = 0. \tag{2.9)(b}$$

Theorem 2.10 (Langlands, Knapp-Zuckerman [16], [15]).

a) Suppose $\gamma_i \in P_f(H_i)$ $(i = 1,2)$. Then $\overline{X}(\gamma_1) \cong \overline{X}(\gamma_2)$ if and only if (H_1,γ_1) is conjugate to (H_2,γ_2) under K.

b) Suppose X is an irreducible $(\mathfrak{g},K)$ module. Then there are a θ-stable Cartan subgroup H, and a $\gamma \in P_f(H)$, such that $X \cong \overline{X}(\gamma)$.

The Knapp-Zuckerman result was formulated differently, and proved only for connected linear groups. This version is a consequence of the results on translation functors in [17] and [20]. Langlands' part of the theorem may also be found in [6].

3. Functoriality and Mackey induction.

The results in this section are all more or less obvious; they are included as motivation for the conjectures which follow.

Definition 3.1 A reductive subgroup $M \subseteq G$ is called real polarizable if one of the following equivalent conditions is satisfied.

a) M is a Levi factor of a parabolic subgroup of G.

b) There is a reductive abelian subalgebra $\mathfrak{a}_0 \subseteq \mathfrak{g}_0$ such that ad(X) has real eigenvalues for $X \in \mathfrak{a}_0$; and M is the centralizer of $\mathfrak{a}_0$ in G.

c) M is conjugate to the centralizer of an abelian subalgebra of p_0 (the -1 eigenspace of θ).

A real polarization of M is a parabolic subgroup P of G with Levi factor M.

Definition 3.2 Suppose M is a θ-stable, real polarizable reductive subgroup of G, and P is a real polarization of it. Let H be a θ-stable Cartan subgroup of M, and suppose

$$\gamma = (\Psi,\Gamma,\bar{\gamma}) \in P^G_{lim}(H).$$

Define

$$\gamma_P = (\Psi,\Gamma,\bar{\gamma}) \in P^M_{lim}(H),$$

that is, γ_P is γ regarded as a limit parameter for M.

Lemma 3.3 In the setting of Definition 3.2,

a) $\gamma \in P^G_f(H) \Longrightarrow \gamma_P \in P^M_f(H)$

b) Every summand of $\overline{X}^G(\gamma)$ occurs in

$$\mathrm{Ind}_P^G(\overline{X}^M(\gamma_P)) \text{ (normalized induction).}$$

Part (a) is trivial. Part (b) is a well-known consequence of either "asymptotics" or "lowest K-types"; see [16] or [19].

Corollary 3.4 Suppose M is a θ-stable, real polarizable reductive subgroup of G, with real polarization P. Suppose $H \subseteq M$ is a θ-stable Cartan subgroup, and $\gamma \in P^G_{lim}(H)$. Assume that $\overline{X}^M(\gamma_P)$ is unitarizable (Definition 3.2). Then $\overline{X}^G(\gamma)$ is as well.

This corollary can be sharpened slightly.

Lemma 3.5 In the setting of Definition 3.2, write H = TA, P = MN. Assume that for every root α of $\mathfrak{a}$ in n,

a) $\mathrm{Im}\langle\alpha,\bar{\gamma}\rangle \neq 0$.

Then if

$$\overline{X}^M(\gamma_P) = \bigoplus_{i=1}^{r} Y_i,$$

we have

$$\overline{X}^G(\gamma) = \bigoplus_i \mathrm{Ind}_P^G(Y_i).$$

This well-known result is due to Harish-Chandra; it gives one version of the desideratum (1.1).

Corollary 3.6 Suppose M is a θ-stable, real polarizable reductive subgroup of G, with real polarization P. Assume that $H \subseteq M$ is a θ-stable Cartan subgroup, and that $\gamma \in P^G_{\text{lim}}(H)$ satisfies (3.5)(a). Then $X^G(\gamma)$ is unitarizable if and only if $\overline{X}^M(\gamma_P)$ is.

This follows from Lemma 3.5 by standard techniques.

4. Functoriality and holomorphic induction.

An important feature of Corollaries 3.4 and 3.6 is that their statements are purely formal. We can therefore hope to formulate conjectural analogues of them without a deep understanding of the analogues of induction which might be necessary for proofs.

Definition 4.1 A reductive subgroup $L \subseteq G$ is called complex polarizable if one of the following equivalent conditions is satisfied.

a) There is a parabolic subalgebra $\mathfrak{q} \subseteq \mathfrak{g}$ (the complexified Lie algebra of G) such that L is the normalizer of $\mathfrak{q}$ in G.
b) There is a reductive abelian subalgebra $\mathfrak{t}_0 \subseteq \mathfrak{g}_0$ such that $\mathrm{ad}(X)$ has purely imaginary eigenvalues for $X \in \mathfrak{t}_0$; and L is the centralizer of $\mathfrak{t}_0$ in G.
c) L is the centralizer of a compact (connected) torus.
d) L is conjugate to the centralizer of an abelian subalgebra of $\mathfrak{k}_0$.

A complex polarization of L is a choice of $\mathfrak{q}$ as in (a). By (d) there is no loss of generality in assuming that L is θ-stable; and in that case, any complex polarization will be θ-stable as well.

Definition 4.2 Suppose L is a θ-stable, complex polarizable reductive subgroup of G, with complex polarization $\mathfrak{q} = \mathfrak{l} + \mathfrak{u}$. Let $H \subseteq L$ be a θ-stable Cartan subgroup, and suppose

$$\gamma = (\Psi, \Gamma, \overline{\gamma}) \in P^G_{\text{lim}}(H).$$

Define the one dimensional representation V of H (trivial on the split component) as in [22], Lemma 8.1.1, using Ψ for $\Delta^+(\mathfrak{u}, \mathfrak{t})$. Put

$$\gamma \quad = (\Psi_{\mathfrak{q}}, \Gamma_{\mathfrak{q}}, \overline{\gamma}_{\mathfrak{q}}) \in P^L_{\text{lim}}(H),$$

with

$$\Psi_{\mathfrak{q}} = \Psi \cap \Delta(\mathfrak{l},\mathfrak{h})$$
$$\bar{\gamma}_{\mathfrak{q}} = \bar{\gamma} - \rho(\mathfrak{u})$$
$$\Gamma_{\mathfrak{q}} = \Gamma \otimes V^*.$$

Here $\rho(\mathfrak{u})$ is half the sum of the roots of $\mathfrak{h}$ in $\mathfrak{u}$. That $\gamma_{\mathfrak{q}}$ belongs to $P^L_{\lim}(H)$ is proved in exactly the same way as Lemma 8.1.2 of [22]. The differential of the character of H on V is determined by this fact, which is why we have not bothered to reproduce the (rather complicated) definition of V. In Definition 3.2, γ_P does not depend on the choice of P. Obviously $\gamma_{\mathfrak{q}}$ does depend on $\mathfrak{q}$, but the dependence is not too serious.

Lemma 4.3 *In the setting of Definition 4.2, let $\mathfrak{q}' = \mathfrak{l} + \mathfrak{u}'$ be another complex polarization of L. Then*

$$\Gamma_{\mathfrak{q}'} = \Gamma_{\mathfrak{q}} \otimes \Lambda^{top}(\mathfrak{u}'/\mathfrak{u}\cap\mathfrak{u}')^*$$

$$\bar{X}^L(\gamma_{\mathfrak{q}'}) \cong X^L(\gamma_{\mathfrak{q}}) \otimes \Lambda^{top}(\mathfrak{u}'/\mathfrak{u}\cap\mathfrak{u}')^*.$$

In particular, $X^L(\gamma_{\mathfrak{q}'})$ is unitarizable if and only if $X^L(\gamma_{\mathfrak{q}})$ is.

Analogy with Corollary 3.4 suggests

Conjecture 4.4 In the setting of Definition 4.2, assume that $\bar{X}^L(\gamma_{\mathfrak{q}})$ is unitarizable and non-zero. Then $\bar{X}^G(\gamma)$ is unitarizable or zero.

The hypothesis that $\bar{X}^L(\gamma_{\mathfrak{q}})$ be non-zero is necessary; $\gamma_{\mathfrak{q}}$ may fail to satisfy F-1) (Definition 2.6) even if γ does. In that case $\bar{X}^L(\gamma_{\mathfrak{q}})$ is zero, and $\bar{X}^G(\gamma)$ may be unitarizable or not. Even as stated, however, this conjecture may be too optimistic; there is little evidence for it except in the following special case.

Conjecture 4.5 In the setting of Definition 4.2, assume that $\bar{X}^G(\gamma)$ is non-zero and irreducible, and that $\bar{\gamma}$ satisfies the positivity hypothesis

(P_0) $\qquad \mathrm{Re}\langle\alpha,\bar{\gamma}\rangle \geq 0$, all $\alpha \in \Delta(\mathfrak{u},\mathfrak{h})$.

Then $\bar{X}^G(\gamma)$ is unitarizable if and only if $\bar{X}^L(\gamma_{\mathfrak{q}})$ is. (A version of (1.1) is established under these hypotheses in [17] -- see Proposition 6.1.) The usefulness of such a conjecture in organizing the unitary dual may be seen in the following result. (Semisimplicity is assumed only to make S finite.)

Proposition 4.6 Assume that G is semisimple. Then there is a finite subset $S \subseteq \hat{K}$ with the following property. Suppose X is an irreducible unitarizable $(\mathfrak{g},K)$ module whose lowest K-types do not belong to the set S. Then there is a complex polarizable, reductive, proper subgroup L of G, with complex polarization $\mathfrak{q}$; a θ-stable Cartan subgroup $H \subseteq L$; and $\gamma \in P_{lim}(H)$, such that $X = \overline{X}^G(\gamma)$, and (P_0) of Conjecture 4.5 holds.

This says that, except for those whose lowest K-type is in S, all unitary representations of G should come from unitary representations of proper subgroups. The proposition is a consequence of the following facts. Suppose $\gamma \in P_{lim}(H)$, and $H = TA$. Given any $C > 0$, we can find a finite set $S_C \subseteq \hat{K}$ such that if no lowest K-type of $\overline{X}^G(\gamma)$ belongs to S_C, then

$$|\overline{\gamma}|_{\mathfrak{t}}| \geq C. \tag{4.7)(a}$$

This is clear from [9]. Next, there is a constant N, depending only on G, such that if $\overline{X}^G(\gamma)$ is unitary (or even if its matrix coefficients are bounded), then

$$|\mathrm{Re}\ \overline{\gamma}|_{\mathfrak{a}}| \leq N. \tag{4.7)(b}$$

This is proved in [6]. If N is fixed, and C is large enough (depending on N), then (4.7) implies the existence of $\mathfrak{q}$ (not equal to $\mathfrak{g}$) satisfying (P_0). We leave the (elementary but not trivial) argument to the reader. This method gives a terrible estimate for S, one which is more or less useless for describing $\hat{G}_u$.

Problem 4.8 Find a description of the best possible set S in Proposition 4.6, even conjecturally.

Using the Dirac inequality, one can (with substantial effort) find the following possibilities for S in some examples.

TABLE 4.9

G	K	S
SO(n,1)	SO(n)	{trivial representation}
SU(n,1)	S(U(n)×U(1))	{trivial representation}
Sp(n,1)	Sp(n)×Sp(1)	$\{\text{trivial} \otimes S^k(\mathbb{C}^2), 0 \leq k \leq n-2\}$
Sp(2,ℂ)	Sp(2)	$\{\text{trivial}, \mathbb{C}^4\}$
SL(3,ℂ)	SU(3)	{trivial}
SL(3,ℝ)	SO(3)	$\{\text{trivial}, \mathbb{C}^3\}$
GL(4,ℝ)	O(4)	$\{\Lambda^i \mathbb{C}^4, 0 \leq i \leq 4\}$

It would not be very difficult to extend this table, finding the smallest set S allowed by the Dirac inequality for some other groups. However, the Dirac inequality does not by itself give the best possible estimate for S in higher rank groups; so such a calculation is not very helpful for Problem 4.8.

We conclude this section with some evidence for Conjecture 4.5. If $\overline{X}^G(\gamma)$ is non-zero, it is tempered if and only if $\Gamma \in \hat{H}$ is unitary ([16],[18]). Consequently, we have

Proposition 4.9 *In the setting of Definition 4.2, assume that $\overline{X}^L(\gamma_{\mathfrak{a}})$ and $\overline{X}^G(\gamma)$ are both non-zero. Then $\overline{X}^G(\gamma)$ is tempered if and only if $\overline{X}^L(\gamma_{\mathfrak{a}})$ is tempered.*

This is evidence even for Conjecture 4.4, since we do not assume (4.5) (P_0). The next result relates the functoriality of Definition 4.2 to unitary induction.

Proposition 4.10 *In the setting of Definition 4.2 assume that L is contained in a real polarizable reductive subgroup M, with real polarization P = MN (Definition 3.1). Then*

a) *$\overline{X}^G(\gamma) \subseteq \mathrm{Ind}_P^G(\overline{X}^M(\gamma))$. If $\overline{\gamma}$ satisfies (4.5)(P_0), then equality holds.*

b) *$\overline{X}^L(\gamma_{\mathfrak{a}})$ differs from $\overline{X}^L(\gamma_{\mathfrak{a}\cap m})$ by tensoring with a unitary character.*

c) *If $\overline{X}^M(\gamma)$ is unitary, so is $\overline{X}^G(\gamma)$. If $\overline{\gamma}$ satisfies (4.5) (P_0), then the converse is also true.*

d) *Conjectures 4.4 and 4.5 for (G,L,γ) follow from their analogues for (M,L,γ).*

The only non-formal assertion is the second part of (a); and that is a consequence of the results in [17]. (This is not obvious, but it is routine.) If G is a complex Lie group, then any complex polarizable subgroup L is automatically real polarizable: take t_0 as in Definition 4.1(b), and let $\mathfrak{a}_0$ be $(\sqrt{-1})(t_0)$. This $\mathfrak{a}_0$ shows that L satisfies Definition 3.1(b). Since Conjectures 4.4 and 4.5 are trivial when L = G, we deduce

Corollary 4.11 *Conjectures 4.4 and 4.5 are true when G is complex.*

For most real groups (for example, whenever rkG = rkK), the only real polarizable subgroup M containing a complex polarizable one is M = G; so Proposition 4.10 is of no use.

Any unitarizable $(\mathfrak{g},K)$ module carries an invariant Hermitian structure; so we can investigate the behaviour of this property under functoriality.

Definition 4.12 Suppose $\gamma = (\Psi,\Gamma,\bar{\gamma}) \in \mathcal{P}_{lim}(H)$, and $H = TA$. Define γ^h, the Hermitian dual of γ, by

$$\gamma^h = (\Psi,\Gamma^h,\bar{\gamma}^h),$$

with

$$\bar{\gamma}^h(x) = \text{complex conjugate of } \bar{\gamma}(x) \qquad (x \in \mathfrak{h}_0),$$

and analogous notation for Γ. For $t \in \mathbb{R}$, define

$$\gamma^t = (\Psi,\bar{\gamma}^t,\Gamma^t),$$

with

$$\begin{aligned} \bar{\gamma}^t\big|_{\mathfrak{t}} &= \bar{\gamma}\big|_{\mathfrak{t}} \\ \operatorname{Im}(\bar{\gamma}^t\big|_{\mathfrak{a}}) &= \operatorname{Im}(\bar{\gamma}\big|_{\mathfrak{a}}) \\ \operatorname{Re}(\bar{\gamma}^t\big|_{\mathfrak{a}}) &= t \cdot \operatorname{Re}(\bar{\gamma}\big|_{\mathfrak{a}}) \\ \Gamma^t\big|_T &= \Gamma\big|_T \\ \Gamma^t\big|_A &= \exp(\bar{\gamma}\big|_{\mathfrak{a}_0}). \end{aligned}$$

Thus $\gamma^1 = \gamma$, and γ^0 is a parameter for a tempered representation.

Lemma 4.13 Suppose $\gamma \in \mathcal{P}^G_{lim}(H)$, and $\bar{X}^G(\gamma) \neq 0$. Then $\bar{X}^G(\gamma)$ admits a non-degenerate invariant Hermitian form if and only if there is an element $w \in W(G,H)$ of order two, taking γ to γ^h (Definition 4.12).

For $\gamma \in \mathcal{P}(H)$, this is well known. The general case is not entirely trivial, but we will not digress to prove it here.

Proposition 4.15 In the setting of Definition 4.2,

a) Assume that $\bar{X}^L(\gamma_{\mathfrak{g}})$ is non-zero and admits a non-degenerate invariant Hermitian form. Then $\bar{X}^G(\gamma)$ admits a non-degenerate invariant Hermitian form.

b) Assume that $\bar{X}^G(\gamma)$ is non-zero, and that $\bar{\gamma}$ satisfies (4.5)(P_0). Then $\bar{X}^G(\gamma)$ admits a non-degenerate invariant Hermitian form if and only if $\bar{X}^L(\gamma_{\mathfrak{g}})$ does.

Part(a) and the "if" part of (b) are more or less obvious from Lemma 4.13. The "only if" part of (b) is a little more subtle; it becomes easy if the inequality in (4.5)(P_0) is required to be strict.

Using complementary series tricks, this can be promoted to a result about unitarity.

Lemma 4.16 Suppose $\overline{X}^G(\gamma)$ admits a non-degenerate invariant Hermitian form, and that $X^G(\gamma^t)$ is irreducible for $0 \le t < 1$ (Definition 4.12). Then $\overline{X}^G(\gamma)$ is unitarizable.

This is well-known.

Lemma 4.17 ([17], Section 4). Under the hypothesis (4.5) (P_0), suppose that $X^L(\gamma_{\mathfrak{g}})$ is irreducible. Then $X^G(\gamma)$ is as well.

Proposition 4.18 In the setting of Definition 4.2, suppose that $\overline{X}^L(\gamma_{\mathfrak{g}})$ is unitarizable because of Lemma 4.16 (with L in place of G), and that $\overline{\gamma}$ satisfies (4.5)(P_0). Then $\overline{X}(\gamma)$ is unitarizable.

Corollary 4.19 The "if" part of Conjecture 4.5 is true whenever every noncompact simple factor of ℓ_0 is $\mathfrak{su}(n,1)$ or $\mathfrak{so}(n,1)$.

5. Life after the conjecture.

In this section, we will try to describe the picture of $\hat{G}_u$ which emerges from Conjecture 4.5 and Problem 4.8. Of course it is inductive in nature, relying on previous knowledge of the unitary duals of certain proper subgroups of G. We may as well assume G is semisimple. Fix a θ-stable Cartan subgroup $H = TA$; a positive system Ψ for t in the imaginary roots of h in $\mathfrak{g}$; and a pair

(5.1)(a) $$(\overline{\lambda},\Lambda) \in t^* \times \hat{T}.$$

Consider the limit characters $\gamma(\nu)$, for $\nu \in \mathfrak{a}^*$, defined by

(5.1)(b) $$\gamma(\nu) = (\Psi,(\overline{\lambda},\nu),(\Lambda \times e^{\nu})).$$

Clearly the condition for $\gamma(\nu)$ to be a limit character is a condition on $(\Psi,\overline{\lambda},\Lambda)$ (see Definition 2.6), and we assume that it is satisfied. To rule out trivialities, we may as well assume that (F-1) is satisfied (see Proposition 2.8). Write

(5.1)(c) $$\overline{X}(\nu) = \overline{X}^G(\gamma(\nu)) \qquad \text{(see 2.9)).}$$
$$X(\nu) = X^G(\gamma(\nu))$$

We want to describe qualitatively the set of ν for which $\overline{X}(\nu)$ is unitary. If $\mathrm{Im}(\nu) \neq 0$, Corollary 3.6 reduces this question to the proper subgroup

$M(\nu) \supseteq H$, defined by

$$\Delta(m(\nu),h) = \{\alpha \in \Delta(\mathfrak{g},h) \mid \mathrm{Im}\langle\alpha,\nu\rangle = 0\}.$$

So we may as well restrict attention to the case $\mathrm{Im}(\nu) = 0$; that is,

(5.1)(d) $$\nu \in \mathfrak{a}_0^* \qquad \text{(the real dual of } \mathfrak{a}_0\text{)}.$$

The set of lowest K-types of $\overline{X}(\nu)$ depends only on $(\Psi,\overline{\lambda},\Lambda)$, and not on ν. If this set does not meet the set S of Proposition 4.6 (and Problem 4.8), then Conjecture 4.5 would reduce the unitarity question for $\overline{X}(\nu)$ to a proper subgroup. So we may assume that the lowest K-types meet S. This can happen for only finitely many choices of $(H,\Psi,\overline{\lambda},\Lambda)$. For each choice, we must describe

(5.2)(a) $$Q = Q(H,\Psi,\overline{\lambda},\Lambda) \subseteq \mathfrak{a}_0^*,$$
$$Q = \{\nu \in \mathfrak{a}_0^* \mid \overline{X}(\nu) \text{ is unitarizable}\}.$$

Fix $w \in W(G,H)$ of order 2, and suppose that

(5.2)(b) $$w(\Psi,\overline{\lambda},\Lambda) = (\Psi,\overline{\lambda},\Lambda).$$

Set

(5.2)(c) $$(\mathfrak{a}_0^*)_w = \{\nu \in \mathfrak{a}_0^* \mid w\nu = -\nu\}$$
$$Q_w = Q \cap (\mathfrak{a}_0^*)_w.$$

By Lemma 4.13

$$Q = \bigcup_{\substack{w \text{ satisfies} \\ (5.2)(b)}} Q_w.$$

We therefore fix w subject to (5.2)(b), and confine our attention to Q_w. We may as well assume that $(\mathfrak{a}_0^*)_w$ is not properly contained in any $(\mathfrak{a}_0^*)_{w'}$. (If (as often happens) we can find w satisfying (5.2)(b), such that

$$w\big|_{\mathfrak{a}_0^*} = -I,$$

then $(\mathfrak{a}_0^*)_w = \mathfrak{a}_0^*$. There is little harm in thinking only of this case.) Write

(5.3) $$\mathcal{H} = \mathcal{H}(H,\Psi,\overline{\lambda},\Lambda)$$
$$= \text{set of maximal affine subspaces of } \mathfrak{a}_0^* \text{ along which } X(\nu) \text{ is reducible.}$$

Every element of H is contained in an affine hyperplane of the form

$$\{\nu \in \mathfrak{a}_0^* \mid \langle \check{\alpha}, \nu \rangle = m/N\}.$$

Here α is a root of h in $\mathfrak{g}$; $m \in \mathbb{Z}$; and N is an integer depending only on G. (If G is linear, then N is 1.) Say that $\nu_1 \sim \nu_2$ if ν_1 and ν_2 lie in exactly the same elements of H. A <u>facet</u> of $(\mathfrak{a}_0^*)_w$ is by definition a connected component of an equivalence class, under this equivalence relation. Each facet is a bounded open subset of an affine subspace of $(\mathfrak{a}_0^*)_w$, defined by strict linear inequalities.

<u>Lemma 5.4 Suppose F is a facet in $(\mathfrak{a}_0^*)_w$, $\nu_1 \in F$, and $\nu_2 \in \overline{F}$. If $\overline{X}(\nu_1)$ is unitarizable, so is $\overline{X}(\nu_2)$.</u>

This is well-known. As a consequence, we get

<u>Proposition 5.5 In the setting just described, the region $Q \subseteq \mathfrak{a}_0^*$ of unitarity is a closed, bounded union of (finitely many) facets. It may be defined by finitely many polynomial inequalities.</u>

To describe $\hat{G}_u$ using Conjecture 4.5, we must therefore supply the following data (for subgroups of G as well): the finite set $S \subseteq \hat{K}$ of Problem 4.8; and for each of the (finitely many) $(H,\Psi,\overline{\lambda},\Lambda)$ as in (5.1) such that the lowest K-types of $\overline{X}(\nu)$ meet S, the unitarity region Q. By Proposition 5.5, this region admits a finite description. To give a <u>nice</u> description of $\hat{G}_u$, we need nice descriptions of S and of the regions Q. (For the latter, one possibility would be to find polynomials as in Proposition 5.5.) If we want to know characters for all elements of $\hat{G}_u$, we need to give them for each facet in every Q; again this is a finite problem.

Here are two examples. Suppose G = SU(n,1), so that S consists only of the trivial representation of K (Table 4.9). We may consider only the maximally split Cartan H = TA, together with any choice of Ψ, and

$$\overline{\lambda} = \rho(\Psi)$$
$$\Lambda = \text{trivial character of } T.$$

Then $\{X(\nu)\}$ is the spherical principal series of SU(n,1). Identify $\mathfrak{a}_0^*$ with $\mathbb{R}$ by sending the unique real root of h to 1. Then the reducibility set of (5.3) is (as computed by Kostant)

$$H = \{\pm(\tfrac{n}{2}+k),\ k = 0,1,2,\ldots\}.$$

The facets are

$$\left(-\frac{n}{2},\frac{n}{2}\right),\quad \pm\left(\frac{n}{2}+k,\frac{n}{2}+k+1\right) \qquad (k = 0,1,2,\ldots)$$

and the points of H. The set Q is

$$Q = [\tfrac{n}{2},\tfrac{n}{2}].$$

The precise structure of $\overline{X}(\nu)$ for $\nu \in Q$ is also easy to find: we have

$$\overline{X}(\nu) = X(\nu),\quad \nu \in \left(-\frac{n}{2},\frac{n}{2}\right)$$

$$\overline{X}\left(\pm\frac{n}{2}\right) \cong \text{trivial representation.}$$

Next, suppose $G = Sp(n,1)$. Again we may consider only the maximally split Cartan $H = TA$. Fix $(\Psi,\overline{\lambda},\Lambda)$ as in (5.1), and suppose that the lowest K-types of $X(\nu)$ meet the set S of Table 4.9. Calculation shows that $X(\nu)$ has a unique lowest K-type, say

$$\text{trivial} \otimes S^r(\mathbb{C}^2),$$

with some r between 0 and n-2. (This r determines $(\Psi,\overline{\lambda},\Lambda)$ up to conjugation.) Again we identify $\mathfrak{a}_0^*$ with $\mathbb{R}$ by sending the real root to 1; for clarity, write $X_r(\nu)$ instead of $X(\nu)$. By [3],

$$H_r = \{\pm[n-\left(\tfrac{r+1}{2}\right)+k],\ k = 0,1,2,\ldots\}.$$

By [1], the unitarity set is

$$Q_r = \left[-\left(n-\left(\tfrac{r+1}{2}\right)\right), n+\left(\tfrac{r+1}{2}\right)\right]$$

if $r \neq 0$; and if $r = 0$,

$$Q_r = \left[-\left(n-\tfrac{1}{2}\right), n-\tfrac{1}{2}\right] \cup \left\{\pm\left(n+\tfrac{1}{2}\right)\right\}.$$

Most of the unitarizable $\overline{X}_r(\nu)$ have simple structure:

$$\overline{X}_r(\nu) = X_r(\nu),\quad \nu \in \text{interior of } Q_r$$

$$\overline{X}_0\left(\pm\left(n+\tfrac{1}{2}\right)\right) \cong \text{trivial representation.}$$

It remains only to describe the representations $\overline{X}_r\left(\pm\left(n-\left(\frac{r+1}{2}\right)\right)\right)$. We will do this in Section 6.

If Conjecture 4.5 is accepted as a natural construction of unitary representations, we see that all the unitary representations of $SU(n,1)$ and $Sp(n,1)$ are obtained as follows. Begin with unitary characters (one dimensional unitary representations), and the mysterious representations $\overline{X}_r\left(\pm\left(n-\left(\frac{r+1}{2}\right)\right)\right)$ of Sp(n,1). Apply as necessary

1) Unitary induction from real parabolic subgroups
2) Holomorphic induction (Conjecture 4.5)
3) Deformation (Lemma 5.4 and obvious generalizations).

6. Mysterious hints.

With a suitable (finite) set of mysterious representations, the picture of $\hat{G}_u$ given for SU(n,1) and Sp(N,1) at the end of the last section seems to persist for many of the (few) known examples. A starry-eyed dreamer might now formulate a conjecture to that effect. (Knapp has informed me of counterexamples to several very simple-minded formulations of the conjecture, but the picture remains.) Without being so brave, we can still consider the obstacles which would remain to finding $\hat{G}_u$ if this hypothetical conjecture were true. The principal practical one is that the application of Lemma 5.4 (deformation) requires a complete knowledge of the reducibility sets of all induced-from-unitary representations. This is provided by the Kazhdan-Lusztig conjecture [22] in the case of linear groups; but the calculations involved even for Sp(n,1) are formidable.

The most interesting obstruction is of course the description of the mysterious representations. These are of several kinds. The first kind consists of dull representations in disguise. To describe them, suppose we are in the setting of Definition 4.2. Recall from [22], Chapter 6, Zuckerman's cohomological parabolic induction functors

$$R^i_{\mathfrak{q}}: \ F(\ell, L\cap K) \to F(\mathfrak{g}, K), \ 0 \le i \le S = \dim \mathfrak{u}\cap\mathfrak{k} \ \text{(notation 2.4)}.$$

Proposition 6.1 In the setting of Definition 4.2, assume that $\overline{\gamma}$ satisfies (4.5)(P_0). Then

$$R^i_{\mathfrak{q}}(\overline{X}^L(\gamma_{\mathfrak{q}})) = \begin{cases} 0 & i \neq S \\ \overline{X}^G(\gamma) & i = S \end{cases}$$

This follows from [17] and technical results in [22]. What we want to do is generalize Conjecture 4.5 in some way that fits well with Proposition 6.1. It is not clear how to do this in maximal generality; but here is a special case, representing joint work with Jeffrey Adams.

Conjecture 6.2 In the setting of Definition 4.2, assume that $\phi \in P^L_{\lim}(H)$; put $\overline{\gamma} = \overline{\phi} + \rho(\mathfrak{u})$. (This ϕ will play the role of $\gamma_{\mathfrak{q}}$, but γ need not actually exist as an element of $P^G_{\lim}$ (H). Assume that

a) $\overline{X}^L(\phi)$ is a one dimensional unitary representation;

b) $\langle\alpha,\overline{\gamma}-\rho(\ell)\rangle \geq 0$, all $\alpha \in \Delta(\mathfrak{u},\mathfrak{h})$;
here $\rho(\ell)$ is half the sum of the positive roots of $\mathfrak{h}$ in ℓ (chosen to make $\overline{\gamma}$ dominant). Then

1) $R^i(\overline{X}^L(\phi)) = 0$, $i \neq S$;

2) $R^S(\overline{X}^L(\phi)) = Y(\mathfrak{q},\phi)$ is completely reducible (or zero). If strict inequality holds in (b), it is irreducible or zero;

3) $Y(\mathfrak{q},\phi)$ is unitarizable.

It should be emphasized that when (4.5)(P_0) fails, there may not be a γ with $\phi = \gamma_{\mathfrak{q}}$; so this conjecture is not a special case even of Conjecture 4.4. Adams has amassed a good deal of evidence for Conjecture 6.2, using the theory of dual reductive pairs; this will appear in a future paper. His work shows that, as ϕ varies subject to (6.2)(a), (b), the representations $Y(\mathfrak{q},\gamma)$ should be regarded as a single family, despite the fact that their Langlands parameters may vary irregularly. (It is in this sense that the Langlands classification is not a perfect context for discussing $\hat{G}_u$.)

Those mysterious representations which are of the form $Y(\mathfrak{q},\phi)$ are the ones we called dull representations in disguise. (We should also include in this class representations arising from some generalization of Conjecture 6.2 in which $\overline{X}^L(\phi)$ need no longer be one dimensional. This requires changing (6.2)(b) in some way which is still unclear.) As an example, let $G = Sp(n,1)$, $L = Sp(n-1,1) \times \mathbb{T}^1$. The choice of is unique up to conjugation. Fix $H = TA \subseteq L$. Up to $W(L,H)$ conjugacy, there are exactly n parameters $\phi \in P^L_{\lim}(H)$ which satisfy (6.2)(a) and (b), but not (4.5)(P_0). Call them $\phi_0,\phi_1,\ldots,\phi_{n-1}$, labelled by the integer $\langle\check{\alpha},\overline{\gamma}-\rho(\ell)\rangle$ for any $\alpha \in \Delta(\mathfrak{u},\mathfrak{h})$. It can be shown that Conjecture 6.2 is true in this case, and that the mysterious representations of Sp(n,1) discussed in Section 5 are

$$\overline{X}_r\left(\pm\left(n-\left(\tfrac{r+1}{2}\right)\right)\right) \cong Y(\mathfrak{q},\phi_{n-r-1}).$$

The second kind of mysterious representation appears in Duflo's work [8] on complex groups of rank 2: he has the two components of the metaplectic representation for $Sp(2,\mathbb{C})$, and three unidentified representations for the complex G_2 as "mysterious representations" in the picture at the end of Section 5. To explain these, we need a more exotic analogue of Conjecture 4.5. It is not clear exactly what that should be, but here is the idea.

<u>Definition 6.3</u> Suppose $H \subseteq G$ is a θ-stable Cartan subgroup.

A reductive group M is called an L-subgroup of G if the following conditions hold.

a) M has a θ-stable Cartan subgroup isomorphic to, and henceforth identified with, H.

b) There is an inclusion $\Delta(m,h) \subseteq \Delta(\mathfrak{g},h)$, preserving the notions of compact and noncompact for imaginary roots.

c) There is a weight $\xi \in h^*$ such that

$$\Delta(m,h) = \{\alpha \in \Delta(\mathfrak{g},h) \mid \langle\check{\alpha},\xi\rangle \in \mathbb{Z}\}.$$

In the setting of algebraic groups, we are essentially requiring (among other things) that ${}^LM^0$ be the centralizer of a semisimple element in ${}^LG^0$. Both real polarizable and complex polarizable subgroups of G are L-subgroups. To give analogues of Conjecture 4.5 and Corollary 3.4, we need a correspondence between $P^G_{\lim}(H)$ and $P^L_{\lim}(H)$. This promises to be hard to define; the "ρ-shift" should be some special representation of H whose differential ξ satisfies (6.3)(c). Here is an example of what is wanted: we will take G complex, M as large as possible, and try to push the trivial representation of M up to G.

So assume G is a complex, connected, simple group, with H = TA a θ-stable Cartan subgroup. The root system $\Delta(\mathfrak{g},h)$ is a union of two disjoint simple subsystems, interchanged by θ:

$$\Delta(\mathfrak{g},h) = \Delta^L(\mathfrak{g},h) \cup \Delta^R(\mathfrak{g},h).$$

Fix a positive system

$$(\Delta^L)^+(\mathfrak{g},h) \subseteq \Delta^L(\mathfrak{g},h),$$

and put

$$\begin{aligned}\Delta^+(\mathfrak{g},h) &= (\Delta^L)^+(\mathfrak{g},h) \cup (-\theta)(\Delta^L)^+(\mathfrak{g},h)\\ &= (\Delta^L)^+ \cup (\Delta^R)^+.\end{aligned}$$

Let β^L be the highest short root in $(\Delta^L)^+$. Fix once and for all a simple root α^L for $(\Delta^L)^+$, and set

$$m = \text{multiplicity of } (\alpha^L)^\vee \text{ in } (\beta^L)^\vee.$$

We may as well assume m > 1 (or all subsequent constructions are trivial). Define $\lambda^L \in h^*$ by

$$\langle\check{\alpha}^L,\lambda^L\rangle = 1$$

$$\langle\check{\varepsilon},\lambda^L\rangle = 0,\ \varepsilon \text{ a simple root} \neq \alpha^L.$$

Define

$$\Delta^L(m,h) = \{\varepsilon \in \Delta(\mathfrak{g},h) \,\Big|\, \tfrac{1}{m}\langle \check{\varepsilon}, \lambda^L\rangle \in \mathbb{Z}\}$$
$$\Delta(m,h) = \Delta^L(m,h) \cup \theta(\Delta^L(m,h))$$
$$\Delta^+(m,h) = \Delta(m,h) \cap \Delta^+(\mathfrak{g},h).$$

There is a complex semisimple group M, with Cartan subgroup H and root system $\Delta(m,h)$. Its trivial representation has a Langlands parameter $\gamma_M \in P^M_{\lim}(H)$ characterized by

$$\overline{\gamma}_M = \rho\big(\Delta^+(m,h)\big) \in \mathfrak{a}^*.$$

Fix an element $w \in W\big(\Delta^L(\mathfrak{g},h)\big)$ such that

$$w\big((\Delta^L)^+(m,h)\big) = (\Delta^L)^+(m,h).$$

Put

$$\lambda^w = \lambda^L - \theta(w\lambda^L) \in h^*$$
$$\overline{\gamma}^w = \overline{\gamma}_M + \frac{1}{m}\lambda^L .$$

It turns out that $\overline{\gamma}^w$ comes from a Langlands parameter

$$\gamma^w \in P^G_{\lim}(H),$$

at least if G is simply connected. The character of $\overline{X}^G(\gamma^w)$ is very nicely related to that of $\overline{X}^M(\gamma_M)$ (the trivial representation of M).

> Conjecture 6.4 Suppose G is a complex simple, simply connected Lie group. Fix a simple root α^L and a Weyl group element w as above, and define γ^w. Then $\overline{X}^G(\gamma^w)$ is unitarizable.

When $G = Sp(n,\mathbb{C})$ and α^L is long, then $M = SO(2n,\mathbb{C})$. There are two choices for w; the representations $\overline{X}(\gamma^w)$ are the two components of the metaplectic representation. For type G_2, one gets a total of three representations; they are Duflo's mysterious representations, and are unitarizable.

Even functoriality from L-subgroups is not enough to produce all mysterious representations, however. Consider for example $G = SO(2n,\mathbb{C})$, with $n \geq 4$. Induce from the one dimensional representation of a real parabolic with Levi factor $SO(2,\mathbb{C}) \times SO(2n-2,\mathbb{C})$. At the end of the interval of irredubility around zero, the subquotient X containing a K-fixed vector is a very singular unitary representation. (It is one of those studied in [21].) Its character is not nicely related to any characters on L-subgroups; apparently it must be regarded as a third kind of mysterious representation. One may perhaps expect to find such representations attached to nilpotent coadjoint orbits which are special in the sense of Lusztig, and which have nothing to do with semisimple orbits in some sense. (For complex groups, the sense is that of Lusztig-Spaltenstein; for real groups the sense is not clear.)

References

1. M.W. Baldoni-Silva, "The unitary dual of Sp(n,1), n ≥ 2", to appear in Duke Math. J.
2. M.W. Baldoni-Silva and D. Barbasch, "The unitary spectrum for real rank one groups", preprint, 1981.
3. M.W. Baldoni-Silva and H. Kraljevic, "Composition factors of the principal series representations of the group Sp(n,1)", Trans. Amer. Math. Soc. 262 (1980), 447-471.
4. V. Bargmann, "Irreducible unitary representations of the Lorentz group", Ann. of Math. 48 (1947), 568-640.
5. A. Beilinson and J. Bernstein, "Localisation de $\mathfrak{g}$-modules", C.R. Acad. Sci. Paris, Série I, t. 292, 15-18.
6. A. Borel and N. Wallach, Continuous Cohomology, Discrete Subgroups, and Representations of Reductive Groups, Princeton University Press, Princeton, New Jersey, 1980.
7. J.L. Brylinski and M. Kashiwara, "Kazhdan-Lusztig conjecture and holonomic systems", Inventiones math. 64 (1981), 387-410.
8. M. Duflo, "Représentations unitaires irréductibles des groupes simples complexes de rang deux", Bull. Soc. math. France 107 (1979), 55-96.
9. M. Flensted-Jensen, "Discrete series for semisimple symmetric spaces" Ann. of Math. 111 (1980), 253-311.
10. Harish-Chandra, "Harmonic analysis on real reductive groups I. The theory of the constant term", J. Functional Analysis 19 (1975), 104-204.
11. J.C. Jantzen, Moduln mit einem Höchsten Gewicht, Lecture Notes in Mathematics 750, Springer-Verlag, Berlin-Heidelberg-New York, 1979.
12. D. Kazhdan and G. Lusztig, "Representations of Coxeter groups and Hecke algebras", Inventiones math. 53 (1979), 165-184.
13. D. Kazhdan and G. Lusztig, "Schubert varieties and Poincaré duality" in Geometry of the Laplace Operator, Proc. Symp. Pure Math. XXXVI American Mathematical Society, Providence, Rhode Island, 1980.
14. A. Knapp and B. Speh, "Status of classification of irreducible unitary representations", in Harmonic Analysis, Proceedings, Minneapolis, 1981, Lecture Notes in Mathematics 908, Springer-Verlag, Berlin-Heidelberg-New York, 1982.

15. A. Knapp and G. Zuckerman, "Classification of irreducible tempered representations of semisimple groups", to appear in Ann. of Math.

16. R. Langlands, "On the classification of irreducible representations of real algebraic groups", mimeographed notes, Institute for Advanced Study, 1973.

17. B. Speh and D. Vogan, "Reducibility of generalized principal series representations", Acta Math. 145 (1980), 227-299.

18. P. Trombi, "The tempered spectrum of a real semisimple Lie group", Amer. J. Math. 99 (1977), 57-75.

19. D. Vogan, "The algebraic structure of the representation of semisimple Lie groups I", Ann. of Math. 109 (1979), 1-60.

20. D. Vogan, "Irreducible characters of semisimple Lie groups I", Duke Math. J. 46 (1979), 61-108.

21. D. Vogan, "Singular unitary representations", in Non-commutative Harmonic Analysis, Proceedings, 1980, Lecture Notes in Mathematics 880, Springer-Verlag, Berlin-Heidelberg-New York, 1981.

22. D. Vogan, Representations of Real Reductive Lie Groups, Birkhäuser, Boston-Basel-Stuttgart, 1981.

ASYMPTOTIC EXPANSIONS OF GENERALIZED MATRIX ENTRIES OF REPRESENTATIONS OF REAL REDUCTIVE GROUPS

Nolan R. Wallach

0. Introduction. In a famous unpublished manuscript Harish-Chandra studied expansions of matrix entries of admissible finitely generated representations of a semi-simple Lie group. Part of that theory appears in Chapter 9 of Warner [25]. Harish-Chandra also proved some much more refined results that do not appear in Warner [25] but which were used in Langlands' proof of his celebrated classification of irreducible admissible representations, [12]. About the same time taking his cue from Jacquet's ideas in the p-adic case, Casselman showed how one can use the expansions (the ones in Warner[25]) to prove an important sharpening of Harish-Chandra's subquotient theorem. Casselman also realized that the more delicate expansions should be related to Deligne's theory of regular singularities. A manuscript containing this material and other results by Casselman and Miličic has finally appeared ([1]).

For some reason the expansions that Harish-Chandra showed to exist have come to be called asymptotic expansions. This is a profound understatement since in the context in which they are gotten they actually converge absolutely and uniformly to the matrix entry.

In this paper we cover similar ground in relation to expansions of matrix entries as do Casselman and Miličic. The main difference is that we derive only asymptotic expansions. But we get these expansions for a much more general class of matrix entries. In particular we only need K-finite on one side and some growth condition on the other side. The precise statements can be found in Sections 5 and 7. Our results are certainly sufficient for the purpose of proving Langlands' classification. Indeed, the subrepresentation theorem and

Theorem 5.8 are all Langlands really needs. We should point out that we show using the main technique of this paper (which we learned in a different context from Harish-Chandra) how the subquotient theorem implies the subrepresentation theorem. Our proof is in a sense the real analogue of the p-adic proof.

Also, due to the general context of the expansions we derive, we actually prove a subrepresentation theorem for the C^∞ vectors of <u>any</u> admissible finitely generated Banach representation of G.

Included in this article is also a substantial part of the work that Bill Casselman and I have been doing since 1977 on C^∞ vectors in admissible representations. Much of this material appears here for the first time.

The work involves the study of how the underlying $(\underline{g},K)$-module determines the topology of admissible representations. It is based on the material of Section 3 which was discovered independently by Casselman and myself. The material of Sections 3, 4 and 6 describe a significant portion of this work (with some simplifications due to the material of Section 5). I thank Bill Casselman for allowing me to include this material in this article.

The material in Sections 5 and 7 should be considered the really new results in this article. These results allow for major simplifications of the results in Section 6. But they go much further than that. First of all the usual reductions to systems of homogeneous linear equations do not work in this context. Even in the case of $\mathbb{R}$-rank 1 the functions to be expanded do not satisfy an ordinary differential equation. (Sometimes if they do it is not regular singular about the point that is being used for the expansion.) The motivation for Theorem 7.2 is a question of Piatetski-Shapiro (we describe the question in 7.3).

As indicated above much of this paper should be construed as joint work with Casselman. I have decided to publish this work in this form because this work has been used by many authors who do not have a reference for it. Here is a reference.

In Section 8 of this paper I derive two results that are fairly easy consequences of our theory. The first is a description of all functions on G that are right K finite and $Z(\underline{g})$-finite and satisfy the growth condition satisfied by automorphic forms. The second result answers a question posed by Piatetski - Shapiro on Whittaker vectors.

1. Some general structural results.

1.1. The class of Lie groups that we will study in this article are the real reductive groups in the Harish-Chandra class (real reductive group for short). These groups can be described as follows:

(1) Let $G_{\mathbb{C}}$ be an affine algebraic subgroup of $GL(n,\mathbb{C})$ defined by $f_1,\ldots,f_m$ polynomials on $M_n(\mathbb{C})$ having the property that

(a) $f_i(M_n(\mathbb{R})) \subset \mathbb{R}$, $i = 1,\ldots,m$.

(b) If $f_i(g) = 0$ then $f_i(g^*) = 0$, $i = 1,\ldots,m$.
($g^* = {}^t\bar{g}$, ${}^tg$ is the transpose of g.)

(2) Let $G_{\mathbb{R}}$ be the group real points of $G_{\mathbb{C}}$ (i.e., $G_{\mathbb{R}} = G_{\mathbb{C}} \cap GL(n,\mathbb{R})$). Let $G^0_{\mathbb{R}}$ denote the identity component of $G_{\mathbb{R}}$.

(3) G a real Lie group is a real reductive group if

(a) G^0 is a finite covering of $G^0_{\mathbb{R}}$.

(b) We identify the Lie algebra of G with $\underline{g}$ the Lie algebra of $G_{\mathbb{R}}$. Then $Ad(G) \subset Aut(\underline{g}_{\mathbb{C}})$ (automorphism group of $\underline{g}_{\mathbb{C}}$) is actually contained in $Ad(G^0_{\mathbb{C}})$.

(c) G has a finite number of connected components.

1.2. Let G be a real reductive group with Lie algebra $\underline{g}$. We carry all of the assumptions in 1.1. In particular we define $\theta(X) = {}^{t}-X$, $X \in \underline{g}$ (see 1.1 (3)(b)). Then θ is an involutive automorphism of $\underline{g}$ called a Cartan involution. We set $B(X,Y) = \mathrm{tr}XY$, $X, Y \in \underline{g}$. Then

(1) $B(\mathrm{Ad}(g)X, \mathrm{Ad}(g)Y) = B(X,Y)$, $g \in G$(1.1(3),(b)).

(2) $-B(X,\theta Y) = \langle X,Y\rangle$ defines a positive, non-degenerate symmetric form on $\underline{g}$.

Let $\underline{k} = \{X \in \underline{g} | \theta X = X\}$. Set $K = \{g \in G | \mathrm{Ad}(g)\underline{k} \subset \underline{k}\}$. Then

(3) K is a maximal compact subgroup of G.

Set $\underline{p} = \{X \in \underline{g} | \theta X = -X\}$.

Then

(4) The map $K \times \underline{p} \to G$, $k, X \mapsto K \exp X$ is a surjective diffeomorphism. We extend θ to G by $\theta(k \exp X) = k \exp(-X)$, $k \in K$, $X \in \underline{g}$.

Let $\underline{a}_0 \subset \underline{p}$ be a subspace such that $[\underline{a}_0, \underline{a}_0] = 0$ and $\underline{a}_0$ is maximal subject to this condition. Observing that if $H \in \underline{a}_0$, $\mathrm{ad}H$ is self adjoint relative to $\langle,\rangle$ we see that

(5) $\underline{g} = \bigoplus_{\lambda \in \underline{a}_0^*} \underline{g}_\lambda$, $\underline{g}_\lambda = \{X \in \underline{g} | \mathrm{ad}H \cdot X = \lambda(H)X, H \in \underline{a}_0\}$.

Set $\Phi(\underline{g},\underline{a}_0) = \{\lambda \in \underline{a}_0^* | \lambda \neq 0,\ \underline{g}_\lambda \neq 0\}$.

Put $\underline{a}_0' = \{H \in \underline{a}_0 | \lambda(H) \neq 0,\ \lambda \in \Phi(\underline{g},\underline{a}_0)\}$.

Choose $H \in \underline{a}_0'$ and put $\Phi^+ = \{\lambda \in \Phi(\underline{g},\underline{a}_0) | \lambda(H) > 0\}$.

Set $\underline{n}_0 = \sum_{\lambda \in \Phi^+} \underline{g}_\lambda$. Then

(6) $\underline{g} = \underline{k} \oplus \underline{a}_0 \oplus \underline{n}_0$.

Take $A_0 = \exp \underline{a}_0$, $N_0 = \exp \underline{n}_0$ (the connected subgroup with Lie algebra $\underline{n}_0$).

(7) The map $K \times A_0 \times N_0 \to G$ $k, a, n \mapsto kan$ is a surjective diffeomorphism.

(6),(7) are called, respectively, Iwasawa decompositions of $\underline{g}$ and G.

1.3. Let $S(G) = \exp(\underline{z}(\underline{g}) \cap \underline{p})$. Then $S(\underline{g})$ is called a standard split component of G. Let $X(G)$ denote the set of all continuous homomorphisms of G into $\mathbb{R}^*$. Set

$$^0G = \bigcap_{\chi \in X(G)} \operatorname{Ker}|\chi|.$$

Then

(1) ^0G has compact center.

(2) $S(G) \times {}^0G \to G$, $s, g \mapsto sg$ defines a Lie isomorphism.

1.4. Set $\underline{p}_0 = \underline{g}_0 \oplus \underline{n}_0$ (note $\underline{a}_0 \subset \underline{g}_0$). Let $M_0 = \{g \in G | Ad(g)|_{\underline{a}_0} = I\}$. Put $P_0 = M_0N_0$. Then P_0 is called a standard minimal parabolic subgroup of G. One checks that

(1) $(M_0 \cap K) \times A_0 \to M_0$ given by $m, a \mapsto ma$ is a surjective Lie group isomorphism.

We note that by its very definition one checks that M_0 is a real reductive group in the Harish-Chandra class, and $M_0 \cap K = {}^0M_0$, $S(M_0) = A_0$. $\underline{p} \subset \underline{g}$ a subalgebra will be called a standard parabolic subalgebra if $\underline{p} \supset \underline{p}_0$. Let $P = \{g \in G | Ad(g)\underline{p} \subset \underline{p}\}$. Then P is called a standard parabolic subgroup. Set $M = \theta(P) \cap P$. Then M is a real reductive group in the Harish-Chandra class. Set $A = S(M)$. Then $A \subset A_0$. Let N be the unipotent radical of P. Then using the Iwasawa decomposition it is not hard to show

(2) $M \times N \to P$, $m, n \mapsto mn$ is a surjective diffeomorphism.

(3) $M = {}^0MA$ as in 1.3 and $P = {}^0MAN$ is called a standard Langlands decomposition.

We note that $M = \{g \in G | Ad(g)|_{\underline{a}} = I\}$, $P = \{g \in G | Ad(g)\underline{n} \subset \underline{n}\}$. We call (P,A) a standard parabolic pair (p-pair for short).

Relative to the action of $\underline{a}$, $\underline{n} = \bigoplus_{\lambda \in \underline{a}^*} n_\lambda$, $n_\lambda = \{X \in \underline{n} | [H,X] = \lambda(H)X, H \in \underline{a}\}$. Set $\Phi(P,A) = \{\lambda \in \underline{a}^* | \underline{n}_\lambda \neq 0\}$. We note that $0 \notin \Phi(P,A)$

and (of course) $\Phi(P,A) \subset \Phi(P_0,A_0)|_{\underline{a}}$.

<u>1.5</u>. We now give a description of all standard p-pairs. Let $\Delta(P_0,A_0) = \Delta$ be the simple roots in $\Phi(P_0,A_0)$. That is, $\lambda \in \Delta$ if and only if it cannot be written as a sum of two or more elements of $\Phi(P_0,A_0)$. Let $F \subset \Delta$ be a subset. Put $\underline{a}_F = \{H \in \underline{a} | \lambda(H) = 0, \lambda \in F\}$. Set $\Phi^F = \{\lambda \in \Phi(P_0,A_0) | \lambda|_{\underline{a}_F} \neq 0\}$. Set $\underline{m}_F = \{X \in \underline{g} | [X,\underline{a}_F] = 0\}$. Put $\underline{p}_F = \underline{m}_F \oplus \underline{a}_F \oplus \bigoplus_{\lambda \in \Phi^F} (\underline{n}_0)_\lambda$. Then $\underline{p}_F$ is clearly a standard parabolic subalgebra of $\underline{g}$. Let (P_F,A_F) be the corresponding standard p-pair. This construction actually gives all standard p-pairs.

1.6. On $G_{\mathbb{R}}$ we put the norm

$$\|g\| = |g \oplus {}^t g^{-1}|$$

where $g \oplus {}^t g^{-1}$ is the operator on $\mathbb{R}^n \oplus \mathbb{R}^n$ given by $(g \oplus {}^t g^{-1})(x,y) = (gx, {}^t g^{-1} y)$, and $|\ldots|$ is the operator norm. Let $\nu : G^0 \to G^0_{\mathbb{R}}$ be the finite covering in 1.1(3)(a). We put for $k \in K$, $g \in G^0$ $\|kg\| = \|\nu(g)\|$. This makes sense since $\|\nu(k)g\| = \|g\|$ for $k \in K$. We have

(1) $\|g\| = \|g^{-1}\|$, $g \in G$.

(2) $\|xy\| \le \|x\|\|y\|$, $x,y \in G$.

(3) $\{x \in G | \|x\| \le r\}$ is compact if $r < \infty$.

(4) $\|k_1 g k_2\| = \|g\|$, $k_1,k_2 \in K$, $g \in G$.

(5) $\|k \exp tX\| = \|\exp X\|^t$, $k \in K$, $X \in \underline{p}$.

2. <u>Some generalities in representation theory</u>.

<u>2.1</u>. Let G be a Lie group. Let V be a Fréchet vector space. Then a <u>representation</u> of G on V is a strongly continuous homomorphism, π, of G into the invertible continuous operators on V. (I.e., $G \times V \to V$, $g,v \mapsto \pi(g)v$ is continuous.) If V is a Banach space then we call (π,V) a <u>Banach representation</u>. If V is a Hilbert space and $\pi(g)$ is unitary for $g \in G$ then (π,V) is called a <u>unitary represen-</u>

tation.

2.2. Let G be a real reductive group as in §1. Let (π, V) be a Banach representation of G.

Lemma. There exists $C > 0$, $r \geq 0$ so that $|\pi(g)| \leq C\|g\|^r$ for $g \in G$. (Here $|A|$ is the operator norm of A, a bounded operator).

Proof. We note that $\|g\| \geq 1$. Set $\sigma(g) = \log\|g\|$. Then $\sigma(x) \geq 0$, $\sigma(xy) \leq \sigma(x) + \sigma(y)$ and $\sigma(x) = \sigma(x^{-1})$. Set $\mu(g) = \log|\pi(g)|$. Then $\mu(xy) \leq \mu(x) + \mu(y)$, $x, y \in G$.

Set $B_r = \{x \in G | \sigma(x) \leq r\}$.

(1) $\mu(x) \leq C_1$ for $x \in B_1$. Indeed, B_1 is compact and $x \mapsto \pi(x)$ is strongly continuous. So (1) is an easy consequence of the uniform boundedness theorem.

(2) $e^{-C_1}|\pi(x)| \leq |\pi(kx)| \leq e^{C_1}|\pi(x)|$, $x \in G$, $k \in K$. Indeed, $K \subset B_1$. So $|\pi(k)| \leq e^{C_1}$ by (1). Since $|\pi(kx)| \leq |\pi(k)||\pi(x)|$ the right inequality is true.

$$|\pi(x)| = |\pi(k^{-1})\pi(k)\pi(x)| \leq |\pi(k^{-1})|\,|\pi(kx)| \leq e^{+C_1}|\pi(kx)|.$$

Thus the left inequality follows.

If $X \in \underline{p}$ then, $\sigma(e^{tX}) \leq t\sigma(e^X)$. Let $j < \sigma(e^X) \leq j + 1$. Then $\sigma|\exp((j+1)^{-1}X) \leq 1$. Hence $\mu(\exp(j+1)^{-1}X) \leq C_1$. But then

$$\mu(\exp X) \leq (j+1)C_1 < C_1(1+\sigma(e^X)).$$

Hence

$$\mu(k \exp X) \leq C_1 + C_1(1+\sigma(e^X)).$$
$$= C_1(2+\sigma(k \exp X)).$$

So $|\pi(g)| \leq e^{2C_1}\|g\|^{C_1}$ as asserted.

This result is obviously a special case of a much more general result (c.f. Warner [24], p.282). Its content is that the matrix entries of a Banach representation grow no faster than those of a finite dimensional representation.

2.3. If (π,V) is a representation of G set $V_\infty = \{v \in V | g \mapsto \pi(g)v$ is of class $C^\infty\}$. If $v \in V_\infty$ and $X \in \underline{g}$ ($\underline{g}$ the Lie algebra of G) define $\pi(X)v = \frac{d}{dt}\pi(\exp tX)v|_{t=0}$. Then

(1) $\pi(X) : V_\infty \to V_\infty$ for $X \in \underline{g}$.

(2) $\pi([X,Y]) = [\pi(X),\pi(Y)]$, $X,Y \in \underline{g}$.

We use the notation π for the corresponding representation of $U(\underline{g}_{\mathbb{C}})$.

Let F be a collection of semi-norms defining the topology of V. (If V is a Banach space $F = \{\|\ldots\|\}$ will do.)

If $g \in U(\underline{g}_{\mathbb{C}})$ and $\nu \in F$ define $\nu_g(v) = \nu(\pi(g)v)$. We topologize V_∞ using $F^\infty = \{\nu_g | \nu \in F, g \in U(\underline{g}_{\mathbb{C}})\}$. Since V is a Fréchet space F can be taken countable and we can take $F^\infty = \{\nu_{g_i} | \nu \in F, \{g_i\}$ a basis of $U(\underline{g}_{\mathbb{C}})\}$. It is easy to prove that V_∞ is complete relative to F^∞. Thus V_∞ is a Fréchet space. It is also not hard to see that

(3) $\pi(g)V_\infty \subset V_\infty$.

(4) (π,V_∞) is a representation of G that is smooth. That is, $g \mapsto \pi(g)v$, $v \in V_\infty$ is C^∞.

2.4. Lemma. There exists $d > 0$ so that $\int_G \|g\|^{-d}\,dg < \infty$. Here dg is some choice of invariant measure on G.

Proof. Let $m_\lambda = \dim \underline{n}_\lambda$, $\lambda \in \Phi(P_0,A_0)$. Set

$$D(a) = \prod_{\lambda \in \Phi(P_0,A_0)} (\sinh \lambda(\log a))^{m_\lambda}.$$

($\log : A_0 \to \underline{a}_0$ is the inverse mapping to $\exp : \underline{a}_0 \to A_0$.) Then up to constants of normalization

$$\int_G f(g)dg = \int_{K\times A_0^+\times K} f(k_1ak_2)D(a)dk_1dadk_2,$$

$$(A_0^+ = \exp \underline{a}_0^+, \underline{a}_0^+ = \{H \in \underline{a}_0 | \lambda(H) > 0, \lambda \in \Phi(P_0,A_0)\}.)$$

(1) If $\lambda \in \Phi(P_0,A_0)$ then $a^\lambda \le \|a\|^2$. Indeed, $\lambda = \varepsilon_i - \varepsilon_j$ for $\varepsilon_i,\varepsilon_j$ weights of the action of A_0 on $\mathbb{R}^n$. Thus $a^\lambda = a^{\varepsilon_i}a^{-\varepsilon_j}$. Since

both ε_i, $-\varepsilon_i$ are weights of A_0 on $\mathbb{R}^n \oplus \mathbb{R}^n$ with the action in 1.6 we see that

$$a^{\varepsilon_i} \leq \|a\|$$

$$a^{-\varepsilon_i} \leq \|a\|.$$

Hence (1).

Thus $D(a) \leq \|a\|^p$ for some p.

Thus to prove the lemma it is enough to show that

$$\int_{A_0^+} \|a\|^{-d}\, da < \infty$$

for some d. But this is clear.

<u>2.5</u>. Let for $f \in C^\infty(G)$, $n \geq 0$, $g \in U(\underline{g}_{\mathbb{C}})$

(1) $\nu_{m,g}(f) = \sup_{x \in G} \|x\|^m |g \cdot f(x)|$. (Here $U(\underline{g}_{\mathbb{C}})$ acts on $C^\infty(G)$ as left invariant differential operators.) Let $S(G)$ denote the space of $f \in C^\infty(G)$ such that $\nu_{m,g}(f) < \infty$ for all m,g. We endow $S(G)$ with the topology given by the semi-norms $\nu_{m,g}$.

The following assertions are not hard to prove.

(2) $S(G)$ is a Fréchet space.

(3) The injection $C_c^\infty(G) \to S(G)$ is continuous with dense image.

(4) If $f \in S(G)$ then $g \cdot f \in L^p(G)$ for all $p > 0$ and all $g \in U(\underline{g}_{\mathbb{C}})$ (see Lemma 2.4).

(5) The map $S(G) \times S(G) \to S(G)$, $f_1, f_2 \mapsto f_1 * f_2$ is continuous.

(6) Let $L_X f(g) = \frac{d}{dt}\Big|_{t=0} f(\exp(-tX)g)$, $X \in \underline{g}$. Then $X \mapsto L_X$ is a representation of $\underline{g}$. Let L_g also denote the extension to $U(\underline{g}_{\mathbb{C}})$. If $f \in S(G)$ then $\sup_{x \in G} \|x\|^m |L_{g_1} g_2 \cdot f(x)| = \nu_{m,g_1,g_2}(f)$ is a continuous semi-norm on $S(G)$. Furthermore the semi-norms $\nu_{m,g,1}$ define the same topology as the semi-norms $\nu_{m,1,g}$ and as the semi-norms ν_{m,g_1,g_2}.

(7) The representation of $G \times G$ on $S(G)$ given by $(x,y) \cdot f(g) = f(x^{-1}gy)$ is a smooth Fréchet representation of G. That is, it is a representation and $S(G)_\infty = S(G)$.

Remark. The notion of $S(G)$ makes sense for a much larger class of groups. Let $G_{\mathbb{R}}$ be the real points of an affine algebraic group G defined over $\mathbb{R}$. Let G be a finite covering of an open subgroup of $G_{\mathbb{R}}$. Let $\nu : G \to G_{\mathbb{R}}$ be the covering map. Let $\mathcal{D}(G_{\mathbb{R}})$ be the algebra of differential operators on $G_{\mathbb{R}}$ with polynomial coefficients. We look at $\mathcal{D}(G_{\mathbb{R}})$ as an algebra of differential operators on G. Let $S(G)$ be the subspace of $C^\infty(G)$ consisting of those f such that

$$\nu_D(f) = \sup_{x\in G}|Df(x)| < \infty, \quad D \in \mathcal{D}(G_{\mathbb{R}}).$$

We topologize $S(G)$ using the semi-norms $\nu_D(f)$.

For example if we look upon $\mathbb{R}^n$ as the unipotent group

$$N = \left\{\left(\begin{array}{c|c} I & x \\ \hline 0 & 1\end{array}\right) \middle| x \in \mathbb{R}^n\right\},$$

then $S(N) = S(\mathbb{R}^n)$.

This is the reason for the notation $S(G)$.

2.6. Let (π,H) be a Banach representation of G. If $f \in S(G)$ we define

(1) $\pi(f)v = \int_G f(g)\pi(g)v\,dg$. Then Lemma 2.2 combined with Lemma 2.4 implies that the integral in (1) converges absolutely. Furthermore

(2) $f \mapsto \pi(f)$ defines a continuous map of $S(G)$ into $L(H,H)$ (with the norm topology) such that $\pi(f_1*f_2) = \pi(f_1)\pi(f_2)$.

(3) If $v \in H$, $f \in S(G)$ then $\pi(f)v \in H_\infty$. Indeed, $\pi(g)\pi(f)v = (L_g f)v$. Since $g \mapsto L_g f$ is C^∞ from G to $S(G)$ (3) follows.

(4) H_∞ is dense in H. Indeed, let $f_j \in C_c^\infty(G)$, $\lim_{j\to\infty} f_j = \delta_1$ (i.e., $\int_G f_j(g)dg = 1$, $f_j \geq 0$ and $\operatorname{supp} f_j \supset \operatorname{supp} f_{j+1}$, $\bigcap \operatorname{supp} f_j = \{1\}$). Then $\lim_{j\to\infty} \pi(f_j)v = v$.

The discussion in Warner [24], 4.4.5 on analytic vectors can also be interpreted in terms of $S(G)$. Let $X_1,\ldots,X_n$ be an orthonormal basis of $\underline{g}$ relative to $\langle,\rangle$. Set $\Delta = \sum X_i^2$. If $f \in C_c^\infty(G)$ let $f_t(x)$

be the square integrable solution to

(5) $$\frac{\partial}{\partial t} f_t(x) = \Delta f_t(x), \ t > 0$$
$$f_0(x) = f(x).$$

Then one sees in Warner [24], 4.4.5 that $f_t \in S(G)$ and $g \mapsto \pi(g)\pi(f_t)v$, is real analytic for $t > 0$, $v \in H$. Furthermore,

(6) $$\lim_{t\to 0} \pi(f_t)v = \pi(f)v.$$

Set H_ω equal to the space of all $v \in H$ such that $g \mapsto \pi(g)v$ is real analytic. Then

(7) (Harish-Chandra [4]). H_ω is dense in H. Clearly H_ω is a $U(\underline{g}_{\mathbb{C}})$-invariant subspace of H.

<u>Lemma</u>. (Harish-Chandra [4]). Let $V \subset H_\omega$ be K-invariant and $U(\underline{g}_{\mathbb{C}})$-invariant. Then $C\ell(V)$ (in H) is G-invariant.

<u>Proof</u>. Let H' be the space of continuous linear functionals on H. Set $V^\perp = \{\lambda \in H' | \lambda(V) = 0\}$. Then the Hahn-Banach theorem implies that $C\ell(V) = \{v \in H | \lambda(v) = 0, \lambda \in V^\perp\}$. If $X \in \underline{g}$ and $v \in V$ then $\pi(\exp tX)v = \sum \frac{t^n}{n!} \pi(X)^n v$ for $|t| < \varepsilon$ and the convergence is in H.

Thus if $\lambda \in V^\perp$, $\lambda(\pi(\exp tX)v) = \sum_{n=0}^{\infty} \frac{t^n}{n!} \lambda(\pi(X)^n v) = 0$ for $|t|<\varepsilon$. So $\lambda(\pi(\exp tX)v) = 0$ for all t.

Let $\pi'(g)\lambda = \lambda \cdot \pi(g)^{-1}$. Then if $\lambda \in V^\perp$ we see that $\pi'(g)\lambda \in V^\perp$ for $g \in G^0$. Since V is K-invariant $\pi'(k)\lambda \in V^\perp$ for $\lambda \in V^\perp$ and $k \in K$. Thus $\pi'(G)V^\perp \subset V^\perp$. Hence $\pi(G)(V^\perp)^\perp \subset (V^\perp)^\perp$. The lemma now follows.

<u>2.7</u>. In Dixmier and Malliavin [26] it has been shown that $\pi(C_c^\infty(G))H = H_\infty$. We conjecture that $\pi(S(G))H_\omega = H_\infty$.

<u>2.8</u>. If $v \in H$ we call v, <u>K-finite</u> if $\pi(K)v$ spans a finite dimensional space. If $\gamma \in E(K)$ ($E(K)$ the equivalence classes of irreducible finite dimensional representations of K) define $\alpha_\gamma = d(\gamma)\overline{\chi}_\gamma$, $d(\gamma)$

the degree of γ, χ_γ the character.

Define $E_\gamma = (\pi|_K)(\alpha_\gamma)$. Then the Peter-Weyl theorem implies that

(1) The algebraic direct sum of the $E_\gamma H$ is dense in H. Furthermore, if $v \in H$ is K-finite then $v \in \sum_{\gamma \in F} E_\gamma H$, $F \subset \mathcal{E}(K)$ a finite set.

(2) Set $H_K = \{v \in H_\omega | v, \text{ K-finite}\}$. Then $H_K = \sum_{\gamma \in \mathcal{E}(K)} E_\gamma H$ and H_K is dense in H_ω. We note that $\pi(U(\underline{g}_{\mathbb{C}}))H_\omega \subset H_\omega$. Thus H_K is a K- and $U(\underline{g}_{\mathbb{C}})$-invariant subspace of H_ω.

This leads to the notion (due to Lepowsky) of a $(\underline{g},K)$-module.

<u>2.9.</u> Let V be a $U(\underline{g}_{\mathbb{C}})$-module that is also a K-module (there is <u>no topology</u> on V). Then V is a $(\underline{g},K)$-module if

(1) If $v \in V$ then $K \cdot v$ spans a finite dimensional subspace W_v.

(2) The map $K \times W_v \to W_v$ is smooth and $\frac{d}{dt} \exp tY \cdot v|_{t=0} = Y \cdot v$ for $Y \in \underline{k}$, $v \in V$.

(3) If $X \in \underline{g}$, $k \in K$ then $k \cdot X \cdot v = (\mathrm{Ad}(k)X) \cdot (k \cdot v)$.

We note that H_K is a $(\underline{g},K)$ module. It is called the <u>underlying</u> $(\underline{g},K)$-module of (π,H).

If V is a $(\underline{g},K)$-module we can define $E_\gamma v = \int_K \alpha_\gamma(k)v\,dk$. Then by (1), (2), $V = \oplus V(\gamma)$ $(V(\gamma) = E_\gamma V)$, algebraic direct sum. Each $V(\gamma)$ is a multiple (possibly infinite) of a representative of γ.

A $(\underline{g},K)$-module is said to be <u>admissible</u> if $\dim V(\gamma) < \infty$ for all $\gamma \in \mathcal{E}(K)$. It is said to be finitely generated if it is finitely generated as a $U(\underline{g}_{\mathbb{C}})$-module.

(π,H) a representation of G on a Banach space is said to be admissible if $\dim E_\gamma H_\omega < \infty$ for all $\gamma \in \mathcal{E}(K)$. Since $E_\gamma H_\omega$ is dense in $E_\gamma H$ we see that

(4) (π,H) is admissible if and only if H_K is admissible. Furthermore if (π,H) is admissible then $E_\gamma H \subset H_\omega$.

2.10. Let $Z(\underline{g})$ denote the center of $U(\underline{g}_{\mathbb{C}})$. Let V be a $(\underline{g},K)$-module. $v \in V$ is said to be $Z(\underline{g})$-finite if $\dim Z(\underline{g}) \cdot v < \infty$.

Lemma. Assume that V is finitely generated. Then V is admissible if and only if every $v \in V$ is $Z(\underline{g})$-finite.

Proof. If V is admissible and $v \in V$ then there is a finite subset, $F \subset E(K)$, such that $v \in \sum_{\gamma \in F} V(\gamma)$. Clearly $Z(\underline{g})V_\gamma \subset V_\gamma$, $\gamma \in E(K)$ so $\dim Z(\underline{g})v \le \sum_{\gamma \in F} \dim V(\gamma) < \infty$. For the converse we need the following important result of Harish-Chandra [4].

Theorem. If W is a finite dimensional (continuous) K-module, define $P(W) = U(\underline{g}_{\mathbb{C}}) \underset{U(\underline{k}_{\mathbb{C}})}{\otimes} W$. Then $P(W)$ is a $(\underline{g},K)$ module with $U(\underline{g}_{\mathbb{C}})$ acting by left translation and K acting by $k \cdot (g \otimes w) = \mathrm{Ad}(k)g \otimes k \cdot w$. Furthermore (and this is the point),

(1) $P(W)(\gamma)$ is a finitely generated as a $Z(\underline{g})$-module for all $\gamma \in E(K)$.

Suppose now that V is $Z(\underline{g})$-finite and finitely generated. Then $V = U(\underline{g}) \cdot W$, $W = \sum_{\gamma \in F} V(\gamma)$, $F \subset E(K)$ a finite set. Thus $P(W) \overset{T}{\to} V$ $g \otimes w \mapsto g \cdot w$ is a surjective $(\underline{g},K)$-module homomorphism. Now $P(W)(\gamma) = \sum_{i=1}^{d} Z(\underline{g})u_i$. Thus $V(\gamma) = \sum_{i=1}^{d} Z(\underline{g})T(u_i)$. So $\dim V(\gamma) \le \sum_{i=1}^{d} \dim(Z(\underline{g})T(u_i)) < \infty$.

2.11. One of the main reasons for the importance of admissible $(\underline{g},K)$-modules is the following theorem of Harish-Chandra [4].

Theorem. If (π,H) is an irreducible unitary representation of G then H_K is admissible.

To prove this one observes that using the spectral theorem there is $\chi : Z(\underline{g}) \to \mathbb{C}$ so that $\pi(z) = \chi(z)I$ on H_∞. Thus if $0 \ne v \in H_K$ then $\mathrm{span}\, \pi(U(\underline{g}_{\mathbb{C}}))\pi(K)v = V$ is a finitely generated admissible $(\underline{g},K)$-module. Now $C\ell(V)$ is G-invariant. So $C\ell(V) = H$. But then $V = H_K$.

2.12. Lemma. (Dixmier's observation). Let V be a countable dimensional vector space over $\mathbb{C}$. Let $T \in \mathrm{End}(V)$. Then there exists $\lambda \in \mathbb{C}$ such that $T - \lambda I$ is not invertible.

Proof. If not we can define $Q(T)$ for Q a rational function on $\mathbb{C}$. The map $Q \mapsto Q(T)v$ is injective for $v \neq 0$. But the space of rational functions on $\mathbb{C}$ has dimension at least the cardinality of $\mathbb{C}$.

2.13. Lemma. An irreducible $(\underline{g},K)$-module is admissible. (Here irreducible means that there are no K and $U(\underline{g}_{\mathbb{C}})$ invariant subspaces other than (0) and the whole space.)

Proof. Let V be irreducible. Let $z \in Z(\underline{g})$; then $z - \lambda$ is not invertible for some $\lambda \in \mathbb{C}$ (V is clearly countable dimensional). If $(z-\lambda)V \neq 0$ then $(z-\lambda)V = V$ by irreducibility. But then if $z - \lambda \neq 0$ it must act bijectively on V. So $(z-\lambda)V = 0$. Hence $\dim Z(\underline{g})v = 1$ for $v \neq 0$. So Lemma 2.10 implies the result.

2.14. If V, W are $(\underline{g},K)$-modules then $\mathrm{Hom}_{\underline{g},K}(V,W)$ is the space of linear maps $T : V \to W$ that commute with the action of K and $\underline{g}$.

2.14. Lemma. (Dixmier's Schur's Lemma). If V is an irreducible $(\underline{g},K)$-module then if $T \in \mathrm{Hom}_{\underline{g},K}(V,V)$, $T = \lambda I$.

Proof. Same proof as Lemma 2.13.

2.15. We now introduce a class of admissible $(\underline{g},K)$-modules that will be very useful in this article. Let (σ,W) be a finite dimensional representation of P_0. We may (and do) assume that $\sigma|_{{}^0M_0}$ is unitary.

Let ${}^{\infty}H^{\sigma}$ denote the space of all $f : G \to W$ of class C^{∞} such that $f(pg) = \sigma(p)f(g)$, $p \in P_0$, $g \in G$. Put on ${}^{\infty}H^{\sigma}$ the inner product

(1) $$\int_K \langle f_1(k), f_2(k)\rangle dk, \quad f_1, f_2 \in {}^{\infty}H^{\sigma}.$$

Define $\pi_{\sigma}(x)f(g) = f(gx)$, $g \in G$, $x \in G$.

Let H^{σ} denote the Hilbert space completion of ${}^{\infty}H^{\sigma}$. Then one checks that $\pi_{\sigma}(g)$ extends to a bounded operator on H^{σ} and $(\pi_{\sigma},H^{\sigma})$

is a representation of G. Furthermore, if we put the C^∞ topology on ${}^\infty H^\sigma$, then $(\pi_\sigma, {}^\infty H^\sigma)$ is a smooth Fréchet representation of G and

(2) $(\pi_\sigma, {}^\infty H^\sigma)$ is topologically equivalent with $(\pi_\sigma, H^\sigma_\infty)$.

(3) $(\pi_\sigma, H^\sigma_\infty)$ is an admissible representation of G.

Indeed, H^σ as a K-representation is just the unitarily induced representation of $\sigma|_{{}^0M_0}$ to K. So (3) follows from Frobenius reciprocity.

<u>2.16</u>. If $\varphi \in S(G)$ we compute $\pi_\sigma(\varphi)$.

$$(\pi_\sigma(\varphi)f)(x) = \int_G f(xg)\varphi(g)dg$$

$$= \int_G f(g)\varphi(x^{-1}g)dg = \int_{P_0\times K} f(p_0k)\varphi(x^{-1}p_0k)d_\ell p_0 dk$$

($d_\ell p_0$ is left invariant measure and the equation is to be understood up to normalization)

$$= \int_{P_0\times K} \sigma(p_0)f(k)\varphi(x^{-1}p_0k)d_\ell p_0 dk.$$

Set $K_{\varphi,\sigma}(x,k) = \int_{P_0} \sigma(p_0)\varphi(x^{-1}p_0k)d_\ell p_0$.

Then

(1) $(\pi_\sigma(\varphi)f)(x) = \int_K K_{\varphi,\sigma}(x,k)f(k)dk.$

Of course, all the calculations are formal. However, the growth conditions on $S(G)$ guarantee that all of the integrals converge absolutely.

Now $P_0 = {}^0M_0 \times A_0 \times N_0$. Define for $\nu \in (\underline{a}_0^*)_{\mathbb{C}}$, $\sigma_\nu(man) = a^\nu\sigma(man)$, $m \in {}^0M_0$, $a \in A_0$, $n \in N_0$. Then H^{σ_ν} can be identified with H^σ $(\sigma_\nu|_{{}^0M_0} = \sigma|_{{}^0M_0})$. So if $f \in S(G)$ $\nu \mapsto \pi_{\sigma_\nu}(f)$ is a function from $(\underline{a}_0)^*_{\mathbb{C}}$ to the continuous endomorphisms of H^σ.

The following lemma is a fairly easy exercise.

2.16. <u>Lemma</u>. If σ is a finite dimensional representation of P_0 and if $f \in S(G)$ then the map $\nu \mapsto \pi_{\sigma_\nu}(f)$ is holomorphic (i.e., complex analytic) on $(\underline{a}_0)^*_{\mathbb{C}}$.

3. The real Jacquet module.

3.1. We begin this section with a general lemma about universal enveloping algebras.

Let $\underline{g}$ be a Lie algebra over $\mathbb{C}$. Let $\underline{n}$ be a nilpotent Lie subalgebra of $U(\underline{g})$. We assume that $[X,\underline{n}] \subset \underline{n}$ for $X \in \underline{g}$.

3.2. Lemma. Let M be a finitely generated $U(\underline{g})$-module. Let $N \subset M$ be a $U(\underline{g})$-submodule. Then there exists r so that

$$(\underline{n}^{k+r}M) \cap N = \underline{n}^r(\underline{n}^k N \cap M).$$

Note. The statement $\underline{n}^{k+r}M \cap N \subset \underline{n}^k N \cap M$ is sometimes called the Artin-Rees property (A-R property). When $\underline{g}$ is nilpotent this is a special case of a general result of McConnell. In this generality it is due to Stafford-Wallach [17] (see that paper for the proof which is fairly simple). Their proof was inspired by an argument in Kostant [10].

We also need the following observation (for a proof cf. [23]).

3.3. Let G be a real reductive group. We carry the notation of Section 2.

Lemma. There exists a finite dimensional subspace E of $U(\underline{g}_{\mathbb{C}})$ so that

$$U(\underline{g}_{\mathbb{C}}) = U((\underline{n}_0)_{\mathbb{C}})Z(\underline{g})EU(\underline{k}_{\mathbb{C}}).$$

3.4. Let $\rho_0(H) = \frac{1}{2}\,\mathrm{tr}(\mathrm{ad}H|_{\underline{n}_0})$ for $H \in \underline{a}_0$. If $\xi \in \mathcal{E}({}^0M_0)$ we extend ξ to P_0 by $\xi(man) = \xi(m)$. We set $H^{\xi,\nu} = H^{\xi}$ and $\pi_{\xi,\nu} = \pi_{\xi_{\nu+\rho_0}}$ (see 2.15). $\{\pi_{\xi,\nu}\}_{\xi\in\mathcal{E}({}^0M_0),\nu\in(\underline{a}_0)^*_{\mathbb{C}}}$ is called the principal series. We note

(1) $\pi_{\xi,\nu}$ is unitary if $\nu \in i(\underline{a}_0)^*$, (cf. [20]).

Theorem. (Harish-Chandra [5], Lepowsky [13], Rader [15]). If V is an irreducible $(\underline{g},K)$-module then there exists $\xi \in E(M)$, $\nu \in (\underline{a}_0)^*_{\mathbb{C}}$ such that V is equivalent with a subquotient (i.e., a quotient of a subrepresentation) of $H_K^{\xi,\nu}$.

An important sharpening of this theorem is:

3.3. Theorem. (Casselman). If $0 \neq V$ is an admissible finitely generated $(\underline{g},K)$-module then $\underline{n}_0 V \neq V$. (This implies that in Theorem 3.3 we can take V to be a sub-module, see Scholium 3.12.)

Proof. (Sketch). We first note that it is enough to prove this result for V irreducible since every finitely generated $(\underline{g},K)$-module has an irreducible quotient ($U(\underline{g}_{\mathbb{C}})$ is noetherian). We prove this result by induction on $\dim \underline{a}_0$. If $\dim \underline{a}_0 = 0$ then $G = K = {}^0M_0$ and the result says that $0 \neq V$. If $G \neq {}^0G$, then V is still admissible as a $({}^0\underline{g},K)$-module. Thus we may assume $G = {}^0G$.

Let now $\alpha \in \Delta(P_0,A_0)$ and $F = \Delta(P_0,A_0) - \{\alpha\}$. Let $\tau(P,A) = (P_F,A_F)$. $P = MN$ as usual. Then

(1) $V/\underline{n}V$ is an admissible finitely generated $(\underline{m}, K\cap M)$-module.

Indeed, let $\mu_{\underline{p}} : Z(\underline{g}) \to Z(\underline{m})$ be defined by $\mu_{\underline{p}}(z) - z \in \underline{n}\,U(\underline{g})$. Then $\mu_{\underline{p}}$ is an algebra homomorphism and $Z(\underline{m})$ is finitely generated as a $\mu_{\underline{p}}(Z(\underline{g}))$-module.

But then every element of $V/\underline{n}V$ is $Z(\underline{m})$-finite. So (1) follows from Lemma 2.10.

Thus if $V/\underline{n}V \neq (0)$ the inductive hypothesis implies that $(\underline{n}_0 \cap \underline{m})(V/\underline{n}V) \neq V/\underline{n}V$. But then $\underline{n}_0 V \neq V$. Hence we are reduced to the case when $\underline{n}^{\alpha} \cdot V = V$ for all $\alpha \in \Delta(P_0,A_0)$; here $\underline{n}^{\alpha} = \underline{n}_F$, $F = \Delta - \{\alpha\}$.

Now there is $\xi \in E({}^0M_0)$, $\nu \in (\underline{a}_0)^*_{\mathbb{C}}$ and $X \subset H_K^{\xi,\nu}$, $Y \subset X$ such that $V \cong X/Y$. Set $H = C\ell(X)/C\ell(Y)$. Let π be the induced action of G on H. Then $H_K \cong V$ (see 2.6).

Let $A_0^+ = \{\exp H \mid H \in \underline{a}_0,\ \lambda(h) > 0$ for $\lambda \in \Phi(P_0,A_0)\}$. Then

$G = KC\ell(A_0^+)K$.

(2) There are K-invariant functions on V, μ_1, μ_2, and there exists $\Lambda \in \underline{a}_0^*$ such that

$$|\langle\pi(a)v,w\rangle| \leq a^\Lambda \mu_1(v)\mu_2(w), \ a \in C\ell(A_0^+), \ v, w \in V .$$

Indeed, $|\pi(g)| \leq C\|g\|^r$ (Lemma 2.2). Let $\Lambda \in \underline{a}_0^*$ be such that $\Lambda(h) \geq \varepsilon(h)$ for any weight, ε, of the action of A_0 on $\mathbb{R}^n \oplus \mathbb{R}^n$ given in 1.6 and any $h \in \underline{a}_0^+ = \{h \in \underline{a}_0 | \lambda(h) > 0, \lambda \in \Phi(P_0,A_0)\}$. Then

$$|\pi(k_1 a k_2)| \leq Ca^{r\Lambda}, \ k_1,k_2 \in K, \ a \in C\ell(A_0^+).$$

Take $\mu_1(v) = C\|v\|$, $\mu_2(w) = \|w\|$.

Let $k \in K$ be such that $Ad(k)\underline{n}_0 = \theta(\underline{n}_0) = \bar{\underline{n}}_0$. Let $\underline{v}^\alpha = Ad(k)\underline{n}^\alpha$ for $\alpha \in \underline{\Delta}(P_0,A_0)$. Then $\underline{v}^\alpha = \theta\underline{n}^{\alpha'}$ with $\alpha' \in \Delta(P_0,A_0)$ and $\alpha \mapsto \alpha'$ is an involution of $\underline{\Delta}(P_0,A_0)$.

Since $\underline{n}^\alpha V = V$ all $\alpha \in \underline{\Delta}(P_0,A_0)$, $\underline{v}^\alpha V = V$ all $\alpha \in \underline{\Delta}(P_0,A_0)$. Hence $\theta(\underline{n}^\alpha)V = V$ all $\alpha \in \Delta(P_0,A_0)$. Let $\bar{\underline{n}}^\alpha = \theta(\underline{n}^\alpha)$.

Let $\bar{X}_1,\ldots,\bar{X}_m$ be a basis of $\bar{\underline{n}}^\alpha$ with $[h,\bar{X}_i] = -\lambda_i(h)\bar{X}_i$ for $h \in \underline{a}_0$. We note that if $h \in \underline{a}_0^+$ then $\lambda_i(h) \geq \alpha(h)$.

Now $V = \sum_{i=1}^{\ell} \pi(\bar{X}_i)v_i$ $(\bar{\underline{n}}^\alpha V = V)$. Thus if $a \in C\ell(A^+)$,

$$\begin{aligned} |\langle\pi(a)v,w\rangle| &\leq \sum|\langle\pi(a)\pi(\bar{X}_i)v_i,w\rangle| \\ &= \sum a^{-\lambda_i}|\langle\pi(\bar{X}_i)\pi(a)v_i,w\rangle| \\ &= \sum a^{-\lambda_i}|\langle\pi(a)v_i,\pi^*(\bar{X}_i)w\rangle| . \end{aligned}$$

(Here π^* is the <u>conjugate</u> <u>dual</u> representation to π, that is, $\pi^*(g) = \pi(g^{-1})^*$)

$$\leq a^{\Lambda-\alpha}\sum\mu_1(v_i)\mu_2(\pi^*(\bar{X}_i)w). \quad \text{Thus}$$

$$|\langle\pi(\alpha)v,w\rangle| \leq a^{\Lambda-\alpha}C(v)C(w)$$

with $C(v)$, $C(w)$ constants depending on v and w. Let $u_1,\ldots,u_d$ and $z_1,\ldots,z_q$ be respectively orthonormal bases of $\mathrm{Span}\{K\cdot v\}$ and $\mathrm{Span}\{K\cdot w\}$. Then if $\mu_1'(v) = \max_{1\leq i\leq d} C(u_i)$, $\mu_1'(w) = \max_{1\leq i\leq q} C(z_i)$, we have

(2) $|<\pi(\alpha)v,w>| \leq a^{\Lambda-\alpha}\mu_1'(v)\mu_2'(w)$

where μ_1' and μ_2' are K-invariant functions on V.

We may thus replace Λ by $\Lambda - \alpha$ and μ_i by μ_i' in (2) and iterate this argument using any $\beta \in \Delta$. We find that

(3) If $Q = \sum_{\alpha\in\Delta} n_\alpha \alpha$, $n_\alpha \geq 0$, $n_\alpha \in \mathbb{Z}$ and if $v, w \in V$ then

$$|<\pi(k_1 a k_2)v,w>| \leq C_{v,w,Q}a^{\Lambda-Q}, \quad a \in C\ell(A^+),$$

with $C_{v,w,Q} > 0$ depending only on v, w, Q.

Now (3) implies that

(3) $f_{v,w} = <\pi(g)v,w>$, $v,w \in V$, $g \in G$ defines an element of $\mathcal{S}(G)$.

Let $f = f_{v,w}$. Then $\nu \mapsto \pi_{\xi,\nu}(\overline{f})$ is a holomorphic function of $\nu \in (\underline{a}_0)^*_{\mathbb{C}}$.

Let $\nu \in i\underline{a}_0^*$. If for some $v,w \in V, \pi_{\xi,\nu}(\overline{f}) \neq 0$, then there must be $x,y \in H_K^{\xi,\nu}$ so that

$$\int_G \overline{f(g)} <\pi_{\xi,\nu}(g)x,y> dg \neq 0.$$

This implies that there is a $(\underline{g},K)$-homomorphism of $\overline{V}$ (the conjugate dual of V) into $(H_K^{\xi,\nu})^-$. Hence there is a non-zero $(\underline{g},K)$ homomorphism of $H_K^{\xi,\nu}$ to V. Since $H^{\xi,\nu}$ is unitary this implies that V is isomorphic with a submodule of $H_K^{\xi,\nu}$. Let $T : V \to H_K^{\xi,\nu}$ be the corresponding $(\underline{g},K)$-homomorphism.

(4) If $X \in \underline{n}_0$ and $\varphi \in H_K^{\xi,\nu}$ then $(\pi_{\xi,\nu}(X)\varphi)(1) = 0$.

Indeed, $(\pi_{\xi,\nu}(X)\varphi)(1) = \frac{d}{dt}\varphi(\exp tX)|_{t=0} = \frac{d}{dt}\varphi(1)|_{t=0} = 0$ ($\varphi(ng) = \varphi(g)$, $n \in N_0$, $g \in G$).

Thus $T(v)(1) = 0$, $v \in V$. But $0 = T(k\cdot v)(1) = T(v)(k)$, $k \in K$. So $T(v) = 0$. Thus $T = 0$. Hence we have a contradiction. So $\pi_{\xi,\nu}(\overline{f}) = 0$, all $\nu \in i\underline{a}_0^*$. Hence $\pi_{\xi,\nu}(\overline{f}) = 0$, all $\nu \in (\underline{a}_0)^*_{\mathbb{C}}$. But f is a matrix entry of $\pi_{\xi,\nu}$ for some $\nu \in (\underline{a}_0)^*_{\mathbb{C}}$. That is $f(g) = <\pi_{\xi,\nu}(g)\varphi,\psi>$ for some $\varphi,\psi \in H_K^{\xi}$. But then

$$0 = \langle \pi_{\xi,\nu}(\bar{f})\varphi, \psi \rangle = \int_G |f(g)|^2 dg.$$

So $f = 0$. But then $V = 0$. Q.E.D.

3.4. This theorem was originally proved by Casselman using the fact that matrix entries have convenient expansions in terms of exponential polynomials. In any event one must have some method of getting matrix entries for $(\underline{g},K)$-modules. The most convenient method is via the subquotient theorem. Recently, a completely algebraic proof of Theorem 3.3 has been given by Bernstein using his theory (with Beilinson) of $\mathcal{D}$-modules. In the case $G = SL(n,\mathbb{R})$ an alternate algebraic proof was given by Stafford-Wallach [17].

3.5. We now introduce two categories of modules. Let $\mathcal{H}$ denote the category of admissible finitely generated $(\underline{g},K)$-modules. Let $\mathcal{V}$ denote the category of $(\underline{g},{}^0M_0)$-modules, V, that are finitely generated as $U(\underline{g}_{\mathbb{C}})$-modules and satisfy the additional condition:

(1) If $v \in V$ then $\dim U((\underline{p}_0)_{\mathbb{C}}) \cdot v < \infty$.

We first look at the class of $(\underline{g},{}^0M_0)$-modules that satisfy (1) (but are not necessarily finitely generated). Let $\mathcal{V}' \supset \mathcal{V}$ be that category.

Let $V \in \mathcal{V}'$. Then if $v \in V$, $\dim U((\underline{a}_0)_{\mathbb{C}})v < \infty$. Hence $U((\underline{a}_0)_{\mathbb{C}}) \cdot v$ splits into generalized eigenspaces relative to $\underline{a}_0$. Thus

(2) $V = \bigoplus_{\lambda \in (\underline{a}_0)^*_{\mathbb{C}}} V_\lambda$ with

$V_\lambda = \{v \in V \mid (h-\lambda(h))^k v = 0$ for $h \in \underline{a}_0$ and some $k > 0$ depending on $v\}$.

Let $V \in \mathcal{V}'$ and suppose that $\dim Z(\underline{g})v < \infty$ for all $v \in V$. Then if $\chi : Z(\underline{g}) \to \mathbb{C}$ is a homomorphism we define $V^\chi = \{v \in V \mid (z-\chi(z))^k v = 0$ for $z \in Z(\underline{g})$ and some $k > 0$ depending on $v\}$. Then $V = \oplus V^\chi$.

3.6. Lemma. A necessary and sufficient condition for $V \in \mathcal{V}'$ to be in $\mathcal{V}$ is that every $v \in V$ is $Z(\underline{g})$-finite, $\dim V_\lambda < \infty$ for each

$\lambda \in (\underline{a}_0)^*_{\mathbb{C}}$ and $V = \bigoplus_{\chi \in F} V^\chi$, F a finite set.

Proof. If $V \in \mathcal{V}$ then $V = \sum_{i=1}^r U(\underline{g}_{\mathbb{C}})v_i$, $v_i \in V$. Each $v_i = \sum (v_i)_\lambda$, $(v_i)_\lambda \in V_\lambda$. Thus we may assume that $v_i \in V_{\lambda_i}$. $\dim U((\underline{p}_0)_{\mathbb{C}})v_i < \infty$. Taking $w_1,\ldots,w_q$ a basis of $\sum U((\underline{p}_0)_{\mathbb{C}})v_i$ with $w_i \in V_{\mu_i}$ we have $V = \sum U((\underline{\bar{n}}_0)_{\mathbb{C}})w_i$ with $\underline{\bar{n}}_0 = \theta(\underline{n}_0)$. (Here we use $\underline{\bar{n}}_0 \oplus \underline{p}_0 = \underline{g}$).

Since as an $\underline{a}_0$-module under ad, $U(\underline{\bar{n}}_0)$ is a sum of weight spaces each being finite dimensional we find

(i) If $V \in \mathcal{V}$ then $V = \bigoplus_{\lambda \in (\underline{a}_0)^*_{\mathbb{C}}} V_\lambda$ and $\dim V_\lambda < \infty$. We also note that we have proved a bit more:

(ii) If $V \in \mathcal{V}$ then there exist $\lambda_1,\ldots,\lambda_r \in (\underline{a}_0)^*_{\mathbb{C}}$ so that $V_{\lambda_i} \neq 0$ $i = 1,\ldots,r$ and if $V_\lambda \neq 0$ then $\lambda = \lambda_i - Q$ with

$$Q = \sum_{\alpha \in \Delta(P_0,A_0)} n_\alpha \alpha,\ n_\alpha \in \mathbb{Z},\ n_\alpha \geq 0.$$

Since $Z(\underline{g})V_\lambda \subset V_\lambda$ and $v_i \in V_{\lambda_i}$ we see that

$$\dim Z(\underline{g})\Big(\sum_{i=1}^r g_i v_i\Big) \leq \sum_{i=1}^r \dim Z(\underline{g})V_{\lambda_i} \leq \sum_{i=1}^r \dim V_{\lambda_i} < \infty.$$

The condition holds if we take for F the set of all χ with $V^\chi \cap V_{\lambda_i} \neq 0$. F is clearly finite.

Now for the converse. Let $V = \bigoplus_{\chi \in F} V^\chi$. $|F| < \infty$. Clearly it is enough to prove the result for $V = V^\chi$. Now $\chi = \chi_\Lambda$ relative to the Harish-Chandra parameterization of homomorphisms of $Z(\underline{g}) \to \mathbb{C}$. We recall this here. Let $\underline{t} \subset {}^0\underline{M}_0$ be a maximal abelian subalgebra. Set $\underline{h} = \underline{t}_{\mathbb{C}} \oplus (\underline{a}_0)_{\mathbb{C}}$. Then $\underline{h}$ is a Cartan subalgebra of $\underline{g}_{\mathbb{C}}$. Let $\Phi(\underline{g}_{\mathbb{C}},\underline{h})$ be the root system of $\underline{g}_{\mathbb{C}}$ relative to $\underline{h}$. Choose $\Phi^+ \subset \Phi(\underline{g}_{\mathbb{C}},\underline{h})$ a system of positive roots such that if $\underline{n}^+ = \sum_{\alpha \in \Phi^+} (\underline{g}_{\mathbb{C}})_\alpha$, $\underline{n}^+ \cap \underline{g} = \underline{n}_0$. Let $\delta = \frac{1}{2}\sum_{\alpha \in \Phi^+} \alpha$. Define $\mu(z)$ to be the element of $U(\underline{h})$ such that $\mu(z) - z \in U(\underline{g}_{\mathbb{C}})\underline{n}^+$. Let $\lambda : U(\underline{h}) \to U(\underline{h})$ be the homomorphism defined by $\lambda(h) = h - \delta(h)1$. Set $\gamma(z) = \lambda \circ \mu(z)$. Then

$\underset{\sim}{\gamma}: Z(\underline{g}) \to U(\underline{h})$ is an algebra isomorphism with image equal to the invariants in $U(\underline{h})$ relative to $W(\underline{g}_{\mathbb{C}},\underline{h})$ (the group of linear automorphisms of $\underline{h}$ generated by the reflections about the hyperplanes $\alpha = 0$, $\alpha \in \Phi(\underline{g}_{\mathbb{C}},\underline{h})$). (For the proofs of these results of Harish-Chandra cf. [2].) We set $\chi_\Lambda(z) = \Lambda(\underset{\sim}{\gamma}(z))$, $\Lambda \in \underline{h}^*$.

Now $\chi = \chi_\Lambda$ and $\chi_\Lambda = \chi_{\Lambda'}$ if and only if $\Lambda' = s\Lambda$ for some $s \in W(\underline{g}_{\mathbb{C}},\underline{h})$. We note that $\underline{b} = \underline{h} + \underline{n}^+ \subset (\underline{p}_0)_{\mathbb{C}}$. Thus if $v \in V$ $\dim U(\underline{b}) \cdot v < \infty$.

(i) If W is a non-zero subquotient of V then $W_{(s\Lambda-\delta)|\underline{a}_0} \neq 0$ for some $s \in W(\underline{g}_{\mathbb{C}},\underline{h})$.

Indeed, $W = W^\chi$, $\chi = \chi_\Lambda$. If $w \in W$, $W \neq 0$ then $\dim U(\underline{b})w < \infty$. Then there must be $0 \neq v \in U(\underline{b})w$ with the property that $\underline{n}^+ v = 0$ and $h \cdot v = \nu(h) \cdot v$, $h \in \underline{h}$, $\nu \in \underline{h}^*$. But then $z \cdot v = \nu(\mu(z))v = \chi_{\nu+\delta}(z)v$. Thus $\nu + \delta = s\Lambda$ for some $s \in W(\underline{g}_{\mathbb{C}},\underline{h})$. Thus $\nu = s\Lambda - \delta$. Hence (i).

Now let $X = \sum_{s \in W(\underline{g}_{\mathbb{C}},\underline{h})} U(\underline{g}_{\mathbb{C}}) V_{(s\Lambda-\delta)|\underline{a}_0}$. Then $(V/X)_{(s\Lambda-\delta)|\underline{a}_0} = 0$ all s. So by (i) $V/X = 0$. So V is finitely generated since $\dim V_{(s\Lambda-\delta)|\underline{a}_0} < \infty$ and $|W(\underline{g}_{\mathbb{C}},\underline{h})| < \infty$.

3.7. Lemma. If $V \in \mathcal{V}$ then V has finite length.

Proof. We may assume $V = V^\chi$ with $\chi = \chi_\Lambda$. Let $V = V_1 \supsetneq V_2 \supsetneq \cdots \supsetneq V_r$ be a chain of submodules of V. Let $W = \bigoplus_{s \in W(\underline{g}_{\mathbb{C}},\underline{h})} V_{(s\Lambda-\delta)|\underline{a}_0}$ (see (i) in 3.6). Then $\oplus(V_i/V_{i+1})_{(s_i\Lambda-\delta)\,\underline{a}_0} \neq 0$ all $i = 1,\ldots,r-1$. Thus $\dim W \geq r - 1$. Thus any decreasing chain of submodules of V can contain at most $\dim W + 1$ elements. So V satisfies both the ascending and descending chain condition.

3.8. Lemma. Let $V \in \mathcal{H}$. We set $j(V) = \{v^* \in V^* | \underline{n}_0^k v^* = 0$ for some $k\}$. Here V^* is the full algebraic dual of V and V^* is an ${}^0M_0$-module under $(m \cdot v^*)(x) = v^*(m^{-1}x)$, $v^* \in V^*$, $m \in {}^0M_0$, $x \in V$. V^* is a

$\underline{g}$-module under $(Xv^*)(v) = v^*(-Xv), X \in \underline{g}, v^* \in V^*, v \in V$.

(1) $j(V)$ is a $\underline{g}$-submodule of V^*.

Indeed, if $X \in \underline{g}$ then $\mathrm{ad}(\underline{n}_0)^p X = 0$, for some fixed p. So if $v^* \in j(V)$, $X \in \underline{g}$ then $\underline{n}_0^k \cdot v^* = 0$ some k. Thus $\underline{n}_0^{k+p} \cdot Xv^* = 0$, for $X \in \underline{g}$.

(2) $j(V) \in \mathcal{V}'$.

Indeed, V is finitely generated as $U(\underline{n}_0)$-module by 3.3. Thus $\dim V/\underline{n}_0^k V < \infty$ for all k. Clearly $j(V) = \bigcup (V/\underline{n}_0^k V)^*$ and $(V/\underline{n}_0^k V)^*$ is a finite dimensional continuous ${}^0M_0$-submodule of V^*. Thus $j(V)$ is a $(\underline{g}, {}^0M_0)$-module. Also $(V/\underline{n}_0^k V)^*$ is a finite dimensional $\underline{p}_0$-submodule of V^*, so $j(V)$ satisfies 3.5(1).

If $V, W \in \mathcal{H}$ and $T \in \mathrm{Hom}_{\underline{g},K}(V,W)$ let $T^* : W^* \to V^*$ be defined by $T^*(w^*) = w^* \circ T$. Then $T^*(j(W)) \subset j(V)$. Set $j(T) = T^*|_{j(W)}$. Then $V \rightsquigarrow j(V)$ is a <u>functor</u> from $\mathcal{H}$ to $\mathcal{V}'$.

<u>3.9</u>. <u>Proposition</u>. (Casselman, Wallach).

(1) If $V \in \mathcal{H}$ then $j(V) \in \mathcal{V}$.

(2) If $0 \to V_1 \overset{\alpha}{\to} V_2 \overset{\beta}{\to} V_3 \to 0$ is an exact sequence in $\mathcal{H}$, then

$$0 \to j(V_3) \overset{j(\beta)}{\to} j(V_2) \overset{j(\alpha)}{\to} j(V_1) \to 0$$

is an exact sequence in $\mathcal{V}$.

<u>Proof</u>. We must show that $j(V) \in \mathcal{V}$. Now if $V \in \mathcal{H}$ then $V = \bigoplus_{\chi \in F} V^\chi$ and F is a finite set. Thus it is enough to prove (1) for $V = V^\chi$. If $v \in V$ then $(z - \chi(z))^k v = 0$ for $z \in \mathcal{Z}(\underline{g})$ and k <u>fixed</u> independent of v. Indeed, V is finitely generated as a $U(\underline{g}_{\mathbb{C}})$-module.

Let $X \mapsto {}^tX$ be defined on $U(\underline{g}_{\mathbb{C}})$ by

(a) ${}^t1 = 1$

(b) ${}^tX = -X, X \in \underline{g}$

(c) ${}^t(XY) = {}^tY\,{}^tX$, $X, Y \in U(\underline{g}_{\mathbb{C}})$.

If $v^* \in V^*$ then

(i) $({}^t z - \chi(z))^k v^* = 0, \ z \in Z(\underline{g})$.

Set ${}^t\chi(z) = \chi({}^t z)$. Then $j(V)' = j(V)^{{}^t\chi}$. Thus to prove that $j(V) \in V$ it is enough to show that $\dim j(V)_\lambda < \infty$ for $\lambda \in (\underline{a}_0)^*_{\mathbb{C}}$.

Let $v^* \in (V/\underline{n}_0^{k+1}V)^*_\nu$, $v^* \notin (V/\underline{n}_0^k V)^*$. Then there exist $X_i \in (\underline{n}_0)_{\lambda_i}$ with $X_1 \ldots X_k \cdot v^* \neq 0$. Clearly $X_1 \ldots X_k v^* \in (V/\underline{n}_0 V)^*_{\nu+\lambda_1+\ldots+\lambda_k}$.

Hence if $v^* \in (V/\underline{n}_0^{k+1}V)_\nu$ and $v^* \notin (V/\underline{n}_0^k V)^*$ then $\nu = \xi - \lambda_1 - \ldots - \lambda_k$ for some $\xi \in (\underline{a}_0)^*_{\mathbb{C}}$ such that $(V/\underline{n}_0 V)_\xi \neq 0$ and $\lambda_i \in \Phi(P_0, A_0)$. Since the set of such ξ is finite and there exists $h_0 \in \underline{a}_0$ so that $\lambda(h_0) \geq 1$ for $\lambda \in \Phi(P_0, A_0)$ we see that if $\nu \in (\underline{a}_0)^*_{\mathbb{C}}$ and $j(V)_\nu \neq 0$ then there exists k so that

$$j(V)_\nu \subset (V/\underline{n}_0^{k+1}V)^*.$$

Hence $\dim j(V)_\nu \leq \dim(V/\underline{n}_0^{k+1}V)^* < \infty$.

To prove (2) we note that the only non-trivial part is to show that $j(\alpha) : j(V_2) \to j(V_1)$ is surjective. We may assume that $V_1 \subset V_2$ is a $(\underline{g}, K)$-submodule. Thus we must show that if $\lambda \in j(V_1)$ there exists $\mu \in j(V_2)$ such that $\mu|_{V_1} = \lambda$.

Now $\lambda \in (V_1/\underline{n}_0^k V_1)^*$. Let r be such that $\underline{n}_0^{k+r} V_2 \cap V_1 \subset \underline{n}_0^k V_1$ (see 3.2 and the note thereafter). Here we use the fact that V is finitely generated as a $U(\underline{n}_0)$-module so 3.2 applies. The natural map

$$V_1/\underline{n}_0^{k+r}\, V_2 \cap V_1 \to V_1/\underline{n}_0^k\, V_1$$

is surjective. Thus every $\lambda \in (V_1/\underline{n}_0^k V_1)^*$ extends to an element of $(V_1/\underline{n}_0^{k+r} V_2 \cap V_1)^*$ which in turn extends to an element of $(V_2/\underline{n}_0^{k+r} V_2)^*$ ($\dim V_2/\underline{n}_0^{k+r} V_2 < \infty$). Let μ be the extension. Then $\mu \in j(V_2)$ and $\mu|V_1 = \lambda$. Q.E.D.

<u>3.10</u>. The main utility of $j(V)$ derives from the following result.

<u>Lemma</u>. If $V \in H$, $V \neq (0)$ then $j(V) \neq (0)$.

<u>Proof</u>. $V/\underline{n}_0 V \neq 0$ by Casselman's Theorem.

Thus $V \rightsquigarrow j(V)$ is an exact, faithful function from the category

$\mathcal{H}$ to the category $\mathcal{V}$.

Let us give an immediate consequence of our theory so far.

3.11. Theorem. If $V \in \mathcal{H}$ then V has finite length.

Proof. Let $V = V_0 \supset V_1 \supset \ldots \supset V_r$ be a chain of submodules of V. Then $j(V/V_1) \subset j(V/V_2) \subset \ldots \subset j(V/V_r) \subset \ldots$. We assert that if $V_i \neq V_{i+1}$ then $j(V/V_i) \neq j(V/V_{i+1})$. Indeed we have the exact sequence in $\mathcal{H}$

$$0 \to V_i/V_{i+1} \xrightarrow{\alpha} V/V_{i+1} \xrightarrow{\beta} V/V_i \to 0$$

α, β the natural maps. This induces

$$0 \to j(V/V_i) \xrightarrow{j(\beta)} j(V/V_{i+1}) \xrightarrow{j(\alpha)} j(V_i/V_{i+1}) \to 0$$

exact in $\mathcal{V}$.

If $j(V/V_i) = j(V/V_{i+1})$ then $j(\alpha) = 0$. But then $j(V_i/V_{i+1}) = 0$ so $V_i/V_{i+1} = 0$. Hence $V_i = V_{i+1}$.

But $j(V/V_i) \subset j(V)$ all i. Thus there must be an r_0 so that $j(V/V_i) = j(V/V_{r_0})$ for $i \geq r_0$. This implies by the above $V_i = V_{r_0}$ for $i \geq r_0$. The result now follows.

We now give another implication of the theory originally due to Casselman.

3.12. Theorem. If $V \in \mathcal{H}$ then there exists an admissible representation (π, H) of G on a Hilbert space such that $V \cong H_K$.

For this we need an auxiliary lemma. Let $V \in \mathcal{H}$. Then $\dim V/\underline{n}_0^k V < \infty$ and $V/\underline{n}_0^k V$ is a $(\underline{p}_0, {}^0M_0)$-module. Since A_0N_0 is simply connected we see that $V/\underline{n}_0^k V$ integrates to a representation $\sigma_{k,V}$ of P_0 (σ_k if V is understood). Let (σ, W) be a finite dimensional representation of P_0. Then there exists $k = k_\sigma$ such that $\sigma(\underline{n}_0^k)W = 0$.

Scholium. $\mathrm{Hom}_{\underline{g},K}(V, H_K^\sigma)$ is naturally isomorphic with

$$\mathrm{Hom}_{\underline{p}_0, {}^0M_0}(V/\underline{n}_0^k V, W)$$

where $k = k_\sigma$.

Proof. Let $T \in \mathrm{Hom}_{\underline{g},K}(V, H_K^\sigma)$; set $\hat{T}(v) = T(v)(1)$. Then $\hat{T}(\underline{n}_0^k V) = 0$. Thus $\hat{T} \in \mathrm{Hom}_{\underline{p}_0, {}^0M_0}(V/\underline{n}_0^k V, W)$.

Let $S \in \mathrm{Hom}_{\underline{p}_0, {}^0M_0}(V/\underline{n}_0^k V, W)$.

Let $q_k : V \to V/\underline{n}_0^k V$ be the natural map. If $v \in V$ set $\check{S}(v)(pk) = \sigma(p) S(q_k(k \cdot v)), p \in P_0, k \in K$. Then $\check{S}(v) \in H_K^\sigma$. $\check{S}(k \cdot v) = k \cdot \check{S}(v)$ is clear. $\check{S}(Xv)(1) = X \cdot \check{S}(v)(1)$, $X \in \underline{p}_0$ is also clear. Since $\underline{g} = \underline{k} + \underline{p}_0$ we see that $\check{S}(Xv)(1) = (X \cdot \check{S}(v))(1)$, $X \in \underline{g}$. But $\check{S}(k \cdot Xv)(1) =$ $\check{S}(\mathrm{Ad}(k)X \cdot (k \cdot v))(1) = ((\mathrm{Ad}(k)X)\check{S}(k \cdot v)(1)) = ((\mathrm{Ad}(k)X \cdot k \cdot \check{S}(v))(1) =$ $(k \cdot X \cdot \check{S}(v))(1) = (X \cdot \check{S}(v))(k)$.

Also $\check{S}(k \cdot Xv)(1) = (k \cdot \check{S}(X \cdot v))(1) = \check{S}(X \cdot v)(k)$. So $\check{S}(Xv) = X\check{S}(v)$ for $X \in \underline{g}$, $v \in V$.

Thus $\check{S} \in \mathrm{Hom}_{\underline{g},K}(V, H_K^\sigma)$. Since $(\hat{T})^\vee = T$ and $(\check{S})^\wedge = S$ the result follows.

We now prove the theorem. Let $T_k = \hat{I}$ where $I : V/\underline{n}_0^k V \to V/\underline{n}_0^k V$ is the identity map. Then $\mathrm{Ker}\, T_k \subset \underline{n}_0^k V$. Clearly $\mathrm{Ker}\, T_k \supset \mathrm{Ker}\, T_{k+1}$. Hence there is k_0 so that $\mathrm{Ker}\, T_k = \mathrm{Ker}\, T_{k_0}$ for $k \geq k_0$ (V has finite length). Thus $\mathrm{Ker}\, T_{k_0} \subset \bigcap_{k \geq k_0} \underline{n}_0^k V$. But there is r such that

$$\underline{n}_0^{k+r} V \cap \mathrm{Ker}\, T_{k_0} \subset \underline{n}_0^k \,\mathrm{Ker}\, T_{k_0}, \; k \geq 0.$$

So $\mathrm{Ker}\, T_{k_0} \subset \underline{n}_0^k \,\mathrm{Ker}\, T_{k_0}$, $k \geq 0$. Now this implies $j(\mathrm{Ker}\, T_{k_0}) = 0$. Hence $\mathrm{Ker}\, T_{k_0} = 0$. Set $\sigma = \sigma_{k_0}$. H^σ is admissible so $T_{k_0}(V) \subset H_\omega^\sigma$. Thus $\mathrm{C\ell}(T_k(V))$ is G-invariant. Clearly $\mathrm{C\ell}(T_k(V))_K = T_k(V)$. So the result follows.

3.13. The following result of [17] is sometimes useful.

Lemma. Let V be a finitely generated $(\underline{g}, K)$-module. Then V is admissible if and only if V is finitely generated as a $U(\underline{n}_0)$-module.

Proof. Suppose that V is finitely generated as a $U(\underline{n}_0)$-module. Consider $j(V)$ (defined the same way as if $V \in \mathcal{H}$). Then $j(V) \in \mathcal{V}'$.

Also $j(V) = U(V/\underline{n}_0^kV)^*$ and $\dim(V/\underline{n}_0^kV)^* < \infty$. $(V/\underline{n}_0^kV)^* = \{\lambda \in j(V) \mid \underline{n}_0^k\lambda = 0\}$. Thus $Z(\underline{g})(V/\underline{n}_0^kV)^* \subset (V/\underline{n}_0^kV)^*$. Hence $(V/\underline{n}_0^kV) = \bigoplus_\chi (V/\underline{n}_0^kV)^\chi$ as a $Z(\underline{g})$-module. Now this implies $j(V) = \oplus j(V)^\chi$. But if $\lambda \in j(V)$, $\underline{n}_0^k\lambda = 0$ for some k. Let $\lambda \neq 0$; then there is k so that $\underline{n}_0^k\lambda \neq 0$, $\underline{n}_0^{k+1}\lambda = 0$. Thus $j(V)^\chi \cap (V/\underline{n}_0V)^* \neq (0)$ if $j(V)^\chi \neq 0$. The argument proving Lemma 3.6 implies that $j(V) \in \mathcal{V}$.

Let $W = \{v \mid j(V)(v) = 0\}$. Then W is a $(\underline{g},K)$-submodule of V. (Indeed it is a $(\underline{g},{}^0M_0)$-module and a $(\underline{g},K^0)$-module. Since $K = {}^0M_0K^0$ it is a $(\underline{g},K)$-submodule.) If $W \neq 0$ then W has an irreducible quotient. So $j(W) \neq 0$ by Lemma 3.10. Hence $j(W) \neq 0$. But the proof of 3.9 (2) only used finite generation relative to $U(\underline{n}_0)$. Thus $j(W) = j(V)|_W$. Hence $W = 0$. Now this implies that $V = \oplus V^{t_\chi}$, the sum over those χ so that $j(V)^\chi \neq 0$. So V is $Z(\underline{g})$-finite. Hence V is admissible. The converse follows from 3.3.

<u>3.14</u>. <u>Corollary</u>. If $V \in \mathcal{H}$ and F is a finite dimensional (g,K)-module then $V \otimes F \in \mathcal{H}$.

<u>Proof</u>. V is finitely generated as a $U(\underline{n}_0)$-module. Hence so is $V \otimes F$. Apply Lemma 3.13.

<u>3.15</u>. <u>Lemma</u>. If $V \in \mathcal{H}$ and if F is a finite dimensional $(\underline{g},K)$-module then $j(V \otimes F) = j(V) \otimes F^*$.

<u>Proof</u>. We note that $(V \otimes F)^* = V^* \otimes F^*$ canonically as a $\underline{g}$ and a ${}^0M$-module. Since $\underline{n}_0^pF^* = 0$ for some p it is clear that $j(V \otimes F) \supset j(V) \otimes F^*$.

Let $f_1, f_2, \ldots, f_d$ be a basis of F^* such that $\underline{n}_0 \cdot f_j \subset \sum_{i>j} \mathbb{C}f_i$. Let $\lambda \in j(V \otimes F)$. Then $\lambda = \sum \mu_i \otimes f_i$. Suppose $\mu_i = 0$ for $i \geq j_0$. Then if $X \in \underline{n}_0$, $X^k\lambda = X^k\mu_{j_0} \otimes f_{j_0} + \sum_{i>j_0} \xi_i \otimes f_i$. Since $X^k\lambda = 0$ for some k, we see that $X^k\mu_{j_0} = 0$ for the same k. Hence $\mu_{j_0} \in j(V)$. Thus $\lambda - \mu_{j_0} \otimes f_{j_0} \in j(V)$. But then $\mu_{j_0-1} \in j(V)$, etc.

<u>3.16</u>. For the remainder of this section we assume that G is a <u>linear</u>

Lie group. That is G is isomorphic with a closed subgroup of $GL(m,\mathbb{R})$ for some m.

Let $\gamma_0 \in \mathcal{E}(K)$ be the class of the trivial representation of K. Let $\xi_0 \in \mathcal{E}({}^0M_0)$ be the class of the trivial representation of ${}^0M_0$. Let H^ν denote $H^{\xi_0,\nu}$.

Lemma. (Lepowsky, Wallach [14]). Let $\xi \in \mathcal{E}({}^0M_0)$, $\nu \in (\underline{a}_0)^*_{\mathbb{C}}$. Then there exists $\mu \in (\underline{a}_0)^*_{\mathbb{C}}$ such that $\operatorname{Re}(\mu,\lambda) > 0$, $\lambda \in \Phi(P_0,A_0)$ and F a finite dimensional $(\underline{g},K)$-module such that $H_K^{\xi,\nu}$ is isomorphic with a quotient of $H_K^\mu \otimes F$.

3.17. We also recall the following result of Kostant [9], Helgason [7]. (Kostant's result is stronger than the statement below.)

Lemma. If $\mu \in (\underline{a}_0)^*_{\mathbb{C}}$ and $\operatorname{Re}(\mu,\lambda) > 0$, $\lambda \in \Phi(P_0,A_0)$ then $H_K^\mu = U(\underline{g}_{\mathbb{C}})H_K^\mu(\gamma_0)$.

3.18. Lemma. If (σ,W) is a finite dimensional representation of P_0 then $H_K^\sigma \in \mathcal{H}$.

Proof. Let $W = W_1 \supset W_2 \supset \dots \supset W_r \supset W_{r+1} = (0)$ be a Jordan-Hölder series for (σ,W). Then the representation of P_0 on W_i/W_{i+1} is equivalent with $(\xi_i)_{\nu_i}$ for some $\xi_i \in \mathcal{E}({}^0M_0)$, $\nu_i \in (\underline{a}_0)^*_{\mathbb{C}}$. Set $V_i = \{f \in H_K^\sigma \mid f(g) \in W_i\}$. Then V_i/V_{i+1} imbeds in $H^{\xi_i,\nu_i-\rho}$.

(One can prove as an easy exercise that in fact

$$(1) \quad V_i/V_{i+1} \cong H_K^{\xi_i,\nu_i-\rho}.)$$

Thus it is enough to show that $H_K^{\xi,\nu}$ is finitely generated as a $U(\underline{g}_{\mathbb{C}})$-module. But this follows from 3.17 and 3.16.

Note. This result is true even if G is not linear. It follows from Harish-Chandra's theorem on characters. One can also give a proof of this result in the same spirit as ours using an analogue of Lemma 3.17. However, for the purposes of these lectures it is Lemmas 3.16 and 3.17 that are important to us. We will use them in the next section.

4. Automatic Continuity Theorems

4.1. We need the following result of Kostant [10]. Let $\underset{\sim}{\lambda} : S(\underline{g}_{\mathbb{C}}) \to U(\underline{g}_{\mathbb{C}})$ be the symmetrization mapping. Let $W(\underline{g},\underline{a}_0) = W$ be the Weyl group of $\underline{g}$ relative to $\underline{a}_0$. That is, $W(\underline{g},\underline{a}_0)$ is the group of isometries of $\underline{a}_0$ generated by the reflections about the hyperplanes $\lambda = 0$ for $\lambda \in \Phi(\underline{g},\underline{a}_0)$. It can also be described as $\mathrm{Ad}(N_G(\underline{a}_0))|_{\underline{a}_0}$ where $N_G(\underline{a}_0) = \{g \in G | \mathrm{Ad}(g)\underline{a}_0 \subset \underline{a}_0\}$.

Theorem 1. (Kostant [10]). There exists an $|W|$-dimensional (graded) subspace of $U(\underline{a}_0)(= S(\underline{a}_0))$, E, such that the map

$$U((\underline{n}_0)_{\mathbb{C}}) \otimes \underset{\sim}{\lambda}(S(\underline{p}_{\mathbb{C}})^K) \otimes E \otimes U(\underline{k}_{\mathbb{C}}) \to U(\underline{g}_{\mathbb{C}})$$

$$n,\ p,\ e,\ k \mapsto npek$$

is a surjective linear isomorphism. Here $S(\underline{p}_{\mathbb{C}})^K$ is the space of $\mathrm{Ad}(K)$-invariants in $S(\underline{p}_{\mathbb{C}})$ $(\underline{p} = \{X \in \underline{g} | \theta X = -X\})$.

This theorem is not too hard to prove using the following theorem of Harish-Chandra. Let $p : U(\underline{g}_{\mathbb{C}}) \to U(\underline{a}_{0\mathbb{C}})$ be the projection corresponding to

$$U(\underline{g}_{\mathbb{C}}) = U(\underline{a}_{0\mathbb{C}}) \oplus (\underline{n}_0 U(\underline{g}_{\mathbb{C}}) + U(\underline{g}_{\mathbb{C}})\underline{k}).$$

Then $p : U(\underline{g}_{\mathbb{C}})^K \to U(\underline{a}_{0\mathbb{C}})$ is an algebra homomorphism. Define the algebra homomorphism $\mu : U((\underline{a}_0)_{\mathbb{C}}) \to U((\underline{a}_0)_{\mathbb{C}})$ by

$$\mu(1) = 1$$

$$\mu(h) = h + \rho_0(h)1.$$

Set $q = \mu \circ p$. Then

Theorem 2. (Harish-Chandra [6]).

(1) $q(U(\underline{g}_{\mathbb{C}})^K) = U((\underline{a}_0)_{\mathbb{C}})^W$

(2) The following sequence is exact

$$0 \to U(\underline{g}_{\mathbb{C}})^K \cap U(\underline{g}_{\mathbb{C}})\underline{k} \to U(\underline{g}_{\mathbb{C}})^K \overset{q}{\to} U((\underline{a}_0)_{\mathbb{C}})^W \to 0.$$

Furthermore

$$\underset{\sim}{\lambda}(S(\underline{p}_{\mathbb{C}})^K) \xrightarrow{q} U((\underline{a}_0)_{\mathbb{C}})^W$$

is bijective.

4.2. We now form some auxiliary modules. Consider the $U(\underline{g}_{\mathbb{C}})$ module $V_0 = U(\underline{g}_{\mathbb{C}})/U(\underline{g}_{\mathbb{C}})\underline{k}$. We make V_0 into a K-module by defining $k \cdot g = \mathrm{Ad}(k) \cdot g$. Then it is easy to see that V_0 is a $(\underline{g},K)$-module. We also note that V_0 is in addition a right $U(\underline{g}_{\mathbb{C}})^K$-module. $(x \in U(\underline{g}_{\mathbb{C}})\underline{k}$, $y \in U(\underline{g}_{\mathbb{C}})^K$ implies $xy \in U(\underline{g}_{\mathbb{C}})\underline{k})$. Clearly $U(\underline{g}_{\mathbb{C}})^K \cap U(\underline{g}_{\mathbb{C}})\underline{k}$ acts by zero on the right. Taking $\underset{\sim}{\lambda}(S(\underline{p}_{\mathbb{C}})^K)$ to have the algebra structure induced by q above, we then see that V_0 is a right $\underset{\sim}{\lambda}(S(\underline{p}_{\mathbb{C}})^K)$-module.

Let X be a finite dimensional $\underset{\sim}{\lambda}(S(\underline{p}_{\mathbb{C}})^K)$ module. Set

$$Y^X = V_0 \underset{\underset{\sim}{\lambda}(S(p_{\mathbb{C}})^K)}{\otimes} X.$$

Then Theorem 4.1.1 implies that

(1) Y^X is a free $U(\underline{n}_0)$-module on $\dim X \cdot |W|$ number of generators.

If $\nu \in (\underline{a}_0)^*_{\mathbb{C}}$ and if $q_\nu(u) = \nu(q(u))$, $u \in U(\underline{g}_{\mathbb{C}})^K$, then if X is the 1-dimensional module with action q_ν we get $Y^X = Y^\nu$.

(2) Clearly $Y^\nu = Y^{s\nu}$, $s \in W$, $\nu \in (\underline{a}_0)^*_{\mathbb{C}}$. Since $Z(\underline{g}_{\mathbb{C}}) \subset U(\underline{g}_{\mathbb{C}})^K$ we see that

(3) $Y^\nu \in \mathcal{H}$.

4.3. Put on $\underline{p}_{\mathbb{C}}$ the Hermitian inner product $\langle , \rangle$, that restricts to $\underline{p}$ to give B. Extend $\langle , \rangle$ to $S(\underline{p}_{\mathbb{C}})$ in the canonical manner. $S(\underline{p}_{\mathbb{C}})^K = \oplus S^j(\underline{p}_{\mathbb{C}}) \cap S(\underline{p}_{\mathbb{C}})^K$ ($S^j(\underline{p}_{\mathbb{C}})$ the homogeneous elements of degree j). Set $Q^j = (S^j(\underline{p}_{\mathbb{C}}) \cap S(\underline{p}_{\mathbb{C}})^K)^\perp \cap S^j(\underline{p}_{\mathbb{C}})$. Let $Q = \oplus Q^j$.

Lemma. The map $T_\nu : Q \to Y^\nu$ given by $T_\nu(g) = \underset{\sim}{\lambda}(\bar{g}) \otimes 1$ ($\bar{g} = g + U(\underline{g}_{\mathbb{C}})\underline{k}$, $g \in U(\underline{g}_{\mathbb{C}})$) defines a K-module isomorphism. (Here K acts on $S(\underline{p}_{\mathbb{C}})$ by the natural extension of $\mathrm{Ad}(K)|_{\underline{p}_{\mathbb{C}}}$.)

This is an elementary exercise that we leave to the reader.

4.3. Let $\underline{t} \subset {}^0\underline{m}_0$ be a maximal abelian subalgebra. Let $\Phi(({}^0\underline{m}_0)_{\mathbb{C}}, \underline{t}_{\mathbb{C}})$ be the root system of $({}^0\underline{m}_0)_{\mathbb{C}}$ relative to $\underline{t}_{\mathbb{C}}$. Fix Φ_0^+ a system of positive roots for $\Phi(({}^0\underline{m}_0)_{\mathbb{C}}, \underline{t}_{\mathbb{C}})$. Set $\delta_0 = \frac{1}{2}\sum_{\alpha\in\Phi_0^+}\alpha$.

Let $\nu \in (\underline{a}_0)^*_{\mathbb{C}}$. We extend ν to $(\underline{p}_0)_{\mathbb{C}}$ by setting $\nu(({}^0\underline{m}_0)_{\mathbb{C}} \oplus (\underline{n}_0)_{\mathbb{C}}) = 0$. Define $M^\nu = U(\underline{g}_{\mathbb{C}}) \otimes_{U((\underline{p}_0)_{\mathbb{C}})} \mathbb{C}_{\nu-\rho_0}$ where $\mathbb{C}_{\nu-\rho_0}$ is the true $(\underline{p}_0)_{\mathbb{C}}$ module $\mathbb{C}$ with action given by $\nu - \rho_0$. Set $\underline{h} = \underline{t}_{\mathbb{C}} \oplus (\underline{a}_0)_{\mathbb{C}}$.

Lemma. Let $\Phi^+ \subset \Phi(\underline{g}_{\mathbb{C}}, \underline{h})$ be a system of positive roots such that $\Phi^+|_{\underline{t}_{\mathbb{C}}} \supset \Phi_0^+$ and $\Phi^+|_{\underline{a}_0} \supset \Phi(P_0, A_0)$; then M^ν is irreducible if $\frac{2(\nu-\delta_0,\alpha)}{(\alpha,\alpha)} \notin \mathbb{N}_+$ for $\alpha \in \Phi^+$ (here $\mathbb{N}_+ = \{1,2,\ldots\}$ and δ_0 is extended to $\underline{h}$ by $\delta_0(\underline{a}_0) = 0$).

For a proof cf. [21].

4.4. Let $\nu \in (\underline{a}_0)^*_{\mathbb{C}}$ and assume that $\mathrm{Re}(\nu,\lambda) > 0$ for $\lambda \in \Phi(P_0, A_0)$. Let $s \in W(\underline{g}_0, \underline{a}_0)$ and let $s^* \in K$ be such that $\mathrm{Ad}(s^*)|_{\underline{a}_0} = s$. We define $A_s(\nu) : H_\infty^\nu \to H_\infty^{s\nu}$ as follows:

(1) $A_s(\nu)f(g) = \int_{N/s^{*-1}Ns^*\cap N} f((s^*)^{-1}ng)\,dn$.

Then if $(\underline{a}_0)^{*+}_{\mathbb{C}} = \{\nu \in (\underline{a}_0)^*_{\mathbb{C}} | \mathrm{Re}(\nu,\lambda) > 0,\ \lambda \in \Phi(P_0,A_0)\}$ we have

(2) The integral defining $A_s(\nu)$ converges absolutely and uniformly for ν in a compactum of $((\underline{a}_0)^*)^+_{\mathbb{C}}$ and g in a compactum of G.

We observe that $H_\infty^\nu = H_\infty^0$ as a K-module. We may look upon $A_s(\nu)$ as an operator on H_∞^0.

(3) $\nu \mapsto A_s(\nu)$ is weakly holomorphic with values in the continuous endomorphisms of H_∞^0.

(4) $A_s(\nu) \in \mathrm{Hom}_G(H_\infty^\nu, H_\infty^{s\nu})$ (here Hom_G denotes the continuous homomorphisms.)

For proofs of (1), (2), (3), (4), cf. [16], [7], [20].

We will call $\nu \in (\underline{a}_0)^*_{\mathbb{C}}$ generic if $\frac{2(s\nu-\delta_0,\alpha)}{(\alpha,\alpha)} \notin \mathbb{N}_+$ for all $\alpha \in \Phi^+$ and all $s \in W(\underline{g}_0,\underline{a}_0)$.

4.5. We note that it is not hard to check (from its very definition) that

(1) $u|_{H_K^\nu(\gamma_0)} = q(u)I$, for $u \in U(\underline{g}_{\mathbb{C}})^K$.

(2) If $f \in H_K^\nu(\gamma_0)$ then $f|_K$ is constant. This makes it also clear that $H_K^\nu(\gamma_0)$ is one dimensional.

(3) Define $c_s(\nu)$ by $A_s(\nu)f|_K = c_s(\nu)f(1)$ for $f \in H_K^\nu(\gamma_0)$. If $\mathrm{Re}(\nu,\lambda) > 0$ for $\lambda \in \Phi(P_0,A_0)$ then $c_s(\nu) \neq 0$. (See [20].)

Define for $f \in H_\infty^\nu$, $\delta_s(\nu)(f) = (A_s(\nu)f)(1)$. Then it is clear that $\delta_s(\nu)|_{H_K^\nu} \in (H_K^\nu/\underline{n}_0 H_K^\nu)^*_{-\nu-\rho_0}$.

Lemma. If $\nu \in ((\underline{a}_0)^*_{\mathbb{C}})^+$ is generic then

(1) The natural map $Y^\nu \to H_K^\nu$ given by $g \mapsto g \cdot 1_\nu$ ($1_\nu \in H_K^\nu$ is the element of $H_K^\nu(\gamma_0)$ which takes the value 1 at 1) defines a surjective $(\underline{g},K)$-module isomorphism.

(2) $j(H_K^\nu) \cong \bigoplus_{s \in W(\underline{g},\underline{a}_0)} M^{-s\nu}$ as a $U(\underline{g}_{\mathbb{C}})$-module.

Proof. Let $H_0 \in \underline{a}_0$ be such that $\alpha(H_0) = 1$ for $\alpha \in \Delta(P_0,A_0)$. Set $U((\underline{n}_0)_{\mathbb{C}})_{(j)}$ equal to the sum of the eigenspaces of $\operatorname{ad} H_0$ in $U((\underline{n}_0)_{\mathbb{C}})$ of eigenvalue less than or equal to j. Then

(i) $U(\underline{n}_0)_{\mathbb{C}} = (U(\underline{n}_0)_{\mathbb{C}})_{(j)} \oplus (\underline{n}_0)^{j+1} U((\underline{n}_0)_{\mathbb{C}})$.

Indeed, $\bigoplus_{\alpha \in \Delta} (\underline{n}_0)_\alpha$ generates $\underline{n}_0$ hence $U((\underline{n}_0)_{\mathbb{C}})$. A monomial $X_1 \dots X_p$ with $X_i \in (\underline{n}_0)_{\alpha_i}$, $\alpha_i \in \Delta$, has eigenvalue p for H_0. Thus $(\underline{n}_0)^{j+1} U((\underline{n}_0)_{\mathbb{C}})$ contains all eigenspaces with eigenvalue bigger than j. It is also clear that any eigenvalue of $\operatorname{ad} H_0$ on $(\underline{n}_0)^{j+1} U((\underline{n}_0)_{\mathbb{C}})$ is bigger than j. Thus (i) follows.

Set $U((\overline{\underline{n}}_0)_{\mathbb{C}})_{(j)}$ equal to the direct sum of the eigenspaces of ad H_0 with eigenvalue greater than or equal to $-j$. Then

(ii) $U((\underline{\bar{n}}_0)_{\mathbb{C}})_{(j)} \oplus (\underline{\bar{n}}_0)^{j+1} U((\underline{\bar{n}}_0)_{\mathbb{C}}) = U((\underline{\bar{n}}_0)_{\mathbb{C}})$.

Let $\gamma_\nu = 1 \otimes 1$ in M^ν.

(iii) Set ${}_jM^\nu = \{m \in M^\nu | \underline{n}_0^j m = 0\}$. Then ${}_jM^\nu \supset U((\underline{\bar{n}}_0)_{\mathbb{C}})_{(j-1)}\gamma_\nu$.

This is clear from weight considerations. Now assume that $\nu \in ((\underline{a}_0)^*_{\mathbb{C}})^+$ is generic. Consider the map $Q_\nu : Y^\nu \to H^\nu_K$, $g \mapsto g \cdot 1_\nu$ as in (1). Then $j(Q_\nu) : j(H^\nu_K) \to j(Y^\nu)$.

$\delta_s(\nu) \in j(H^\nu_K)$ for $s \in W$. Thus $U(\underline{g}_{\mathbb{C}})\delta_s(\nu) \subset j(H^\nu{}_K)$. Since ν is generic $U(\underline{g}_{\mathbb{C}})\delta_s(\nu) \cong M^{-s\nu}$. Furthermore $\delta_s(\nu)(1_\nu) \neq 0$. Thus $j(Q_\nu)(\delta_s(\nu)) \neq 0$. Hence we have $j(Q_\nu) : U(\underline{g}_{\mathbb{C}})\delta_s(\nu) \to j(Y^\nu)$ is injective. Since $M^{-s\nu}$ and $M^{-s\nu'}$ are inequivalent for $s \neq s'$ we see that $j(Q_\nu) : \bigoplus_{s \in W} U(\underline{g}_{\mathbb{C}})\delta_s(\nu) \to j(Y^\nu)$ is injective.

By (iii), $(Y^\nu/\underline{n}_0^{j+1}Y^\nu)^* \supset \bigoplus_{s \in W} U((\underline{\bar{n}}_0)_{\mathbb{C}})_{(j)} j(Q_\nu)\delta_s(\nu)$. But

$$\dim(Y^\nu/\underline{n}_0^{j+1}Y^\nu)^* = \dim(Y^\nu/\underline{n}_0^{j+1}Y^\nu) =$$
$$|W| \dim U((\underline{n}_0)_{\mathbb{C}})/\underline{n}_0^{j+1}U((\underline{n}_0)_{\mathbb{C}}) =$$
$$|W| \dim U((\underline{n}_0)_{\mathbb{C}})_{(j)} = |W| \dim U((\underline{\bar{n}}_0)_{\mathbb{C}})_{(j)}.$$

Hence

(iv) $(Y^\nu/\underline{n}_0^{j+1}Y^\nu)^* = \bigoplus_{s \in W} U((\underline{\bar{n}}_0)_{\mathbb{C}})_{(j)} j(Q_\nu)\delta_s(\nu)$.

But (iv) implies that

(v) $j(Q_\nu) : j(H^\nu_K) \to j(Y^\nu)$ is surjective.

Since $U(\underline{g}_{\mathbb{C}})1_\nu = H^\nu_K$ we see that Q_ν is surjective. Hence we have $0 \to \operatorname{Ker} Q_\nu \to Y^\nu \xrightarrow{Q_\nu} H^\nu_K \to 0$ exact in H gives

$$0 \to j(H^\nu_K) \xrightarrow{j(Q_\nu)} j(Y^\nu) \to j(\operatorname{Ker} Q_\nu) \to 0$$

exact in V. Since $j(Q_\nu)$ is surjective this implies that $j(\operatorname{Ker} Q_\nu) = 0$. So $\operatorname{Ker} Q_\nu = 0$. This proves (1).

(2) follows since by (iv)

$$j(Y^{\nu}) = \bigoplus_{s\in W} j(Q_{\nu})(U(\underline{g}_{\mathbb{C}})\delta_s(\nu)) \cong \bigoplus_{s\in W} M^{-s\nu}.$$

Since $H_K^{\nu} \cong Y^{\nu}$ this implies (2).

4.6. Corollary. (Kostant, Rallis [11]). Let Q be as in 4.3; then as a K-module $Q = \bigoplus_{\gamma\in E(K)} Q(\gamma)$ and $\dim Q(\gamma) = d(\gamma)\dim V_{\gamma}^{{}^0M_0}$ ($V_{\gamma} \in \gamma$ and $V_{\gamma}^{{}^0M_0}$ is the space of ${}^0M_0$ invariants in V_{γ}).

Proof. Let $\nu \in ((\underline{a}_0)^*_{\mathbb{C}})^+$ be generic; then $Q_{\nu} \circ T_{\nu} : Q \to H_K^{\nu}$ is a bijective K-module isomorphism. The result now follows from Frobenius reciprocity.

4.7. Proposition. If $\nu \in ((\underline{a}_0)^*_{\mathbb{C}})^+$ then $Q_{\nu} : Y^{\nu} \to H_K^{\nu}$ is a bijective $(\underline{g},K)$-module isomorphism.

Proof. Q_{ν} is surjective and $\dim Y^{\nu}(\gamma) = \dim H_K^{\nu}(\gamma)$ for $\gamma \in E(K)$ by (4.6). Hence Q_{ν} is injective.

Note. This result for ν satisfying the weaker hypothesis $\mathrm{Re}(\nu,\lambda) \geq 0$, $\lambda \in \Phi(P_0,A_0)$ is due to Kostant.

4.8. We now come to the main result of this section. The remainder of the section will be devoted to its proof.

Theorem. (Casselman, Casselman-Wallach). We assume that G is linear. Let σ be a finite dimensional representation of P_0. If $\lambda \in j(H_K^{\sigma})$ then λ extends to a continuous functional on H_{∞}^{σ}. (Since H_K^{σ} is dense in H_{∞}^{σ} this extension is unique.)

The proof consists of proving the result for a special case and then reducing the general case to the special case. This proof follows the broad lines of the proof of the analogous result for Whittaker vectors in [3].

Let σ be a fixed finite dimensional unipotent representation of A_0 on X. Here unipotent means $\sigma(a)-I$ is nilpotent for $a \in A_0$. Let $\nu \in (\underline{a}_0)^*_{\mathbb{C}}$. Define $\sigma_{\nu}(man) = a^{\nu+\rho}\sigma(a)$ for $m \in {}^0M_0$, $a \in A_0$,

$n \in N_0$. If $s \in W$ define $s\sigma(a) = \sigma(s^{-1}a)$ on X. Thus $(s\sigma)_{s\nu}$ makes sense.

We define for $s \in W$ and $\nu \in ((\underline{a}_0)^*_{\mathbb{C}})^+$,

(1) $A_s(\nu) : H_\infty^{\sigma_\nu} \to H_\infty^{(s\sigma)_{s\nu}}$

by

$$A_s(\nu)f(g) = \int_{N/s^{*-1}Ns^* \cap N} f(s^{*-1}ng)\,dn.$$

As before $H_\infty^{\sigma_\nu} = H_\infty^{\sigma_0} = H_\infty^{s\sigma_0}$ as a K-representation and the estimates that were used to prove the results above about $A_s(\nu)$ for σ the trivial representation apply also to this case. One has

(2) $\nu \mapsto A_s(\nu)$ is weakly holomorphic from $((\underline{a}_0)^*_{\mathbb{C}})^+$ to the continuous endomorphisms of $H_\infty^{\sigma_0}$.

Let for $\lambda \in X^*$, $\delta_{s,\lambda}(\nu)(f) = \lambda(A_s(\nu)f(1))$. Then

(3) $\nu \mapsto \delta_{s,\lambda}(\nu)$ is weakly holomorphic from $((\underline{a}_0)^*_{\mathbb{C}})^+$ to $(H_\infty^{\sigma_0})'$ (continuous dual), for $s \in W$, $\lambda \in X^*$. Furthermore $\delta_{s,\lambda}(\nu)\big|_{H_K^{\sigma_\nu}} \in j(H_K^{\sigma_\nu})$.

We come to the first special case of the theorem. Let $\lambda_1, \ldots, \lambda_d$ be a basis of X^*.

<u>4.9</u>. <u>Lemma</u>. Let $\nu \in ((\underline{a}_0)^*_{\mathbb{C}})^+$ be generic; then $j(H_K^{\sigma_\nu}) = \bigoplus_{\substack{s \in W \\ 1 \le i \le d}} U((\bar{\underline{n}}_0)_{\mathbb{C}})(\delta_{s,\lambda_i}(\nu))$.

<u>Proof</u>. We first note that for general $\nu \in (\underline{a}_0)^*_{\mathbb{C}}$ that

(1) $H_K^{\sigma_\nu} = V_1 \supset V_2 \supset \ldots \supset V_d \supset V_{d+1} = (0)$ a chain of $(\underline{g}, K)$-submodules and $V_1/V_{i+1} \cong H_K^\nu$.

This follows from the exactness of the inducing functor.

Now we look at the case at hand.

(1) implies in particular that as a $U(\underline{n}_0)$-module $H_K^{\sigma_\nu}$ is free on $(\dim X)|W|$ generators. Since ν is generic one checks that

(2) $U(\underline{g}_{\mathbb{C}})(\sum_{i=1}^{d} \mathbb{C}\, \delta_{s,\lambda_i}(\nu)) = M_1 \supset M_2 \supset \ldots \supset M_d \supset M_{d+1} = 0$

with $M_i/M_{i+1} \cong M^{-s\nu}$.

The result now follows from the same counting argument as in 4.5.

4.10. Lemma. If $\nu \in ((\underline{a}_0)^*_{\mathbb{C}})^+$ and if $\lambda \in j(H_K^{\sigma_\nu})$ then λ extends to a continuous functional on $H_\infty^{\sigma_\nu}$.

Proof. We look at $H_K^{\sigma_\nu}$ as $H_K^{\sigma_0}$ with action depending on ν. Let us call the actions π_ν.

(1) The map $U((\underline{n}_0)_{\mathbb{C}}) \otimes E \otimes H_K^{\sigma_0}(\gamma_0) \xrightarrow{T_\nu} H_K^{\sigma_\nu}$ given by $T_\nu(n \otimes e \otimes \varphi) = \pi_\nu(n)\pi_\nu(e)\varphi$ defines a linear bijection.

This is proved by induction on $\dim X$ using the already known result for $\dim X = 1$ and 4.9(1).

Using (1) it is clear that

$$H_K^{\sigma_\nu} = T_\nu(U((\underline{n}_0)_{\mathbb{C}})_{(j)} \otimes E \otimes H_K^{\sigma_0}(\gamma_0)) \oplus \underline{n}_0^{j+1} H_K^{\sigma_\nu}.$$

Fix $w_1,\ldots,w_p$ a basis of $U((\underline{n}_0)_{\mathbb{C}})j) \otimes E \otimes H_K^{\sigma_0}(\gamma_0)$. Define

$$u_i(\nu)(f) = \delta_{ik} \quad \text{if} \quad f \in T_\nu(w_k) + \underline{n}_0^{j+1} H_K^{\sigma_\nu}.$$

Then, clearly, $u_1(\nu),\ldots,u_p(\nu)$ is a basis of $(H_K^{\sigma_\nu}/\underline{n}_0^{j+1}H_k^{\sigma_\nu})^*$.

(2) If $f \in H_K^{\sigma_\nu}$ then $\nu \mapsto u_i(\nu)(f)$ is a rational mapping of $((\underline{a}_0)^*_{\mathbb{C}})^+$ to $\mathbb{C}$.

This is proved in exactly the same way as Lemma 5.11 p. 249 in [3].

Let $\varepsilon > 0$ be such that $\nu + z\rho_0$ is generic for $0 < |z| < \varepsilon$. Let $\bar{n}_1,\ldots,\bar{n}_q$ be a basis of $U((\bar{\underline{n}}_0)_{\mathbb{C}})_{(j)}$. Set $\delta_{i,s,\lambda_j}(\nu+z\rho_0) = \bar{n}_i \cdot \delta_{s_i,\lambda_j}(\nu+z\rho)$. Then $\delta_{i,s,\lambda_k}(\nu+z\rho)$ is a basis of $(H_K^{\sigma_\nu+z\rho_0}/\underline{n}_0^{j+1}H_K^{\sigma_\nu+z\rho_0})^*$ for $0 < |z| < \varepsilon$. Hence

$$u_i(\nu+a\rho_0) = \sum a_{\ell,s,k;i}(z)\delta_{\ell,s,\lambda_k}(\nu+z\rho)$$

with $a_{\ell,s,k;i}(z)$ holomorphic on $0 < |z| < \varepsilon$ and having at worst a pole at $z = 0$. On the other hand

$$z \mapsto u_i(\nu + z\rho_0)$$

is holomorphic on $|z| < \varepsilon$.

Now one can argue using the maximal principle (as in [3], p. 250) to see that $u_i(\nu)$ extends to a continuous functional on $H_\infty^{\sigma_\nu}$.

4.11. Lemma. Let σ be a finite dimensional representation of P_0. Then there exist $\sigma_1, \ldots, \sigma_r$ unipotent representations of A_0, $\nu_i \in ((\underline{a}_0)^*_{\mathbb{C}})^+$, $i = 1, \ldots, r$ and F_i finite dimensional representations of G so that H_∞^σ is topologically isomorphic as a G-representation with a subquotient of $\oplus(H_\infty^{(\sigma_i)_{\nu_i}} \otimes F_i)$.

Let us show how this lemma implies Theorem 4.8.

$$j(H_K^{(\sigma_i)_{\nu_i}} \otimes F_i) = j(H_K^{(\sigma_i)_{\nu_i}}) \otimes F_i^*.$$

Clearly $(H_\infty^{(\sigma_i)_{\nu_i}})' \otimes F_i^* \subset (H_\infty^{(\sigma_i)_{\nu_i}} \otimes F_i)'$ (' denotes continuous dual). We have seen that $j(H_K^{(\sigma_i)_{\nu_i}}) \subset (H_\infty^{(\sigma_i)_{\nu_i}})'$. Thus $j(H_K^{(\sigma_i)_{\nu_i}} \otimes F_i) \subset (H^{(\sigma_i)_{\nu_i}} \otimes F_i)'_\infty$. By exactness of j we see that if $\lambda \in j(H_K^\sigma)$ then λ extends to an element, μ, of $\oplus j(H_K^{(\sigma_i)_{\nu_i}} \otimes F_i)$. But then $\mu \in \oplus(H^{(\sigma_i)_{\nu_i}} \otimes F_i)'_\infty$. By restriction we see that $\lambda \in H_\infty^\sigma$.

To prove the lemma we note that by exactness of the induction functor it is enough to show that σ is equivalent with a subquotient of $\oplus((\sigma_i)_{\nu_i} \otimes F_i)$ for appropriate σ_i, ν_i and F_i.

We prove the result by a series of reductions. As an M_0-module,

$$H = \bigoplus_{\substack{\lambda \in (\underline{a}_0)^*_{\mathbb{C}} \\ \xi \in \mathcal{E}({}^0M_0)}} H_\lambda(\xi)$$

where $H_\lambda(\xi)$ is the ξ-isotypic component of H_λ, the λth generalized weight space relative to the action of $\underline{a}_0$. Thus

$$\oplus U(\underline{p}_0) \cdot H_\lambda(\xi) \to H \to 0$$

is an exact sequence of $(\underline{p}_0, {}^0M_0)$-modules.

Hence we may assume that $H = U(\underline{p}_0)H_\lambda(\xi)$ for fixed $\lambda \in (\underline{a}_0)^*_{\mathbb{C}}$, $\xi \in \mathcal{E}({}^0M_0)$.

Now $H_\lambda(\xi)$ is $\underline{m}_0$-invariant. Thus $H = U(\underline{n}_0)H_\lambda(\xi)$. But $\underline{n}_0^k \cdot H_\lambda(\xi) = 0$ for some k. Hence H is a quotient of the $\underline{p}_0$-representation

(1) $(U((\underline{n}_0)_{\mathbb{C}})/\underline{n}_0^k U((\underline{n}_0)_{\mathbb{C}})) \otimes H_\lambda(\xi)$, with $\underline{n}_0$ acting trivially on $H_\lambda(\xi)$.

Let $\alpha_1,\ldots,\alpha_\ell$ be the simple roots in $\Phi^+ = \Phi^+(\underline{g}_{\mathbb{C}},\underline{h})$. Let $\Lambda \in \underline{h}^*$ be Φ^+ dominant integral. Let F^Λ be the irreducible finite dimensional $\underline{g}_{\mathbb{C}}$-representation with lowest weight $-\Lambda$. Let $e_i \in (\underline{g}_{\mathbb{C}})_{\alpha_i} - \{0\}$, $i = 1,\ldots,\ell$. Then as a $\underline{b} = \underline{n}^+ \oplus \underline{h}$ module

(2) $F^\Lambda = (U(\underline{n}^+)/\sum U(\underline{n}^+)e_i^{m_i}) \otimes \mathbb{C}_{-\Lambda}$ ($\mathbb{C}_{-\Lambda}$ the $\underline{b}$-module $\mathbb{C}$ with $\underline{n}^+ \cdot 1 = 0$ and $\underline{h}$ acting by Λ). Here $m_i = \dfrac{2(\Lambda,\alpha_i)}{(\alpha_i,\alpha_i)} + 1$. (Cf. Dixmier [2], Chapter 7.)

Let $\alpha_1,\ldots,\alpha_r$ be the α_i such that $(\underline{g}_{\mathbb{C}})_{\alpha_i} \subset (\underline{m}_0)_{\mathbb{C}}$. Assume that $(\Lambda,\alpha_i) = 0$. Then by taking the m_i large we see that F^Λ has as a quotient

$$U((\underline{n}_0)_{\mathbb{C}})/\underline{n}_0^k U((\underline{n}_0)_{\mathbb{C}}) \otimes \mathbb{C}_{-\Lambda}.$$

Thus taking H in the form (1) as we may we have

(3) H is a quotient of $F \otimes H_{\lambda'}(\xi')$ for λ', $(\underline{a}_0)^*_{\mathbb{C}}$ and $\xi \in \mathcal{E}({}^0M_0)$, and F a finite dimensional G-module all appropriately chosen.

Now there is F' a finite dimensional G-module such that $H_{\lambda'}(\xi')$ is a subquotient of $H_\nu(\xi_0) \otimes F'$ with $\nu \in (\underline{a}_0)^*_{\mathbb{C}}$, ξ_0 the trivial ${}^0M_0$-module. (See Lepowsky-Wallach [14].)

But $H_\nu(\xi_0) \equiv \sigma_\nu$ for σ a unipotent representation. So the result follows.

5. Some asymptotic expansions.

5.1. Let (π, H) be a Banach representation of G. Let H_∞ be the space of C^∞ vectors of H. Let H'_∞ be the space of continuous linear functionals on H_∞.

Lemma. If $\lambda \in H'_\infty$ then there exists $r \geq 0$ and a continuous functional, ν, on H_∞ so that

$$|\lambda(\pi(g)v)| \leq \|g\|^r \nu(v), \quad v \in H_\infty.$$

Proof. There exist $x_1, \ldots, x_k \in U(\underline{g}_{\mathbb{C}})$ such that

$$|\lambda(v)| \leq \sum_{i=1}^{k} \|\pi(x_i)v\|, \quad v \in H_\infty$$

by the definition of the topology on H_∞.

Now
$$\|\pi(x_i)\pi(g)v\| = \|\pi(g)\pi(\mathrm{Ad}(g)^{-1}x_i)v\|$$
$$\leq C\|g\|^p \|\pi(\mathrm{Ad}(g)^{-1}x_i)v\|.$$

$(|\pi(g)| \leq C\|g\|^p$; see 2.2.) Let $V = \operatorname{Span}_{g \in G} \mathrm{Ad}(g)x_i$. Then $\dim V < \infty$ and V is a representation of G. Put on V any Banach structure. Then $|\mathrm{Ad}(g)|_V| \leq C_1\|g\|^q$. Now $\mathrm{Ad}(g)^{-1}x_i = \sum g_{ji}z_j$ with z_j a basis of V. Clearly $|g_{ji}| \leq C_2\|g\|^q$. Hence

$$\|\pi(x_i)\pi(g)v\| \leq C \cdot C_2\|g\|^{p+q}\sum\|\pi(z_j)v\|.$$

Arguing the same way for each i gives the result.

5.2. Let V be an admissible $(\underline{g},K)$-module. Define $\tilde{V}$ to be the space of all $v^* \in V^*$ such that $K \cdot v^*$ spans a finite dimensional space. Then $\tilde{V}$ is a $(\underline{g},K)$-module which is easily seen to be admissible.

Lemma. If $V \in \mathcal{H}$ then $\tilde{V} \in \mathcal{H}$.

Proof. If $\tilde{V} \subset \tilde{V}_2 \subset \ldots \subset \tilde{V}_r \subset \tilde{V}_{r+1} \subset \ldots$ is a chain of submodules of $\tilde{V}$, set $V_i = \{v \in V | \tilde{V}_i(v) = 0\}$. Then $V_1 \supset V_2 \supset \ldots$ is a chain of $(\underline{g},K)$-submodules of V. Since V has finite length there exists

r_0 so that $V_{r_0} = V_j$ for all $j \geq r_0$. But $\tilde{V}_j = \{\tilde{v} \in \tilde{V} | \tilde{v}(V_j) = 0\}$. Hence $\tilde{V}_j = \tilde{V}_{r_0}$, $j \geq r_0$.

Lemma 2. Let V be an admissible $(\underline{g},K)$-module. Let W be a $(\underline{g},K)$-module and let $\beta : V \times W \to \mathbb{C}$ be a complex bilinear map such that

(1) $\beta(Xv,w) = -\beta(v,Xw)$, $X \in \underline{g}$, $v \in V$, $w \in W$

(2) $\beta(kv,w) = \beta(v,k^{-1}w)$, $k \in K$, $v \in V$, $w \in W$.

Then there is a $(\underline{g},K)$ homomorphism $T : W \to \tilde{V}$ such that $\beta(v,w) = T(w)(v)$. Furthermore, if β is non-degenerate ($\beta(v,W) = 0$ implies $v = 0$ and $\beta(V,w) = 0$ implies $w = 0$), then $T : W \to \tilde{V}$ is a surjective isomorphism of $(\underline{g},K)$-modules.

Proof. Define $T(w)(v) = \beta(v,w)$. This proves the first assertion. As for the second, if $T(w) = 0$ then $w = 0$ so T is injective. Let $W = \bigoplus_{\gamma \in \mathcal{E}(K)} W(\gamma^*)$; then $T : W(\gamma^*) \to \tilde{V}(\gamma^*)$. ($\gamma^*$ is the class of the dual representation of any element in the class γ). By assumption $T(W(\gamma^*))$ pairs non-degenerately with $V(\gamma)$. Thus $\dim W(\gamma^*) \geq \dim V(\gamma)^* = \dim \tilde{V}(\gamma)$. Since T is injective this implies $T(W(\gamma^*)) = V(\gamma^*)$. So T is surjective.

5.3. Let (π,H) be an admissible Bánach representation of G. Set $(H'_\infty)_K$ equal to the space of K-finite vectors in H'_∞.

Lemma. $(H'_\infty)_K \big|_{H_K} = \tilde{H}_K$.

Proof. The Hahn-Banach theorem implies that the pairing between H'_K and H_K is non-degenerate. Clearly $H'_K \subset (H'_\infty)_K$.

5.4. Lemma. Let (π,H) be an admissible finitely generated Banach representation. Then there exists $\Lambda \in \underline{a}_0^*$ such that if $\lambda \in (H'_\infty)_K$ then there exists a continuous semi-norm ν on H_∞ such that

$$|\lambda(\pi(k_1ak_2)v)| \leq a^\Lambda \nu(v)$$

for $v \in H_\infty$, $k_1,k_2 \in K$ and $a \in C\ell(A_0^+)$.

Proof. $(H'_\infty)_K = (H_K)^\sim$. Thus since H_K is finitely generated and admissible so is $\tilde{H}_K = (H'_\infty)_K$ by Lemma 5.2.1.

Thus in particular $(H'_\infty)_K$ is finitely generated as a $U(\underline{n}_0)$-module. So $(H'_\infty)_K = \sum_{i=1}^{p} U(\underline{n}_0)\lambda_i$.

Now $\|a\| \leq Ca^\mu$, $a \in C\ell(A_0^+)$ for some $\mu \in \underline{a}_0^*$. Thus

$$|\lambda_i(\pi(a)v)| \leq a^{r_i\mu}\nu_i(v) \quad \text{for} \quad a \in C\ell(A_0^+)$$

and $v \in H_\infty$. (r_i is the r for λ_i in Lemma 5.1.) Now if $\lambda \in (H'_\infty)_K$ then $\lambda = \sum n_i\lambda_i$ with $n_i \in U(\underline{n}_0)$. Thus

$$\begin{aligned}\lambda(\pi(a)v) &= \sum \lambda_i(\pi({}^tn_i)\pi(a)v) \\ &= \sum \lambda_i(\pi(a)\pi(\mathrm{Ad}\,(a)^{-1}\,{}^tn_i)v).\end{aligned}$$

Let $X_1,\ldots,X_r$ be a basis in $U(\underline{n}_0)$ of the $\mathrm{Ad}(A_0)$-span of the ${}^tn_i$. We may assume $\mathrm{Ad}(a)X_j = a^{\mu_j}X_j$. We note that $a^{\mu_j} \geq 1$ for $a \in C\ell(A_0^+)$ for such μ_j. Thus if ${}^tn_i = \sum c_jX_j$

$$\begin{aligned}|\lambda(\pi(a)v)| &= |\sum a^{-\mu_j}c_j\lambda_i(\pi(a)\pi(X_j)v)| \\ &\leq \sum |c_j|Ca^{r_i\mu}\nu_i(\pi(X_j)v).\end{aligned}$$

Now let $\mu_1,\ldots,\mu_r$ be a basis for the span of $K\cdot\lambda$. Then

$$|\lambda(\pi(k)\pi(a)v)| \leq \sum c'_j|\mu_j(\pi(a)v)|.$$

Thus we see that $(\Lambda = (\max r_i)\mu)$

$$|\lambda(\pi(k)\pi(a)v)| \leq a^\Lambda\xi(v),\ k \in K,\ a \in C\ell(A_0^+),\ v \in H_\infty,$$

with ξ a continuous semi-norm on H_∞. Finally set $\nu(v) = \sup_{k\in K}\xi(k\cdot v)$. This gives the result.

5.5. Let $V \in \mathcal{H}$. Let (P,A) be a standard p-pair. Let $P = {}^0MAN$ a standard Langlands decomposition. Then $\underline{n} \subset \underline{n}_0$ is normal.

Lemma. $V/\underline{n}^kV$ is an admissible finitely generated $(\underline{m}, M\cap K)$-module.

Proof. V is finitely generated as a $U(\underline{n}_0)$-module. So $V/\underline{n}^k V$ is finitely generated as an $U(\underline{n}_0/\underline{n}) = U(\underline{n} \cap \underline{m})$-module. That $V/\underline{n}^k V$ is an $(\underline{m}, M \cap K)$-module is clear. The result now follows from 3.13.

In particular $V/\underline{n}V = \bigoplus_{\xi \in \underline{a}^*_{\mathbb{C}}} (V/\underline{n}V)_\xi$ with $(V/\underline{n}V)_\xi = \{\bar{v} \in V/\underline{n}V \mid (H-\xi(H))^d \bar{v} = 0 \ \forall H \in \underline{a}$ for some $d\}$. Furthermore since $V/\underline{n}V$ is finitely generated and $\underline{a} \subset \underline{m}$ is central there is a fixed d such that $(H-\xi(H))^d (V/\underline{n}V)_\xi = 0$.

Set $E(P,V) = \{\xi \in \underline{a}^*_{\mathbb{C}} \mid (V/\underline{n}V)_\xi \neq 0\}$. Then since $(V/\underline{n}V)/(\underline{n}_0 \cap \underline{m})(V/\underline{n}V) = V/\underline{n}_0 V$ and $(n_0 \cap \underline{m})(V/\underline{n}V)_\xi \neq (V/\underline{n}V)_\xi$, $\xi \in E(P,V)$ (Casselman's theorem) we see

(1) $E(P,V) = \{\xi|_{\underline{a}} \mid \xi \in E(P_0,V)\}$.

5.6. We now assume that $G = {}^0G$. Let $\Delta(P_0,A_0) = \{\alpha_1,\dots,\alpha_\ell\}$. Let $H_i \in \underline{a}_0$ be defined by $\alpha_i(H_j) = \delta_{ij}$. If $V \in \mathcal{H}$ define $\Lambda_V \in \underline{a}^*_0$ by

$$\Lambda_V(H_i) = \max_{\xi \in E(P_0,\tilde{V})} - \xi(H_i).$$

(Since $G = {}^0G$, $\alpha_1,\dots,\alpha_\ell$ is a basis of $\underline{a}^*_0$.)

Theorem. Let (π,H) be an admissible, finitely generated Banach representation of G. Set $V = H_K$. Then there exists $d \geq 0$ such that if $\lambda \in (H'_\infty)_K$ $(\cong\tilde{V})$ then there exists a continuous semi-norm, ν_λ on H_∞ such that

$$|\lambda(\pi(a)v)| \leq (1 + \log\|a\|)^d a^\Lambda \ \nu_\lambda(v)$$

for $v \in H_\infty$ and $a \in C\ell(A_0^+)$, and $\Lambda = \Lambda_V$.

Note. The analogue of this result for $v \in H_K$ is a famous unpublished theorem of Harish-Chandra. See also Casselman, Miličic [1].

The proof of this result uses a standard technique in the theory of automorphic forms modified to apply to our case. We will be using this technique several more times in this article. We give the full details in this proof and we will be more sketchy later.

By Lemma 5.4 there exists $\Lambda \in \underline{a}_0^*$ so that if $\lambda \in (H_\infty')_K$ then

(1) $|\lambda(\pi(a)v)| \leq a^\Lambda \mu_\lambda(v)$, $a \in C\ell(A_0^+)$, $v \in H_\infty$ with μ_λ a continuous semi-norm on H_∞.

The idea of the proof is to show that if $\Lambda(H_i) > \Lambda_V(H_i)$ then we can replace Λ in (1) by $\Lambda - m\alpha_i$ with $m = \min\{\frac{1}{2}, \Lambda(H_i) - \Lambda_V(H_i)\}$ at the cost of possibly putting in a term $(1+\log\|a\|)^d$. Of course, it will also be necessary to change the semi-norm μ_λ.

We begin with α_1. Let $F = \Delta - \{\alpha_1\}$. Let (P,A) be the standard p-pair corresponding to F. Then $\underline{a} = \mathbb{R}H_1$. If $a \in C\ell(A_0^+)$ then $a = a' \cdot \exp tH_1$ with $a' = \exp(\sum_{i>1} x_i H_i)$, $x_i \in \mathbb{R}$, $x_i \geq 0$ and $t \geq 0$.

Let $q : \tilde{V} \to \tilde{V}/\underline{n}\tilde{V}$ be the canonical projection.

(2) If $q(\lambda) = 0$ then there exists ν_λ' a continuous semi-norm on H_∞ such that

$$|\lambda(\pi(a)v)| \leq a^{\Lambda-\alpha_1}\nu_\lambda'(v), \quad a \in C\ell(A_0^+), \; v \in H_\infty.$$

Indeed, let $X_1,\ldots,X_r$ be a basis of $\underline{n}$ so that $[H_i,X_i] = \beta_i(H)X_i$, $\beta_i \in \Phi(P,A)$. Then $\lambda = \sum X_i\lambda_i$ with $\lambda_i \in \tilde{V}$. So

$$\begin{aligned}
|\lambda(\pi(a)v)| &= |\textstyle\sum X_i\lambda_i(\pi(a)v)| \\
&= |-\textstyle\sum \lambda_i(\pi(X_i)\pi(a)v)| \\
&\leq \textstyle\sum|\lambda_i(\pi(a)\pi(\mathrm{Ad}(a)^{-1}X_i)v)| \\
&= \textstyle\sum a^{-\beta_i}|\lambda_i(\pi(a)\pi(X_i)v)| \\
&\leq \textstyle\sum a^{-\Lambda-\beta_i}\mu_{\lambda_i}(\pi(X_i)v).
\end{aligned}$$

Now $a^\lambda \geq a^{\alpha_1}$ for $\lambda \in \Phi(P,A)$ and $a \in C\ell(A_0^+)$. So take $\nu_\lambda'(v) = \sum \mu_{\lambda_i}(\pi(X_i)v)$.

Let $P_\xi : \tilde{V}/\underline{n}\tilde{V} \to (\tilde{V}/\underline{n}\tilde{V})_\xi$ be the projection onto the ξ-generalized eigenspace for H_1. Let $\lambda \in \tilde{V}$; then if $q(\lambda) \neq 0$ then $q(\lambda) = \sum P_\xi q(\lambda)$. Let $\lambda_\xi \in \tilde{V}$ be such that $q(\lambda_\xi) = P_\xi q(\lambda)$. Then $\lambda - \sum\lambda_\xi \in \underline{n}\tilde{V}$.

We estimate $\lambda_\xi(\pi(a)v)$ for $\xi \in E(P,A)$. Let $\mu = \lambda_\xi$. Let $\bar{\mu}_1,\ldots,\bar{\mu}_p$ be a basis of $U(\underline{a})q(\mu)$. We assume $\bar{\mu}_1 = q(\mu)$. Let $\mu_i \in \tilde{V}$ be such that $q(\mu_i) = \bar{\mu}_i$, $i \geq 2$. Now

$$H_1 \cdot \bar{\mu}_i = \sum b_{ij}\bar{\mu}_j$$

and $B = [b_{ij}]$ has the property that

$$(3) \quad (B-\xi I)^p = 0.$$

Also

$$(4) \quad \gamma_i = H_1 \cdot \mu_i - \sum b_{ij}\mu_j \in \underline{n}\tilde{V}.$$

Set $a_t = \exp tH_1$, $a' \in C\ell(A_0^+)$, $(a')^{\alpha_1} = 1$.

Also set, $$F(t,a';v) = \begin{bmatrix} \mu_1(\pi(a_t a')v) \\ \vdots \\ \mu_p(\pi(a_t a')v) \end{bmatrix}$$

and, $$G(t,a';v) = \begin{bmatrix} \gamma_1(\pi(a_t a')v) \\ \vdots \\ \gamma_p(\pi(a_t a')v) \end{bmatrix}.$$

Then

$$(5) \quad \frac{d}{dt} F(t,a';v) = -BF(t,a';v) - G(t,a';v).$$

Hence

$$(6) \quad F(t,a';v) = e^{tB}F(0,a';v) - e^{-tB}\int_0^t e^{tB}G(t,a';v)dt.$$

We now estimate the terms in this equation.

(7) $\|F(0;a';v)\| \leq (a')^{\Lambda}\alpha(v)$ with α a continuous semi-norm on H_∞.

This is clear from (1).

(8) $\|G(t,a';v)\| \leq e^{(\Lambda(H_1)-1)t}(a')^{\Lambda}\beta(v)$ with β a continuous semi-norm on H_∞.

This follows directly from (2).

(9) $\|e^{sB}\| \leq C(1+|s|)^{p-1}e^{s\mathrm{Re}\xi}$ for $s \in \mathbb{R}$, $p \leq d$, d fixed.

Indeed, $(B-\xi I)^p = 0$. d is fixed as observed in 5.5.

Using these estimates we have

$$\|F(t,a';v)\| \le C(1+t)^{p-1} e^{-t\mathrm{Re}\xi(H_1)} \cdot (a')^{\Lambda}\alpha(v)$$

$$+ C^2(1+t)^{p-1} e^{-t\mathrm{Re}\xi(H_1)} \int_0^t (1+\tau)^{p-1} e^{\tau(\mathrm{Re}\xi(H_1)+\Lambda(H_1)-1)} d\tau \cdot (a')^{\Lambda}\beta(v).$$

We observe that $(1+\tau)^{p-1} e^{-\varepsilon\tau} \le C_\varepsilon$ for each $\varepsilon > 0$, and all $\tau > 0$. Thus we get

$$\text{(10)}\quad \|F(t,a';v)\| \le C(1+t)^{p-1} e^{-t\mathrm{Re}\xi(H_1)} (a')^{\Lambda}\alpha(v)$$

$$+ C^2(1+t)^{p-1} e^{-t\mathrm{Re}\xi(H_1)} (a')^{\Lambda}\beta(v)$$

$$+ C_1(1+t)^{p-1} e^{t(\Lambda_1(H_1)-\frac{2}{3})} (a')^{\Lambda}\beta(v).$$

<u>Case I</u>. $\Lambda_1(H_1) - \frac{2}{3} \le -\mathrm{Re}\xi(H_1)$; then we have

$$\|F(t, a';v)\| \le (1+t)^{p-1} e^{-t\mathrm{Re}\xi(H_1)} (a')^{\Lambda}\gamma(v)$$

with γ a continuous semi-norm on H_∞.

<u>Case II</u>. $\Lambda(H_1) - \frac{2}{3} > -\mathrm{Re}\xi(H_1)$; then in (1) we may replace Λ by $\Lambda - \frac{1}{2}\alpha_1$. Iterating the argument leading to (10) we are eventually in Case I.

We do this argument for $\alpha_2,\ldots,\alpha_\ell$ and get after a finite number of steps:

$$|\lambda(\pi(\exp(\textstyle\sum t_i H_i))v)| \le (1+t_1)^{p_1}(1+t_2)^{p_2}\ldots(1+t_\ell)^{p_\ell} \cdot$$

$$e^{\Sigma t_i \Lambda_v(H_i)} \nu_\lambda(v)$$

for $t_i \ge 0$ and ν_λ a continuous semi-norm on H. We note that each $p_i \le d$ with d fixed. This is clearly the desired estimate.

<u>5.7</u>. We now use exactly the same technique as in the proof of Theorem 5.6 to derive some asymptotic expansions. By a <u>formal</u> <u>exponential</u> <u>polynomial</u> <u>series</u> we mean a formal sum

$$(1)\quad \sum_{i=1}^{p} e^{z_i t} \sum_{n=0}^{\infty} p_{i,n}(t)e^{-nt},$$

such that $p_{i,n}$ is a polynomial in t for each i,n.

If f is a function on $\mathbb{R}$ we say that f is asymptotic to a formal exponential series as $t \to +\infty$, denoted

$$f(t) \sim \sum_{i=1}^{p} e^{z_i t} \sum_{n=0}^{\infty} p_{i,n}(t)e^{-nt},$$

if when we rearrange the terms of the series in the form

$$\sum_{i=1}^{\infty} e^{\xi_i t} p_{\xi_i}(t)$$

with $\xi_i \in \{z_i - n \mid 1 \le i \le p,\ n \ge 0,\ n \in \mathbb{Z}\}$ and $\mathrm{Re}\xi_1 \ge \mathrm{Re}\xi_2 \ge \ldots$, and p_{ξ_i} is gotten by adding together the $p_{i,n}$ with $z_j - n = \xi_i$, then

$$(2)\quad |f(t) - \sum_{i=1}^{N} e^{\xi_i t} p_{\xi_i}(t)| \le Ce^{(\mathrm{Re}\xi_N - \varepsilon)t} \quad \text{for } t \ge 1 \text{ and some}$$

$\varepsilon > 0$. Here N is such that $\mathrm{Re}\xi_N > \mathrm{Re}\xi_{N+1}$.

This is a slight modification of Poincaré's original definition of asymptotic series.

Notice in particular if $\mathrm{Re}\xi_N > \mathrm{Re}\xi_{N+1}$ then

$$(3)\quad \lim_{t\to+\infty} e^{-t\mathrm{Re}\xi_N} |f(t) - \sum_{i=1}^{N} e^{\xi_i t} p_{\xi_i}(t)| = 0.$$

__Lemma__. If $\sum_{i=1}^{p} e^{z_i t} \sum_{n=0}^{\infty} p_{i,n}(t)e^{-nt}$ and $\sum_{i=1}^{q} e^{w_i t} \sum_{n=0}^{\infty} q_{i,n}(t)e^{-nt}$ are formal exponential polynomial series such that $z_i - z_j \notin \{1,2,\ldots\}$, $i \ne j$, $p_{1,0} \ne 0$, $i = 1,\ldots,p$ and $w_i - w_j \notin \{1,2,\ldots\}$, $i \ne j$, $q_{i,0} \ne 0$, $i = 1,\ldots,q$ then if $f(t)$ is asymptotic to both series as $t \to +\infty$ then $p = q$ and after relabeling $w_i = z_i$, $p_{i,n} = q_{i,n}$.

__Proof__. We use

(1) If $\lim_{t\to+\infty} \sum_{i=1}^{n} e^{u_i t} p_i(t) = 0$ with $u_i \in \mathbb{C}$, $\mathrm{Re}u_i = 0$, p_i polynomials in t and $u_i \ne u_j$ for $i \ne j$, then $p_i = 0$ for all

$i = 1,\ldots,n$.

The proof is an interesting exercise (cf. [25], Appendix.)

Let $\operatorname{Re} z_1 \geq \ldots \geq \operatorname{Re} z_r > \operatorname{Re} z_{r+1} \geq \ldots \geq \operatorname{Re} z_p$ and let $\operatorname{Re} w_1 \geq \ldots \geq \operatorname{Re} w_s > \operatorname{Re} w_{s+1} \geq \ldots \geq \operatorname{Re} w_q$. Then

$$|f(t) - \sum_{i=1}^{r} e^{z_i t} p_{i,0}(t)| \leq Ce^{(\operatorname{Re} z_1 - \varepsilon)t}$$

for some $\varepsilon > 0$ and $t \geq 1$,

$$|f(t) - \sum_{i=1}^{s} e^{w_i t} q_{i,0}(t)| \leq Ce^{(\operatorname{Re} w_1 - \varepsilon)t}$$

for some $\varepsilon > 0$, and $t \geq 1$.

Suppose that $\operatorname{Re} z_1 > \operatorname{Re} w_1$. Then

$$|f(t)| \leq C_1(1+t)^{d''} e^{\operatorname{Re} w_1 t}, \quad t \geq 1.$$

Hence

$$|\sum_{i=1}^{r} e^{z_i t} p_{i,0}(t)| \leq C_2 e^{(\operatorname{Re} z_1 - \eta)t}, \quad t \geq 1.$$

Thus

$$\lim_{t\to+\infty} |\sum_{i=1}^{r} e^{(z_i - z_1)t} p_{i,0}(t)| = 0.$$

So (1) implies $p_{i,0} = 0$, $i = 1,\ldots,r$. This is contrary to our assumptions. Hence $\operatorname{Re} z_1 \leq \operatorname{Re} w_1$. By symmetry $\operatorname{Re} w_1 = \operatorname{Re} z_1$. Hence

$$\lim_{t\to+\infty} |\sum e^{(z_i - z_1)t} p_{i,0}(t) - \sum e^{(w_i - z_1)t} q_{i,0}(t)| = 0.$$

Now (1) implies that $r = s$ and $w_i = z_i$, $i = 1,\ldots,r$ after rearrangement.

Let $\operatorname{Re}\xi_1 \geq \operatorname{Re}\xi_2 \geq \ldots$ be the $\{z_i - n \mid$ with $p_{i,n} \neq 0$ $i = 1,\ldots,p$, $n = 0,1,\ldots\}$ in decreasing order. Set $p_\xi(t) = p_{i,n}(t)$ if $\xi = z_i - n$ (i,n are unique). Similarly let $\operatorname{Re}\eta_1 \geq \operatorname{Re}\eta_2 \geq \ldots$ be the $\{w_j - n \mid$ with $q_{j,n} = 0$ $i = 1,\ldots,q$, $n = 0,1,\ldots\}$ in decreasing order and let $q_\eta(t)$ be defined as above. Let $\operatorname{Re}\xi_{N_j} > \operatorname{Re}\xi_{N_j+1}$, $\operatorname{Re}\eta_{M_j} >$

$\mathrm{Re}\eta_{M_j+1}$. At this point we have shown $N_1 = M_1$ and after relabeling $\xi_j = \eta_j$, $1 \le j \le N_1$ and $p_{\xi_i}(t) = q_{\eta_i}(t)$, $1 \le i \le N_1$. Suppose that we have shown that $N_1 = M_1, \ldots, N_j = M_j$ and after relabeling $\xi_i = \eta_i$, $i \le N_j$, $p_{\xi_i}(t) = q_{\eta_i}(t)$, $i \le N_j$. Then

$$|f(t) - \sum_{i=1}^{N_{j+1}} e^{\xi_i t} p_{\xi_i}(t)| \le Ce^{(\mathrm{Re}\xi_{N_j+1}-\varepsilon)t}, \quad t \ge 1$$

$$|f(t) - \sum_{i=1}^{M_{j+1}} e^{\eta_i t} q_{\eta_i}(t)| \le Ce^{(\mathrm{Re}\xi_{M_{j+1}}-\varepsilon)t}, \quad t \ge 1.$$

Replacing $f(t)$ by $f(t) - \sum_{i=1}^{N_j} e^{\xi_i t} p_{\xi_i}(t) = f(t) - \sum_{i=1}^{M_j} e^{\eta_i t} q_{\eta_i}(t)$ we can argue as in the first step to see that $M_{j+1} = N_{j+1}$ and after relabeling that $\xi_i = \eta_i$ for $i \le N_{j+1}$.

Now we have shown that

$$\{z_i - n | p_{i,n} \neq 0,\ i = 1,\ldots,p,\ n = 0,1,\ldots\} =$$
$$\{w_i - n | q_{i,n} \neq 0,\ i = 1,\ldots,q,\ n = 0,\ldots\},$$

for each i,n such that $p_{i,n} \neq 0$ there is i',n' so that $w_{i'} - n' = z_i - n$ and $q_{i',n'} = p_{i,n}$ and for each j,m with $q_{j,m} \neq 0$, there is j'',m'' so that $p_{j'',m''} = q_{j,m}$ and $z_{j''} - m'' = w_j - m$. In particular, $z_i = w_{i'} - 0'$ and $w_j = z_{j''} - 0''$ all i,j. Hence if $w_{i'} = z_\alpha - m$ we have $z_i = z_\alpha - n' - m$. But then $n' = m = 0$ and $\alpha = i$. Thus $z_i = w_{i'}$, $w_j = z_{j''}$. So after the above relabeling $p = q$ and $z_i = w_i$. But then if $p_{i,n} \neq 0$, $p_{i,n} = q_{i,n}$ and if $p_{i,n} = 0$ then $q_{i,n} = 0$. So the result follows.

5.8. Let (π,H) be an admissible finitely generated Banach representation of G. We assume as before that $G = {}^0G$. Let $\Delta(P_0,A_0) = \{\alpha_1,\ldots,\alpha_\ell\}$ as above and let H_i be as above. We look at $(P,A) = (P_F,A_F)$ with $F = \{\alpha_2,\ldots,\alpha_\ell\}$. Set $\widetilde{V} = (H_\infty')_K$ as above. Let $E(P,\widetilde{V}) = \{\xi_1,\ldots,\xi_r\}$. Let $z_1,\ldots,z_p$ be a subset of $\{-\xi_1(H_1),\ldots,$

$-\xi_r(H_1)\}$ such that

(i) Each $\xi_i(H_1)$ is of the form $z_j - m$ for some $m = 0,1,2,\ldots$.

(ii) If $z_i - z_j \in \{1,\ldots\}$ then $i = j$.

Theorem. Let $\lambda \in \tilde{V}$ $(=(H_\infty')_K)$ and $v \in H_\infty$.
Then there exist polynomials

$$t \mapsto P_{i,n}(t;\lambda,v),\ 1 \le i \le p,\ n = 0,1,\ldots$$

such that

(1) $\deg_t P_{i,n}(t;\lambda,v) \le d_{i,n}$ for some $d_{i,n}$ fixed independent of λ and v.

(2) The map $\mathbb{R} \times H_\infty \to \mathbb{C}$, $t,v \mapsto P_{i,n}(t;\lambda,v)$ is continuous and linear in v.

(3) $\lambda(\pi\ (\exp tH_1)v) \sim \sum_{i=1}^{p} e^{tz_i} \sum_{n=0}^{\infty} e^{-nt} P_{i,n}(t;\lambda,v)$ as $t \to +\infty$ for $v \in H_\infty$.

(4) $|P_{i,n}(t;\lambda,\pi(a')v)| \le (1+t)^d(1+\log\|a'\|)^p \cdot (a')^{\Lambda_V} \mu_{i,n}(v)$

for $a' \in (\mathrm{Ker}\ \alpha_1 \cap \mathrm{C\ell}(A_0^+))$. Here $\mu_{i,n}$ is a continuous semi-norm on H_∞.

This result is essentially Harish-Chandra's famous unpublished result on the "asymptotic expansions along the walls." (See Casselman-Miličić [1].) However, Harish-Chandra's result is for $v \in H_K$ and the expansion he proves to exist converges to the matrix entry.

Before we go into the proof of our theorem we point out to the reader that in the proof more precise information about the expansions will become apparent.

We now begin the proof. First of all let us consider the $(\underline{m}, K\cap M)$-modules $\underline{n}^k\tilde{V}/\underline{n}^{k+1}\tilde{V}$. Let E_k denote the set of weights of $\underline{a}$ on $\underline{n}^k\tilde{V}/\underline{n}^{k+1}\tilde{V}$.

(1) $E_{k+1} \subset \{\xi+\alpha \mid \alpha \in \Phi(P,A),\ \xi \in E_k\}$.

Indeed, we consider

$$\underline{n} \otimes (\underline{n}^k\tilde{V}/\underline{n}^{k+1}\tilde{V}) \xrightarrow{\gamma_k} (\underline{n}^{k+1}\tilde{V}/\underline{n}^{k+2}\tilde{V})$$

$$x \otimes (u+\underline{n}^{k+1}\tilde{V}) \overset{\gamma_k}{\longmapsto} (xu+\underline{n}^{k+2}\tilde{V}).$$

Then γ_k is a surjective $\underline{a}$-module homomorphism. This clearly proves (1).

Set $E = \bigcup_{k=1}^{\infty} E_k$. Then $E \subset \{\xi+\alpha \mid \xi \in E(P,\tilde{V}),\ \alpha$ a sum of elements of $\Phi(P,A)\} = \{\xi+n\alpha_1|_{\underline{a}} \mid \xi \in E(P,\tilde{V}),\ n = 0,1,2,\ldots\}$. So identifying $\nu \in \underline{a}^*_{\mathbb{C}}$ with $\nu(H_1) \in \mathbb{C}$ we have

(2) $-E = \{-\lambda \mid \lambda \in E\} \subset \{z_i-n \mid i = 1,\ldots,r,\ n = 0,1,2,\ldots$. $-E_k \subset \{z_i-n \mid i=1,\ldots,r,\ n \geq k\}$.

(3) Given $\xi \in \{z_i-n \mid i = 1,\ldots,r,\ n = 0,1,2,\ldots\}$, there exists $k_\xi \geq 0$ such that $\Lambda_V(H_1) - k_\xi < \mathrm{Re}\,\xi-1$. This is clear.

Now let $\{z_i-n \mid i = 1,\ldots,r,\ n = 0,1,2,\ldots\} = \{\xi_1,\xi_2,\ldots\}$ with $\mathrm{Re}\,\xi_1 \geq \mathrm{Re}\,\xi_2 \geq \ldots$. Let $\mathrm{Re}\,\xi_{N_i} > \mathrm{Re}\,\xi_{N_{i+1}}$ and let $k_i = k_\xi$, $\xi = \xi_{N_i}$.

We now proceed as in the proof of Theorem 5.6. Set $a_t = \exp tH_1$. Fix $0 \neq \lambda \in (H'_\infty)_K = \tilde{V}$. Let $q_k : \tilde{V} \to \tilde{V}/\underline{n}^k\tilde{V}$ be the natural projection.

(i) If $q_{k_i}(\lambda) = 0$ then there exists $\varepsilon > 0$ so that

$$|\lambda(\pi(a_t a')v)| \leq e^{(\mathrm{Re}\,\xi_{N_i}-\varepsilon)t}(1+\log\|a\|)^p (a')^{\Lambda_V} \nu'_{\lambda_{j_i}}(v)$$

for $t \geq 1$, $a' \in \mathrm{Ker}\,\alpha_1 \cap \mathrm{C}\ell(A_0^+)$; here $\nu'_{\lambda,i}$ is a continuous semi-norm on H_∞.

Indeed, $\lambda \in \underline{n}^k\tilde{V}$. Thus if x_i is a basis of $\underline{n}$ such that $[h,x_i] = \beta_i(h)x_i$, $\beta_i \in \Phi(P,A)$ for $h \in \underline{a}_0$ then $\lambda = \sum x_{i_1}\ldots x_{i_k}\lambda_{i_1\ldots i_k}$, $\lambda_{i_1\ldots i_k} \in \tilde{V}$. Thus if $a \in \mathrm{C}\ell(A_0^+)$

$$|\lambda(\pi(a)v)| = |\sum \lambda_{i_1\ldots i_k}(\pi(x_{i_1}\ldots x_{i_k})\pi(a)v)|$$

$$\leq \sum a^{-(\beta_{i_1}+\cdots+\beta_{i_k})} |\lambda(\pi(a)\pi(x_{i_1}\cdots x_{i_k})v)|$$

$$\leq (1+\log\|a\|)^p a^{-k\alpha_1} a^{\Lambda_V} \sum \nu_{\lambda_{i_1}\cdots i_k}(\pi(x_{i_1}\cdots x_{i_k})v).$$

This clearly implies (i).

Suppose now that $q_{k_i}(\lambda) \neq 0$. Let $\overline{\lambda} = q_{k_i}(\lambda)$. Let $\overline{\lambda}_1 = \overline{\lambda}, \overline{\lambda}_2, \ldots, \overline{\lambda}_q$, be a basis of $U(\underline{a}) \cdot \overline{\lambda} \subset \widetilde{V}/\underline{n}^k\widetilde{V}$ $(k=k_i)$. Then $H_1 \cdot \overline{\lambda}_i = \sum b_{ij}\overline{\lambda}_j$. Then $B = (b_{i_j})$ has eigenvalues contained in $\bigcup_{j\leq k} E_j$. Let $\lambda_1 = \lambda$, $\lambda_2, \ldots, \lambda_q \in \widetilde{V}$ be such that $q_k(\lambda_i) = \overline{\lambda}_i$. Then

$$H_i \cdot \lambda_i = \sum b_{ij}\lambda_j + \gamma_i$$

with $\gamma_i \in \underline{n}^k\widetilde{V}$.

Define

$$F(t,a';v) = \begin{bmatrix} \lambda_1(\pi(a_t a')v) \\ \vdots \\ \lambda_q(\pi(a_t a')v) \end{bmatrix}$$

$$G(t,a';v) = \begin{bmatrix} \gamma_1(\pi(a_t a')v) \\ \vdots \\ \gamma_q(\pi(a_t a')v) \end{bmatrix}.$$

Then as usual

$$\text{(ii)} \quad \frac{d}{dt} F(t,a';v) = -BF(t,a';v) - G(t,a';v).$$

So

$$\text{(ii')} \quad F(t,a';v) = e^{-tB}F(0,a';v) - e^{-tB}\int_0^\tau e^{\tau B}G(t,a';v)d\tau.$$

Let $\mu_1, \ldots, \mu_s$ be the eigenvalues of $-B$ on $\mathbb{C}^q$ arranged so that $\mathrm{Re}\mu_1 \geq \ldots \geq \mathrm{Re}\mu_s$. Now $s \geq N_i$ and we may assume that $\mu_i = \xi_i$, $i \leq N_i$. Let P_i be the projection onto the μ_i-generalized eigenspace of B on $\mathbb{C}^q$. Let $Q = \sum_{i>N_i} P_i$. Then it is clear that if $t \geq 1$,

$a' \in \operatorname{Ker} \alpha_1 \cap C\ell(A_0^+)$

(iii) $\|Q(e^{-tB}F(0,a';v) - e^{-tB}\int_0^\tau e^{\tau B}G(\tau,a';v)d\tau)\|$

$$\leq e^{(\operatorname{Re}\xi_{N_i}-\varepsilon)t}(1+\log\|a'\|)^p(a')^{\Lambda_V}\mu(v)$$

with μ a continuous semi-norm on H_∞.

We now consider $R = I - Q$. $R = \sum_{i\leq N_i} P_i$.

(iv) Let $i \leq N_i$.

$$|P_i(e^{tB}G(t,a'v))| \leq (1+t)^d e^{-t\operatorname{Re}\xi_i}(1+t)^p(a')^{\Lambda_V}\cdot e^{(\Lambda_V(H_1)-k)t}\delta(v)$$

where δ is a continuous semi-norm on H_∞; here $t \geq 1$, $a' \in \operatorname{Ker}\alpha_1 \cap C\ell(A_0^+)$.

(iv) implies that

$$\int_0^\infty Re^{\tau B}G(\tau,a';v)d\tau$$

converges absolutely.

Set $F^0(t,a';v) \;=\; Re^{tB}F(0,a';v) - Re^{-tB}\int_0^\infty e^{\tau B}G(\tau,a';v)d\tau.$

Then

$$RF(t,a';v) - F^0(t,a';v) \;=\; -Re^{-tB}\int_t^\infty e^{\tau B}G(\tau,a';v)d\tau.$$

Now

$$\|Re^{-tB}\int_t^\infty e^{\tau B}G(\tau,a';v)d\tau\|$$

$$\leq \sum_{i\leq N_i}(1+t)^d e^{+t\operatorname{Re}\xi_i}\left(\int_t^\infty (1+\tau)^{p+d} e^{-\tau\operatorname{Re}\xi_i} e^{(\Lambda_V(H_1)-k)\tau}d\tau\right).$$

$$(1+\log\|a'\|)^p(a')^{\Lambda_V}\delta(v)$$

$$\leq C\sum_{i\leq N_i} e^{t\operatorname{Re}\xi_i}\int_t^\infty e^{(-\operatorname{Re}\xi_i+\Lambda_V(H_1)-k+1)\tau}d\tau\;\cdot$$

$$(1+t)^d(1+\log\|a'\|)^p(a')^{\Lambda_V}\delta(v)$$

$$\leq C'(1+t)^d(1+\log\|a'\|)^p(a')^{\Lambda_V}\cdot\delta(v)\cdot e^{t(\Lambda_V(H_1)-k+1)}.$$

Set $\psi_i(t,a';v)$ equal to the first component of $F^0(t,a';v)$. Combining all of the material above we have

$$\text{(v)}\quad |\lambda(\pi(a_t a')v) - \psi_i(t,a';v)| \le e^{(\mathrm{Re}\,\xi_{N_i} - \varepsilon)t} (a')^{\Lambda_V} \cdot (1 + \log\|a'\|)^{p} \sigma_i(v)$$

for $t \ge 1$, $a' \in \mathrm{Ker}\,\alpha_1 \cap C\ell(A_0^+)$, where σ_i is a continuous semi-norm on H_∞.

Now $\psi_i(t,a';v) = \sum_{j=1}^{N_i} e^{t\xi_j} P_j^i(t,a';v)$ with $t \mapsto P_j^i(t,a';v)$ a polynomial in t. Furthermore the map

$$a' \mapsto P_j^i(t,a';v)$$

is C^∞ on $\mathrm{Ker}\,\alpha_1 \cap (A_0^+)$ and continuous on the closure. Finally

$$t,a',v \longmapsto P_j^i(t,a';v)$$

is continuous on $\mathbb{R} \times \mathrm{Ker}\,\alpha_1 \cap C\ell(A_0^+) \times H_\infty$.

$$\text{(vi)}\quad P_j^i(t,a',v) = P_j^{i-1}(t,a';v) \quad \text{if } j \le N_{i-1}.$$

This is clear from the estimate (v) and the proof of Lemma 5.7.

Define $P_{\xi_i}(t;v) = P_j^i(t,a';v)$ for $j \le N_i$. If $q_{k_i}(\lambda) \ne 0$ $P_{\xi_j}(t;v) = 0$ for $j \le N_i$. Then we have shown

$$\text{(vii)}\quad \lambda(\pi(a_t)v) \sim \sum_{i=1}^{\infty} e^{\xi_i t} P_{\xi_i}(t;v)$$

with the properties asserted in (1),(2),(3).

The uniqueness of the expansion gives

$$P_{\xi_i}(t;\pi(a')v) = P_j^i(t,a';v) \quad \text{for } j \le N_i,$$

which implies (4). This completes the proof of the theorem.

5.9. We now note that the material of this section applies in slightly greater generality. Let (π,V) be a Fréchet representation of G.

Let V_K denote the K-finite C^∞ vectors in V. If $V_K \in H$ we say that (π,V) is admissible and finitely generated. Assume that $V_K \in H$. We set $(V_\infty)_K$ equal to the K-finite continuous functionals on V_∞. Then $(V_\infty)'_K = (V_K)^\sim$ as above.

We say that (π,V) is of moderate growth if for each $\lambda \in (V_\infty)'_K$ there exists ν a continuous functional on V_∞ and $r \geq 0$ such that

$$|\lambda(\pi(g)v)| \leq \|g\|^r \nu(v).$$

Lemma 5.4 now follows without change. Let $\Lambda_V = \Lambda_{V_K}$. Then Theorem 5.6 and Theorem 5.8 also go through in this generality. We will have to use this mild generalization of Theorem 5.6 and Theorem 5.7 in the proof of the next result.

5.10. Theorem. Let (π,H) be an admissible finitely generated Banach representation of G. Then there exists (σ,W) a finite dimensional representation of P_0 and an injective, continuous G-module homomorphism on T of H_∞ into H_∞^σ.

Proof. We prove this theorem by induction on $\mathrm{rk}_{\mathbb{R}}\, {}^0G$. If $\mathrm{rk}_{\mathbb{R}}\, {}^0G = 0$ then we take T equal to the identity map. We assume that we have proved the result for all M with $\mathrm{rk}_{\mathbb{R}}\, {}^0M < \mathrm{rk}_{\mathbb{R}}\, {}^0G$. Let $\Delta = \Delta(P_0,A_0) = \{\alpha_1,\ldots,\alpha_\ell\}$ as above. The usual reductions allow us to assume $G = {}^0G$.

Set $F_i = \Delta - \{\alpha_i\}$ and let $(P_i,A_i) = (P_{F_i},A_{F_i})$. Then $\underline{a}_i = \mathbb{R}H_i$.

Let for $i = 1,\ldots,\ell$, z_j^i be as in Theorem 5.8 for α_i (rather than α_1) and $P_{j,n}^i$ be as in Theorem 5.8 for α_i.

Lemma. If $v \in C\ell(\underline{n}_i^{-k} H_\infty) \subset H_\infty$ then $P_{j,n}^i(t;\lambda,v) \equiv 0$ for $\mathrm{Re}\, z_j^i - n > \Lambda_V(H_i) - k$ and all $\lambda \in (H'_\infty)_K$.

Proof. Since $v \mapsto P_{j,n}^i(t,v)$ is continuous on H we may assume $v \in \underline{n}_i^{-k}H_\infty$. Let $x_1,\ldots,x_r$ be as in the proof of 5.8(i). Set $\psi_i = \theta x_i$. If $v \in \underline{n}_i^{-k} H_\infty$ then

$$v = \sum \psi_{i_1} \cdots \psi_{i_k} v_{i_1 \ldots i_k}, \quad v_{i_1 \ldots i_k} \in H_\infty.$$

Thus if $t \geq 0$ $|\lambda(\pi(\exp tH_i)v)|$

$$= |\sum \lambda(\pi(\exp tH_i)\psi_{i_1} \cdots \psi_{i_k} v_{i_1 \ldots i_k})|$$

$$\leq \sum e^{-kt} |(\psi_{i_k} \cdots \psi_{i_1} \lambda)(\pi(\exp tH_i) v_{i_1 \ldots i_k})|$$

$$\leq Ce^{(\Lambda_V(H_i)-k)t}(1+t)^p.$$

Now use the definition of the asymptotic expansions.

For each $1 \leq i \leq \ell$ and $k \geq 1$ consider the continuous $\bar{P}_i$-module $(\bar{P}_i = M_i \theta(N_i))$

$$H_\infty / C\ell(\underline{\bar{n}}_i^k H_\infty) = Z_{i,k}.$$

Here we give $Z_{i,k}$ the quotient topology. One checks that $Z_{i,k}$ is a Fréchet space and as an M_i-module it is admissible, finitely generated and of moderate growth.

Let $I(Z_{i,k})$ denote the representation of G smoothly induced from $Z_{i,k}$. That is: $I(Z_{i,k})$ is the space of all C^∞ functions $f : G \to Z_{i,k}$ such that

(a) $f(\bar{p}g) = \bar{p}f(g)$, $\bar{p} \in \bar{P}_i$.

We set $(\pi_{i,k}(g)f)(x) = f(xg)$. We put on $I(Z_{i,k})$ the C^∞ topology.

Define $T_{i,k} : H_\infty \to I(Z_{i,k})$ by $T_{i,k}(v)(g) = q_{i,k}(\pi(g)v)$; here $q_{i,k}\colon H_\infty \to H_\infty/C\ell(\underline{\bar{n}}_i^k H_\infty)$.

Let $W = \bigcap_{i,k} \operatorname{Ker} T_{i,k}$. Then W is a G-submodule of H_∞. So $W_K = W \cap H_K$ is a $(\underline{g},K)$-submodule of H_K.

<u>Lemma 2</u>. $W = (0)$.

<u>Proof</u>. By Lemma 1 we see that if $w \in W$, $\lambda \in (H_\infty')_K$ then

$$\lambda(\pi(\exp t H_i)v) \sim 0, \quad t \to +\infty$$

for all i.

Combining this with Theorem 5.8 and the technique in the proof of Theorem 5.6 one finds that if $\lambda \in (H_\infty)'_K$ and $w \in W_K$ then

$$g \mapsto \lambda(\pi(g)w) \in \mathcal{S}(G) \quad \text{(see 2.5)}.$$

But then if f is this function then $f \in \mathcal{S}(G)$ and f is right and left K-finite and is also $Z(\underline{g})$-finite.

We assert that if $f \in \mathcal{S}(G)$ and f is $Z(\underline{g})$-finite then $\phi(m) = \int_{N_0} \sigma(n)\overline{f}(nm)dn = 0$, $m \in M_0$, σ a finite dimensional representation of P_0.

Indeed it is clear that $\phi \in \mathcal{S}(M_0)$. But it is also easy to see that ϕ is $Z(\underline{m}_0)$-finite, hence $\dim U(\underline{a}_0)\cdot\phi < \infty$. This is clearly impossible unless $\phi = 0$. Now using the material in 2.16 we see that $\pi_\sigma(\overline{f}) = 0$ for all σ a finite dimensional representation of P_0. But f is a matrix coefficient of some π_σ (see the proof of Theorem 3.12). Thus $f = 0$.

But then $\lambda(w) = 0$ for $\lambda \in (H'_\infty)_K$ and $w \in W$. Hence $W = (0)$ as asserted.

Now for each i

$$((\mathrm{Ker}\ T_{i,k}))_K \supset (\mathrm{Ker}(T_{i,k+1}))_K \supset \cdots$$

is a chain of $(\underline{g},K)$-modules in H_K. Hence there is k_0 so that $\mathrm{Ker}(T_{i,k_0}))_K = (\mathrm{Ker}(T_{i,k}))_K$ for all i and all $k \ge k_0$.

We therefore see that $\mathrm{Ker}\ T_{i,k_0} = \mathrm{Ker}\ T_{i,k}$ for $k \ge k_0$ and all i.

Now let $Z_{i,k_0} \xrightarrow{S_i} {}_iH_\infty^{\sigma_i}$ be a continuous injective P_i-module homomorphism. Here (σ_i, W_i) is a finite dimensional representation of $\overline{P}_0 (= \theta(P_0) = M_0\theta(N_0))$ and ${}_iH_\infty^{\sigma_i}$ is the space of C^∞ functions on $\overline{P}_i$ with values in w_i such that

$$f(\overline{p}x) = \sigma_i(\overline{p})f(x),\ \overline{p} \in \overline{P}_0,\ x \in \overline{P}_i.$$

$\overline{P}_i$ acts by

$$(y\cdot f)(x) = f(xy).$$

We give ${}_\infty H_\infty^{\sigma_i}$ the C^∞ topology. Such an S_i exists by the inductive hypothesis. (Obviously we must extend the class of (π,H) to H being a Fréchet space and (π,H) admissible finitely generated and of moderate growth.)

Let $Q_i : I(Z_{i,k_0}) \to I({}_iH_\infty^{\sigma_i})$ be given by $Q_i f(g) = S_i(f(g))$. Set $\tilde{T} = \oplus Q_i \circ T_{i,k_0}$. Then $\tilde{T}$ is injective by our construction.

We note that $I({}_iH_\infty^{\sigma_i}) = \bar{H}_\infty^{\sigma_i}$ with $\bar{H}^{\sigma_i}$ defined for $\bar{P}^0$ in the same way as H^σ is defined for P_0. Finally, let $y \in K$ be such that $y^{-1}P_0 y = \bar{P}_0$. Set $\sigma = \oplus\, \sigma_i \circ \mathrm{Ad}(y)$. Let $\beta(f)(g) = f(yg)$ for $f \in \oplus\, H_\infty^{-\sigma_i} = H_\infty^{\oplus_{i=1}^{\ell}\sigma_i}$. Then $\beta : \oplus_i \bar{H}_\infty^{\sigma_i} \to H_\infty^\sigma$ is a G-isomorphism. Set $T = \beta \circ \tilde{T}$. This completes the proof.

<u>5.11. Corollary</u>. Assume that G is linear. Let (π,H) be an admissible finitely generated Banach representation of G. If $\lambda \in j(H_K)$ then λ extends to a continuous functional on H_∞.

<u>Proof</u>. Let σ be a finite dimensional representation of P_0 such that there exists $T : H_\infty \to H_\infty^\sigma$ a continuous injective G-homomorphism. Then $T : H_K \to H_K^\sigma$ is injective. Let $\lambda \in j(H_K)$; then there exists $\mu \in j(H_K^\sigma)$ such that $\mu|_{T(H_K)} \circ T = \lambda$ (3.9.(2)). But μ extends to an element of $(H_\infty^\sigma)'$ (Theorem 4.8). Thus $\mu|_{T(H_\infty)} \circ T$ is the desired extension.

<u>5.12</u>. In the next section we show how these results lead to an intrinsic topology on objects in $\mathcal{H}$.

6. Topologies on $(\underline{g},K)$-modules.

In this section we assume that G is linear.

6.1. Let for $V \in \mathcal{H}$, V' be the subset of V^* consisting of those λ such that if $v \in V$ then there exists $f_{\lambda,v}$ real analytic on G such that

(i) $x \cdot f_{\lambda,v}(k) = \lambda(k \cdot x \cdot v)$, $k \in K$, $x \in U(\underline{g})$ and

(ii) $|f_{\lambda,v}(g)| \le C_{\lambda;v} \|g\|^{d_\lambda}$, $g \in G$.

here $C_{\lambda,v}$ depends on λ and v, d_λ depends only on λ.

We note that Lemma 5.1 implies that

Lemma. If (π,H) is a realization of V as a Banach representation (that is, $H_K \cong V$), then $(H_\infty)'|_V \subset V'$.

6.2. If $V \in \mathcal{H}$ let $\mathcal{B}(V)$ denote the set of all K-invariant pre-Hilbert space structures, $\langle , \rangle$ on V such that if H is the completion of V relative to $\langle , \rangle$ then there is a representation of G on H, π, such that $H_K = V$ and the $K - C^\infty$ vectors of (π,H) is the same as the space of $G - C^\infty$ vectors of (π,H) (with the same topology). If $b \in \mathcal{B}(V)$ we use the notation, $b(v)$, for $b(v,v)^{1/2}$.

If $b \in \mathcal{B}(V)$ let (π_b,H_b) denote the corresponding representation of G.

Lemma. Let (σ,W) be a finite dimensional representation of P_0 so that there exists $T : V \to H_K^\sigma$ a $(\underline{g},K)$-module injective homomorphism. If $b \in \mathcal{B}(V)$ then T extends to a continuous G-module homomorphism of $(H_b)_\infty$ into $C\ell(T(V)) \subset H_\infty^\sigma$.

Proof. This is just a restatement of Corollary 5.11.

6.3. Lemma. Let (σ,W), T be as in Lemma 6.2. Let us identify $T(V)$ with V and set $\bar{V}_\sigma$ equal to $C\ell(T(V))$ in H_∞^σ. If $(\mu,\mathcal{W})$, S also satisfy the conditions of Lemma 6.2 then the identity map $I : V \to V$ extends to a continuous, surjective G-module isomorphism of $\bar{V}_\sigma$ onto $\bar{V}_\mu$.

Proof. This is clear from Lemma 6.2.

We set $\bar{V} = \bar{V}_\sigma$ for any $(\sigma,W),T$ as in Lemma 6.2. Then $\bar{V}$ is a smooth, Fréchet representation of G. Furthermore, if $b \in \mathcal{B}(V)$ then $(H_b)_\infty \subset \bar{V}$.

We will call $\bar{V}$ the maximal completion of V.

We now define a minimal completion of V. Let for $b \in \mathcal{B}(V)$, $x \in U(\underline{g})$,

$$\nu_{b,x}(v) = b(x\cdot v).$$

Let $\bar{\bar{V}}$ be the completion of V relative to all of the semi-norms $\nu_{b,x}$, $b \in \mathcal{B}(V)$, $x \in U(\underline{g})$. It is by no means obvious that V is a Fréchet space.

If W is a topological vector space we denote by W' the space of all continuous functionals on W. What is clear about $\bar{\bar{V}}$ is

6.4. Lemma. $\bar{\bar{V}}'|_V = V'$.

Proof. This is another statement of Lemma 6.1.

We note that if (σ,W) is a finite dimensional representation of P_0 then $(H_K^\sigma)^\sim = H_K^{\tilde{\sigma}\otimes e^{2\rho_0}}$ under the pairing

$$\int_G \langle f(k),g(k)\rangle dk$$

where $\langle,\rangle$ is the pairing of σ and $\tilde{\sigma}$ (the dual or contragredient module to (σ,W)).

Thus if $V \in \mathcal{H}$ and $\tilde{V} \hookrightarrow H_K^\sigma$ then $H_K^{\tilde{\sigma}\otimes e^{2\rho_0}} \to V \to 0$ is exact in $\mathcal{H}$. That is

(1) If $V \in \mathcal{H}$ there exists (σ,W) a finite dimensional representation of P_0 and $T : H_K^\sigma \to V$ a surjective $(\underline{g},K)$-module homomorphism.

6.5. Proposition. Let $V \in \mathcal{H}$ and let $(\sigma,W),T$ be as in 6.4 (1). We identify V with $H_K^\sigma/\text{Ker } T$. Then the identity map $I : V \to V$ extends to a continuous bijection of $\bar{\bar{V}}$ onto $H_\infty^\sigma/C\ell(\text{Ker } T)$. Here $C\ell(\text{Ker } T)$ is taken in H_∞^σ.

Proof. Let $Z = C\ell(\mathrm{Ker}\, T)$ in H^{σ}. We consider (π, H) the representation of G in the Hilbert space H^{σ}/Z. (Z is G-invariant since $\mathrm{Ker}\, T \subset H^{\sigma}_{K} \subset H^{\sigma}_{\omega}$.) Then $H_K = V$ under the obvious identification. Let $\hat{\sigma} = \tilde{\sigma} \otimes e^{2\rho_0}$. Then we have $\tilde{V} \overset{S}{\to} H^{\hat{\sigma}}_{K}$ relative to the pairing above between H^{σ} and $H^{\hat{\sigma}}$. Let $\hat{H} = C\ell(S(\tilde{V}))$ in $H^{\hat{\sigma}}$. Let $\hat{\pi}$ be the action of G on $\hat{H}$. Then we have a G-invariant pairing between H and $\hat{H}$. Clearly $\hat{H}_K = \tilde{V}$ relative to this pairing.

Let $\lambda \in V'$. Let $d \geq 0$ be so large that there exists $C > 0$ such that

(i) $\|\hat{\pi}(g)\| \leq C\|g\|^{d}$, $g \in G$.

(ii) $|f_{\lambda,v}(g)| \leq C_{\lambda,v}\|g\|^{d}$, $g \in G$, $v \in V$.

Let $v_1,\ldots,v_m$ generate V as a $U(\underline{g})$-module. Set for $w \in \hat{H}$,

$$\|w\|_1^2 = \sum_{i=1}^{m} \int_G |\langle v_i, \hat{\pi}(g)w\rangle|^2 \|g\|^{-2d-d_0}\, dg.$$

Here d_0 is chosen so that $\int_G \|g\|^{-d_0} dg < \infty$.

(1) If $g \in G$ and $w \in \hat{H}$ then

$$\|\hat{\pi}(g)w\|_1 \leq \|g\|^{d+d_0/2}\|w\|_1.$$

Indeed,

$$\|\hat{\pi}(g)w\|_1^2 = \sum_{i=1}^{m} \int_G |\langle v_i, \hat{\pi}(xg)w\rangle|^2 \|x\|^{-2d-d_0}\, dx$$

$$= \sum_{i=1}^{m} \int_G |\langle v_i, \hat{\pi}(x)w\rangle|^2 \|xg^{-1}\|^{-2d-d_0}\, dx.$$

Now $\|xg^{-1}\| \geq \|x\|\,\|g\|^{-1}$. This implies (1).

Let H_1 denote the Hilbert space completion of $\hat{H}$ relative to $\|\ldots\|_1$. Then (1) implies that $\hat{\pi}(g)$ extends to a bounded operator, $\pi(g)$, on H_1 for $g \in G$.

(2) (π_1, H_1) is a representation of G.

We must show that if $w \in \hat{H}$ then

$$\lim_{g\to 1}\|\hat{\pi}(g)w-w\|_1 = 0.$$

This follows from the uniform continuity on compacta of the matrix entries.

(3) $\hat{H}_\infty \subset (H_1)_\infty$ and the action of $(\underline{g},K)$ on $\hat{H}_\infty$ induced from $(H_1)_\infty$ is the original action.

Let $X \in \underline{g}$. Then

$$\langle v_i,\hat{\pi}(g \exp tX)v\rangle = \langle v_i,\hat{\pi}(g)v\rangle + t\langle v_i,\hat{\pi}(g)\hat{\pi}(X)v\rangle + t^2\langle v_i,\hat{\pi}(g \exp cX)\hat{\pi}(X^2)v\rangle/2$$

with c between 0 and t of course depending on g. Hence

$$\left\| \frac{\hat{\pi}(\exp tX)w-w}{t} - \hat{\pi}(X)w\right\|_1^2$$

$$= \sum_i \int_G \left|\left\langle v_i, \frac{\hat{\pi}(g \exp tX)w-\hat{\pi}(g)w}{t} - \hat{\pi}(g)\hat{\pi}(X)w\right\rangle\right|^2 \|g\|^{-2d-d_0}dg$$

$$= \sum_i \int_G |\langle v_i,t\hat{\pi}(g \exp cX)\hat{\pi}(X^2)w\rangle|^2\|g\|^{-2d-d_0}dg/2$$

c between 0 and t (depending on g),

$$\le C|t|^2\int_G \|g\|^{-d_0} dg \quad (C = m \max\|v_i\|^2 \cdot \|\hat{\pi}(X^2)w\|^2 \max_{|t|\le 1} \|\exp tX\|^{2d}\cdot(\text{const in (i)})^2/2).$$

Hence

$$\lim_{t\to 0}\left\| \frac{\hat{\pi}(\exp tX)w-w}{t} - \hat{\pi}(X)w\right\|_1 = 0.$$

So if $w \in \hat{H}_\infty$ then

$$g \mapsto \hat{\pi}(g)w$$

is of class C^1 with values in H_1.

By iterating this we have (3).

Using (3) we see that if $w \in (H_1)_\infty$ then $\dim Z(\underline{g})w < \infty$. Hence the K-finite vectors in $(H_1)_\infty$ are analytic vectors.

(4) If $w \in \hat{H}$ then

$$\|w\|_1 \leq C_1\|w\|.$$

Indeed

$$\|w\|_1^2 = \sum_i \int_G |\langle v_i, \hat{\pi}(g)w\rangle|^2 \|g\|^{-2d-d_0} dg$$

$$\leq \sum_i \int_G C_1\|v_i\|^2\|g\|^{2d}\|w\|^2\|g\|^{-2d-d_0} dg$$

$$\leq C_1 \cdot (\sum\|v_i\|^2)\|w\|^2 \int_G \|g\|^{-d_0} dg.$$

Hence(4).

Thus the imbedding $i : \hat{H} \to H_1$ given by $i(w) = w$ is continuous.

Since $\hat{H}_K$ is dense in $\hat{H}$ (relative to the topology of $\hat{H}$) and $\hat{H}$ is dense in H_1 (relative to $\|\ldots\|_1$) we see that $\hat{H}_K$ is dense in H_1. Hence

(5) $(H_1)_K = \hat{H}_K$.

The upshot is that (π_1, H_1) is a realization of $\tilde{V}$.

If $\gamma \in \mathcal{E}(K)$ define $\lambda_\gamma = \lambda \cdot E_\gamma$. Then $\lambda_\gamma \in \hat{H}(\gamma)$. We assert that

(6) $\sum\lambda_\gamma \in H_1$. Indeed,

$$f_{\lambda,v_i}(g) = \sum_{\gamma\in\mathcal{E}(K)} f_{\lambda_\gamma,v_i}(g)$$

is the K-Fourier series (relative to the left action of K) of f_{λ,v_i}.

$$\infty > \sum_i \int_G |f_{\lambda,v_i}(g)|^2\|g\|^{-2d_0-d_0} dg$$

$$= \sum_{i,\gamma} \int_G |\langle\pi(g)v_i, \lambda_\gamma\rangle|^2\|g\|^{-2d-d_0} dg$$

$$= \sum_{i,\gamma} \int_G |\langle v_i, \hat{\pi}(g)^{-1}\lambda_\gamma\rangle|^2\|g\|^{-2d-d_0} dg$$

$$= \sum_{i,\gamma} \int_G |\langle v_i, \hat{\pi}(g)\lambda_\gamma\rangle|^2\|g\|^{-2d-d_0} dg$$

$$= \sum\|\lambda_\gamma\|_1^2.$$

Thus $\sum \lambda_\gamma \in H_1$ as asserted.

Now by (4) the imbedding $i : \hat{H} \to H_1$ is continuous. On the other hand $(\hat{H}_K)^- = \hat{H}_\infty$ Thus the identity map of $\hat{H}_K \to \hat{H}_K$ extends to a continuous G-module imbedding

$$j : (H_1)_\infty \to \hat{H}_\infty.$$

But, $ij = I$ and $ji = I$.

In particular, the $K - C^\infty$ vectors of H_1 are the same as the $G - C^\infty$ vectors of H_1 (since this is true for $\hat{H}$). Let C_K be the Casimir operator of K corresponding to $B|_{\underline{k}\times\underline{k}}$. Then we have

(7) $\|(1+C_K)^{-d}v\| \le C\|v\|_1$ for some fixed $d \ge 0$, $C > 0$ and all $v \in \hat{H}_K = (H_1)_K$ (say). But then

(8) $$\sum_{\gamma\in E(K)} \|(1+C_K)^{-d}\lambda_\gamma\|^2 < \infty.$$

By (8) we see that λ extends to a continuous functional on H_∞.

We have thus proved that

(9) $H'_\infty|_V = V'$.

Let $b \in \mathcal{B}(V)$. Then by (9) $(H_b)'_\infty|_V \subset (H'_\infty)|_V$. Thus we see that we have a continuous imbedding

$$H_\infty \hookrightarrow (H_b)_\infty$$

by Lemma A.6.2.

But then the semi-norms $\nu_{b,\infty}$ in 6.3 are all continuous on H_∞. Hence $H_\infty \subset \bar{\bar{V}}$. Since $\bar{\bar{V}} \subset H_\infty$ is clear the result follows.

<u>6.6</u>. We therefore see in particular that $\bar{\bar{V}} \cong (H_b)_\infty$ for some $b \in \mathcal{B}(V)$ and one has a compatible G-module structure on $\bar{\bar{V}}$ (compatible with the $\underline{g}$ and K-actions) which is unique.

We will call $\bar{\bar{V}}$ the <u>minimal</u> completion of V.

We have

(1) If $b \in \mathcal{B}(V)$ then

$$\bar{\bar{V}} \subset (H_b)_\infty \subset \bar{V};$$

the inclusions are as smooth Fréchet G-modules.

<u>6.7.</u> <u>Lemma.</u> If σ is a finite dimensional representation of P_0 then $\overline{(H_K^\sigma)} = \overline{\overline{(H_K^\sigma)}} = H_\infty^\sigma$.

<u>Proof.</u> This follows from Lemma 6.3 and Proposition 6.5.

We now describe a larger class of cases of $V \in \mathcal{H}$ such that $\bar{\bar{V}} = \bar{V}$. Let (P,A) be a standard p-pair $P = MN$, $M = {}^0MA$ as usual. Let W be an $(\underline{m}, M \cap K)$-module that is admissible and finitely generated. Let (σ, U) be a realization of W in $\mathcal{B}(W)$.

We may form the induced representation $I_{P,\sigma}$ at three levels. The first is the C^∞ version $I_{P,\sigma}^\infty$, the space of $f : G \to U_\infty$, f of class C^∞ and $f(pg) = \sigma(p)f(g)$. We put on $I_{P,\sigma}^\infty$ the C^∞ topology. If we set $(x \cdot f)(g) = f(gx)$, $x, g \in G$ then $I_{P,\sigma}^\infty$ is a smooth Fréchet module. We can also take the completion $I_{P,\sigma}$ of $I_{P,\sigma}^\infty$ relative to

$$\langle f_1, f_2 \rangle = \int_K \langle f_1(k), f_2(k) \rangle dk.$$

We then get a representation of G on a Hilbert space. It is standard that $(I_{P,\sigma})_\infty = I_{P,\sigma}^\infty$ and (in light of the results in Section 3), $(I_{P,\sigma})_K \in \mathcal{H}$.

<u>6.8.</u> <u>Lemma.</u> If $\bar{W} = \bar{\bar{W}}$ (for $(\underline{m}, K \cap M)$) then $\overline{((I_{P,\sigma})_K)} = \overline{\overline{((I_{P,\sigma})_K)}} = I_{P,\sigma}^\infty$.

<u>Proof.</u> If $\bar{W} = \bar{\bar{W}}$ then there are finite dimensional representations μ_1, μ_2 of $P_0 \cap M$ such that if ${}_MH^{\mu_1}$, ${}_MH^{\mu_2}$ are the induced representations from $P_0 \cap M$ then there exist S, T, M continuous module homomorphisms such that

$${}_MH_\infty^{\mu_1} \xrightarrow{S} U_\infty$$

$$U_\infty \xrightarrow{T} {}_MH_\infty^{\mu_2}$$

with S surjective and T injective with closed image.

We extend μ_1, μ_2 to P_0 by setting them equal to 1 on N. Then using S and T we have

$$I^\infty_{P,\sigma} \xrightarrow{\hat{T}} H^{\mu_2}_\infty$$

$$H^{\mu_1}_\infty \xrightarrow{\hat{S}} I^\infty_{P,\sigma}$$

with $\hat{T}$ injective with closed image and $\hat{S}$ surjective with both continuous G-module homomorphisms. The lemma now follows from 6.3 and 6.5.

6.9. Lemma. If (π, H) is an irreducible square integrable representation of G (we assume $G = {}^0G$) then $(H_K)^- = (H_K)^= = H_\infty$.

Proof. Let σ be a finite dimensional representation of P_0 such that there exists $T : H_K \to H^\sigma_K$ an injective $(\underline{g}, K)$-module homomorphism. Then T extends to $T : H_\infty \to H^\sigma_\infty$.

Also $\overline{H}_K = C\ell(T(H_K))$ in H^σ_∞.

Let $H_1 = C\ell(T(H_K))$ in H^σ; then $\pi_\sigma(g)H_1 \subset H_1$, $g \in G$. Set $\pi_1(g) = \pi_\sigma(g)|_{H_1}$. Then (π_1, H_1) is a realization of H_K. Let $\lambda \in ((H_1)'_\infty)_K$, $\lambda \neq 0$. Then

$$|\lambda(\pi_1(a)v)| \le (1 + \log\|a\|)^d a^{\Lambda_V} \nu_\lambda(v)$$

for $a \in C\ell(A_0^+)$, ν_λ a continuous semi-norm on $(H_1)_\infty$.

Take $\overline{\nu}_\lambda(v) = \sup_K \nu_\lambda(\pi_1(k)v)$. Let $\lambda_1, \ldots, \lambda_r$ be a basis of the span of $K \cdot \lambda$ and let $C = \sup|k_{ij}|$ where $k \cdot \lambda_i = \sum k_{ji}\lambda_j$, $k \in K$. Put $\mu_\lambda(v) = C \max_{1 \le i \le r} \overline{\nu}_{\lambda_i}(v)$. Then

$$|\lambda(\pi_1(k_1 a k_2)v)| \le (1 + \log\|a\|)^d a^{\Lambda_V} \mu_\lambda(v)$$

for $k_1, k_2 \in K$, $a \in C\ell(A_0^+)$.

But then if we set $S(v)(g) = \lambda(\pi_1(g)v)$ for $v \in (H_1)_\infty$ then $S : (H_1)_\infty \to L^2(G)$ is a G-module homomorphism. Clearly $C\ell(\operatorname{Im} S)$

in $L^2(G)$ is equivalent with (π,H). We thus have $S:(H_1)_\infty \to H_\infty$ a continuous G-module homomorphism. Clearly $S\circ T = I$, $T\circ S = I$. So $\overline{H}_K = (H_1)_\infty = H_\infty$. But the conjugate dual of (π,H) is (π,H).

We need

<u>Scholium.</u> If $V \in \mathcal{H}$ and $b \in \mathcal{B}(V)$ is such that $(H_b)_\infty = \overline{V}$ and if $H_b^\vee$ is the conjugate dual to H_b then $\overline{\overline{(H_b^\vee)}}_K = (H_b^\vee)_\infty$.

<u>Proof</u>. $(H_b)_\infty$ is a subrepresentation of H_∞^σ some σ a finite dimensional representation of P_0. So $(H_b^\vee)_\infty$ is a quotient of H_∞^σ for an appropriate $\sigma^\vee$.

In light of the Scholium the lemma follows.

6.10. <u>Proposition</u>. If $V \in \mathcal{H}$ is a tempered $(\underline{g},K)$-module (i.e. $(\Lambda_V+\rho)(H_i) \leq 0 \quad i = 1,\dots,\ell)$ then $\overline{V} = \overline{\overline{V}}$.

<u>Proof</u>. We prove the result by induction on the length of V. If V is irreducible then there is (P,A) a standard p-pair, $\sigma \in \mathcal{E}_2({}^0M)$ and $\nu \in \underline{a}^*$ such that V is a summand of $(I_{P,\sigma_{i\nu+\rho_P}})_K$ (Trombi [19], Langlands [12]). $(\sigma_{i\nu+\rho_P}(ma) = a^{i\nu+\rho_P}\sigma(m) \quad a \in A,\ m \in {}^0M.)$ Thus in this case the result follows from 6.8 and 6.9.

Suppose that the result has been proved for all tempered V of length $\leq r$, $r \geq 1$ and that V is tempered of length $= r+1$. Let $b \in \mathcal{B}(V)$ and let $\pi = \pi_b$, $H = H_b$. Let $W \subset V$ be a non-zero irreducible $(\underline{g},K)$- submodule. Set $H_1 = C\ell(W) \subset H$. Then H_1 is $\pi(g)$ invariant for $g \in G$. By the inductive hypothesis $\overline{(H_1)}_K = \overline{\overline{(H_1)}}_K = (H_1)_\infty$ and $\overline{(H/H_1)}_K = \overline{\overline{(H/H_1)}}_K = (H/H_1)_\infty$. Let $\lambda \in V'$. Then $\lambda|_W \in W'$.

Thus $\lambda|_W$ extends to an element μ of $(H_1)_\infty'$. But as a Fréchet space $(H_1)_\infty$ is a summand of H_∞ (the $K-C^\infty$ vectors equal the $G-C^\infty$ vectors). Let ξ be an extension of μ to an element of H_∞'. Then $\xi|_V \in V'$. Consider $\gamma = \lambda - \xi|_V$; then $\gamma \in V'$ and $\gamma|_W = 0$. Thus $\gamma \in (V/W)'$, hence γ extends to an element of $(H/H_1)_\infty'$. Thus λ extends to an element of H_∞'. This implies $H_\infty'|_V = V'$. Now applying Lemma A.6.2 we see $H_\infty = \overline{V} = \overline{\overline{V}}$.

Note that the above argument actually proves

6.11. Lemma. If $V \in \mathcal{H}$ and if every irreducible subquotient, W, of V satisfies $\overline{W} = \overline{\overline{W}}$ then $\overline{V} = \overline{\overline{V}}$.

We also note the following implications of $\overline{V}$ equalling $\overline{\overline{V}}$.

6.12. Proposition. If $V \in \mathcal{H}$ and $\overline{V} = \overline{\overline{V}}$ and if (π, H) is a Banach representation of G such that $H_K = V$ then $H_\infty = \overline{V} = \overline{\overline{V}}$.

Proof. Let σ be a finite dimensional representation of P_0 such that there exists $T : V \to H_K^\sigma$ an injective $(\underline{g}, K)$-module homomorphism. Then $C\ell(T(V))$ in H_∞^σ is $\overline{V}$, hence it is also $\overline{\overline{V}}$ by assumption. Now Corollary 5.11 implies $T : V \to H_K^\sigma$ extends to a continuous homomorphism $\overline{T} : H_\infty \to H_\infty^\sigma$. Clearly $\overline{T}(H_\infty) \subset C\ell(T(V))$. Thus we have $\overline{T} : H_\infty \to \overline{V} = \overline{\overline{V}}$. Hence ${}^t\overline{T}(\overline{\overline{V}}') \subset H_\infty'$. But ${}^t\overline{T}(\overline{\overline{V}}')|_V = V'$. So $H_\infty'|_V \supset V'$. But $H_\infty'|_V \subset V'$ (Lemma 6.1). So ${}^t\overline{T}(\overline{\overline{V}}')|_V = H_\infty'|_V$. Since V is dense in H_∞ we see that ${}^t\overline{T}(\overline{\overline{V}}') = H_\infty'$. Thus ${}^tT$ is bijective. The result now follows from Corollary A.6.4.

6.13. Theorem. Let $V \in \mathcal{H}$ and assume that (π, H) is a Banach representation of G such that $H_K = V$ and $H_\infty'|_V = V'$. Let $W \subset V$ be a finite dimensional subspace of V such that $\text{span}\{\pi(U(\underline{g}))\pi(K)\cdot W\} = V$. Then $\pi(\mathcal{S}(G))W = H_\infty$. (Here the important thing to note is that we are not taking the closure of $\pi(\mathcal{S}(G))W$.)

Proof. We topologize $\mathcal{S}(G) \otimes W$ as the product of $\dim W$ copies of $\mathcal{S}(G)$. The map $T : \mathcal{S}(G) \otimes W \to H_\infty$ is continuous.

(1) If $v \in V$ then there exists $f \in \mathcal{S}(G)$ such that $\pi(f)v = v$.

Consider for $\gamma \in \mathcal{E}(K)$, $\pi(\alpha_\gamma \underset{K}{*} \mathcal{S}(G) \underset{K}{*} \alpha_{\gamma^*})$. Let f_j be a δ-sequence in $\mathcal{S}(G)$. Then $\lim\limits_{j\to\infty} \pi(\alpha_\gamma \underset{K}{*} f_j \underset{K}{*} \alpha_{\gamma^*}) = E_\gamma$. Hence $\pi(\alpha_\gamma \underset{K}{*} \mathcal{S}(G) \underset{K}{*} \alpha_{\gamma^*})$ contains E_γ in its closure. But $\dim \pi(\alpha_\gamma \underset{K}{*} \mathcal{S}(G) \underset{K}{*} E_{\gamma^*}) < \infty$. Thus $E_\gamma \in \pi(\alpha_\gamma \underset{K}{*} \mathcal{S}(G) \underset{K}{*} \alpha_{\gamma^*})$. Fix $f_\gamma \in \mathcal{S}(G)$ such that $\pi(f_\gamma) = E_\gamma$. Then if $\sum_{\gamma \in F} E_\gamma v = v$, F a finite set, $\sum_\gamma \pi(f_\gamma)v = v$. This proves (1).

(2) $\pi(\mathcal{S}(G))W \supset V$.

Indeed $\pi(S(G))W \supset W$ by (1). Clearly $\pi(L(k)f) = \pi(k)\pi(f)$, $\pi(L(X)f) = \pi(X)\pi(f)$, $k \in K$, $X \in U(\underline{g})$. Thus $\pi(S(G))W \supset V$ by the hypothesis on W.

But then ${}^tT : H'_\infty \to (S(G) \otimes W)'$ is injective. To prove the theorem we must only prove that ${}^tT(H'_\infty)$ is weakly closed in $(S(G) \otimes W)'$ (see A.6.3.)

Let λ_α be a net in H'_∞ and assume that ${}^tT(\lambda_\alpha) \to \mu$ in $(S(G) \otimes W)'$ weakly.

Then if $u \in \operatorname{Ker} T$, $\mu(u) = \lim_\alpha \lambda_\alpha(T(u)) = 0$. Thus $\mu(\operatorname{Ker} T) = 0$. So μ is defined on $\pi(S(G)W)$. In particular μ defines an element of V^*. We must show that $\mu|_V \in V'$. Let $v \in V$ and let $u \in S(G) \otimes W$ be such that $T(u) = v$. Define $f_{\mu,v}(g) = \mu(L(g)u)$. Then $g \mapsto \mu(L(g)u)$ is C^∞. $f_{\mu,v}$ satisfies 6.1 (i). We must show that it satisfies 6.1 (ii). Now $(S(G) \otimes W)' = S(G)' \otimes W^*$. Thus we must show that if $\gamma \in S(G)'$ then

$$(*) \qquad |\gamma(L(g)f)| \le C_{\gamma,f}\|g\|^d \nu(f)$$

ν a continuous semi-norm on $S(G)$. But $\gamma(f) \le \nu(f)$, ν a continuous semi-norm on $S(G)$. If ν is a continuous semi-norm on $S(G)$ then

$$\nu \le \sum \nu_{g_i,d_i}$$

for some $g_i \in U(\underline{g})$, $d_i \ge 0$. Each ν_{g_i,d_i} satisfies

$$\nu_{g_i,d_i}(L(g)f) \le C_{i,d_i}\|g\|^{\mu_i}\nu_{g_i,d_i}(f).$$

Thus γ satisfies (*). Hence $f_{\mu,v}$ satisfies 6.1(ii). But then $\mu|_V \in V'$. So $\mu|_V$ extends to an element, $\bar{\mu}$, of H'_∞. Clearly $\bar{\mu}|_{\pi(S(G))W} = \mu$. Hence ${}^tT(H'_\infty)$ is weakly closed and the result follows from A.6.3.

<u>6.14. Corollary</u>. If (π,H) is an admissible irreducible representation of G on a Banach space such that $H'_\infty|_V = V'$ then H_∞ is algebraically

irreducible as a representation of $\mathcal{S}(G)$.

Proof. Let $v \in H_\infty$, $v \neq 0$. Then there is $f \in \mathcal{S}(G)$ such that $\pi(f)v \neq 0$. Thus there is $\gamma \in \mathcal{E}(K)$ such that if $f_\gamma = \alpha_\gamma *_K f$ then $\pi(f_\gamma)v \neq 0$. Clearly $z = \pi(f_\gamma)v = E_\gamma \pi(f)v \in V(\gamma) - \{0\}$. Since (π,H) is irreducible $V^\bullet$ is irreducible as a $(\underline{g},K)$-module. Hence span $\pi(U(\underline{g}))\pi(K)z = V$. Thus by Theorem 6.13, $H_\infty = \pi(\mathcal{S}(G))z \subset \pi(\mathcal{S}(G))v \subset H_\infty$. Hence the result.

6.15. We note that 6.14 (in light of 6.10) is an analogue for reductive groups of a theorem of Howe for nilpotent groups. A version of 6.14 was suggested to us by J. Bernstein as a possible interpretation of the space V', $V \in \mathcal{H}$.

6.16. This is as far as one can go without developing more theory. One can actually prove that if $V \in \mathcal{H}$ then $\overline{V} = \overline{\overline{V}}$. This is a relatively deep theorem whose proof uses versions of the above results and Casselman's theory of the "fattened principal series." The proof of the above mentioned result in its full generality is due to Casselman and should appear in some form in the not too distant future.

Appendix to Section 6. A Lemma on sequence spaces

A.6.1. Let V be a vector space over $\mathbb{C}$ that has a gradation $V = \bigoplus_{j=1}^{\infty} V(j)$ with $\dim V(j) < \infty$ $j = 1,2,\ldots$. Let $\{a_j\}_{j=1}^{\infty}$ be a sequence of positive real numbers such that $\sum \dim V(j) a_j^{-p} < \infty$ for some $p > 0$.

Let $\langle,\rangle$ and $(,)$ be pre-Hilbert space structures on V so that $V(j)$ and $V(k)$ are orthogonal for $j \neq k$.

If $v \in V$ let $v_j \in V(j)$ be its components relative to the above direct sum decompositions.

If $v \in V$ set $\|v\|_d^2 = \sum_j a_j^d \langle v_j, v_j\rangle$ and set $|v|_d^2 = \sum_j a_j^d (v_j, v_j)$.

Let X denote the completion of V relative to the norms $\|\ldots\|_d$ and let Y denote the completion of V relative to the norms $|\ldots|_d$.

If U is a topological vector space let U' denote the space of continuous functionals on U.

A.6.2. Lemma. If $X'|_V \subset Y'|_V$ then the identity map $I : V \to V$ extends to a continuous map of Y into X.

Proof. Let $(v,w) = \langle C_j v, w\rangle$ for $v,w \in V(j)$. Then $C_j : V(j) \to V(j)$ is self adjoint relative to $\langle,\rangle|_{V(j)}$ and invertible. We may thus find a basis orthonormal relative to $\langle,\rangle$, u_i, of V with $u_i \in V(j)$ for some j and if $u_i \in V(j)$ then $C_j u_i = \lambda_{ij} u_i, \lambda_{ij} > 0$ $\lambda_{ij} \in \mathbb{R}$.

By setting $b_i = a_j$ if $u_i \in V(j)$ we see that we may look upon X as the set of all sequences $\{X_j\}$ with $\sum b_j^d |X_j|^2 < \infty$ for all $d \geq 0$. We may look at Y as the set of all sequences $\{X_j\}$ with $\sum b_j^d \lambda_j |X_j|^2 < \infty$ for all $d \geq 0$.

The assertion of the lemma is now equivalent to

(1) $\lambda_j \geq C b_j^k$ for some $C > 0$, and some $k \in \mathbb{R}$, all $j = 1,2,\ldots$.

Indeed if (1) is true then if $\{X_j\} \in Y$ then $\sum \lambda_j b_j^d |X_j|^2 < \infty$ all d implies $\sum b_j^{d+k} |X_j|^2 < \infty$ all d hence $\{X_j\} \in X$. If we set $\|\{X_j\}\|_d^2 = \sum b_j^d |X_j|^2$ then $\|\{X_j\}\|_d^2 \leq C \sum b_j^{d-k} \lambda_j |X_j|^2$. So the lemma follows if (1) is proved.

We assume that (1) is false. Then for each $n = 1,2,\ldots,$ $k = 1,2,\ldots$ there is $j(n,k)$ such that

$$\lambda_{j(n,k)} \le \frac{1}{n} b_{j(n,k)}^{-k}.$$

Set $j(k) = j(1,k)$. Then $\lambda_{j(k)} \le b_{j(k)}^{-k}$. Since $\lambda_j > 0$ all j we may take $j(k) \neq j(r)$ if $k \neq r$.

Now the hypothesis says that if $\{c_j\}$ satisfies the condition that $|c_j| \le Cb_j^r$ then $\sum c_j X_j$ converges for all $\{X_j\} \in Y$. But if we set $X_j = \begin{cases} 1 & \text{if } j = j(k) \\ 0 & \text{if } j \neq j(k) \end{cases}$ then $\sum \lambda_j b_j^d |X_j|^2 < \infty$ for all d. Hence $\{X_j\} \in Y$. Take $c_j = b_j$; then $\infty > \sum c_j X_j = \sum b_{j(k)}$. This clearly contradicts the assumption that $\sum b_j^{-p} < \infty$.

<u>A.6.3</u>. We also record here a general theorem, due to Bánach, about surjections of Fréchet spaces. For a proof see Treves [18], Theorem 37.2, p. 382.

Let V and W be Fréchet spaces. Let $u : V \to W$ be continuous and let ${}^t u : W' \to V'$ be given by ${}^t u(\lambda) = \lambda \circ u$, $u \in W'$.

<u>Theorem</u>. u is surjective if ${}^t u$ is injective and ${}^t u(W')$ is weakly closed in V'.

This result has as its immediate corollary:

<u>A.6.4</u>. <u>Corollary</u>. u is bijective if and only if ${}^t u$ is bijective.

<u>Proof</u>. If ${}^t u$ is surjective then $V'|_{\text{Ker } u} = 0$. This implies that $\text{Ker } u = 0$. ${}^t u(W') = V'$ hence ${}^t u(W')$ is certainly weakly closed in V'. Hence u is surjective by Theorem A.6.3.

7. More asymptotic expansions

7.1. We now return to the general situation of G a real reductive Lie group in the Harish-Chandra class. We retain the notation of the previous sections. We assume $G = {}^0G$.

Let (P,A) be a standard parabolic subgroup of V such that $\dim A = 1$. Let $H \in \underline{a}$ be such that the smallest eigenvalue of $\operatorname{ad} H$ on $\underline{n}$ $(P = MN)$ is 1. Let $\bar{\underline{n}} = \theta(\underline{n})$ as usual.

Let $V \in \mathcal{H}$. Let (π,H) be a representation of G on a Banach space such that $H_K = V$. Let $V_1 = \sum_{g\in G} \pi(g)V$. Then V_1 is a G-module and a $\underline{g}$-module with the obvious compatibility

(1) $g \cdot X \cdot v = (\operatorname{Ad}(g)X)g \cdot v$, $X \in \underline{g}$, $g \in G$ and $v \in V$.

Let $\lambda \in V_1^*$ (algebraic dual, there is no topology on V_1). Then λ is said to be of <u>moderate growth in the direction P</u> if the following two conditions are satisfied:

(2) If $v \in V$ then $g \mapsto \lambda(g\cdot v)$ is C^∞. Also $\frac{d}{dt}\lambda(g \exp tX\cdot v)\big|_{t=0} = \lambda(gXv)$, $g \in G$, $X \in \underline{g}$, $v \in V$.

(3) There exists $\mu \in \mathbb{R}$ such that if $\bar{n} \in U(\bar{\underline{n}})$ and $v \in V$ then

$$|\bar{\underline{n}}\cdot\lambda(\exp tH\cdot v)| \le C_{v,\bar{n}}e^{t\mu},\ t \ge 1.$$

Notice that if $\lambda \in V_1^*$ satisfies (2) then $g\cdot\lambda$ satisfies (2) for all $g \in U(\underline{g})$.

7.2. Let $V \in \mathcal{H}$; then

$$V/\bar{\underline{n}}^kV$$

is an admissible finitely generated $(\underline{m}, M\cap K)$-module. Thus in particular $V/\bar{\underline{n}}^kV = \bigoplus_\xi (V/\bar{\underline{n}}^kV)_\xi$ under the action of H. Here $(V/\bar{\underline{n}}^kV)_\xi = \{u \in V/\bar{\underline{n}}^kV \mid (H-\xi)^du = 0$ for some $d\}$. Let $E_k(\bar{P},V) = \{\xi \in \mathbb{C} \mid (V/\bar{\underline{n}}^kV)_\xi \ne 0\}$. Then $E_k(\bar{P},V)$ is a finite subset. Furthermore if $\xi \in E_k(\bar{P},V)$ then there is a <u>fixed</u> $d = d_{\xi,k}$ so that $(H-\xi)^d\big|_{(V/\bar{\underline{n}}^kV)_\xi} = 0$.

One finds just as in 5.8 that

(1) $E_{k+1}(\bar{P},V) - E_k(\bar{P},V) \subset \{\xi-\alpha(H) \mid \xi \in E_k(\bar{P},V),\ \alpha \in \Phi(P,A)\}$.

Let $z_1,\ldots,z_d \in E_1(\bar{P},V)$ be such that if $z_i - z_j$ is a non-zero integer then $i = j$ and such that if $\xi \in E_1(\bar{P},V)$ then there is $1 \le i \le d$, $n = 0,1,\ldots$ such that $\xi = z_i - n$.

Fix a realization (π,H) of G.

Theorem. If $\lambda \in V_1^*$ is of moderate growth in the direction P then there exist for $v \in V$, $P_{i,n}(t;v)$, polynomials on $\mathbb{R}$ of degree $d_{i,n}$ (independent of λ and V) such that

$$\lambda(\pi(\exp tH)v) \sim \sum_{i=1}^{d}\sum_{n=0}^{\infty} e^{(z_i-n)t} P_{i,n}(t;v),\ t \to +\infty.$$

Proof. The proof of this result follows the same line as the proof of Theorem 5.8. Let $\mu \in \mathbb{R}$ be such that

$$|(\underline{\bar{n}}\cdot\lambda)(\pi(\exp tH)v)| \le C_{v,\bar{n}}e^{\mu t} \quad \text{for } t \ge 1.$$

The first step is to "bring μ down to" the maximum of the $\operatorname{Re} z_i$.

(i) If $v \in \underline{\bar{n}}V$ then

$$|\bar{n}\lambda(\pi(\exp tH)v)| \le C'_{v,\bar{n}}e^{(\mu-1)t},\ \bar{n} \in U(\underline{\bar{n}}),\ t \ge 1.$$

Indeed, let $\bar{X}_1,\ldots,\bar{X}_r$ be a basis of $\underline{\bar{n}}$ such that $[H,\bar{X}_i] = -\alpha_i\bar{X}_i$. ($\alpha_i \ge 1$ by the choice of H).

Then $v = \sum \bar{X}_i v_i$, $v_i \in V$. So

$$\begin{aligned}
|(\bar{n}\lambda)(\pi(\exp tH)v)| &\le \sum_i |\bar{n}\cdot\lambda(\pi(\exp tH)\bar{X}_i\cdot v_i)| \\
&= \sum_i e^{-\alpha_i t}|\bar{n}\cdot\lambda(\pi(\bar{X}_i)\pi(\exp tH)v_i)| \\
&= \sum_i e^{-\alpha_i t}|\bar{X}_i\bar{n}\lambda(\pi(\exp tH)v_i)| \\
&\le \left(\sum C_{v_i,\bar{X}_i\bar{n}}\right)e^{(\mu-1)t},\ t \ge 1.
\end{aligned}$$

This proves (i).

Let $q_k : V \to (V/\bar{\underline{n}}^k V)$ be the natural map. Let $\gamma_\xi : V/\bar{\underline{n}}^k V \to (V/\bar{n}^k V)_\xi$ be the projection corresponding to the direct sum decomposition of $V/\bar{\underline{n}}^k V$ above. Set $q_{k,\xi} = \gamma_\xi \circ q_k$.

If $v \in V$ and $q_1(v) \neq 0$ then $(E_1 = E_1(\bar{P},V))q_1(v) = \sum_{\xi \in E_1} \gamma_\xi(q_1(v))$. We may thus choose $v_\xi \in V$ for each $\xi \in E_1$ such that $q_1(v_\xi) = \gamma_\xi(q_1(v))$ and $\sum v_\xi = v$. Thus for estimation purposes we may assume $v = v_\xi$.

Let $\bar{v}_1 = q_1(v), \bar{v}_2, \ldots, \bar{v}_d \in V/\bar{\underline{n}}^1 V$ be a basis of $U(\underline{a}) \cdot q_1(v)$. Let $v_1 = v$, $v_2, \ldots, v_d$ be such that $q_1(v_i) = \bar{v}_i$. Then $H \cdot \bar{v}_i = \sum b_{ij} \bar{v}_j$ and $(B-\xi I)^d = 0$. $H \cdot v_i = \sum b_{ij} v_j + w_i$ with $w_i \in \underline{\bar{n}}V$.

$$\text{Set } F_{\bar{n}}(t,v) = \begin{bmatrix} \bar{n}\lambda(\pi(\exp tH)v_1) \\ \vdots \\ \bar{n}\lambda(\pi(\exp tH)v_d) \end{bmatrix}$$

$$G_{\bar{n}}(t,v) = \begin{bmatrix} \bar{n}\lambda(\pi(\exp tH)w_1) \\ \vdots \\ \bar{n}\lambda(\pi(\exp tH)w_d) \end{bmatrix}.$$

Then

$$\frac{d}{dt} F_{\bar{n}}(t,v) = BF_{\bar{n}}(t,v) + G_{\bar{n}}(t,v).$$

Now one argues in exactly the same way as we did in the proof of Theorem 5.6 to find that if $\operatorname{Re} z_1 \geq \operatorname{Re} z_2 \geq \ldots \geq \operatorname{Re} z_r$

$$(1) \quad |\bar{n}\lambda(\pi(\exp tH)v)| \leq C'_{\bar{n},v}(1+t)^d e^{\operatorname{Re} z_1 t}, \quad t \geq 1.$$

It is now obvious how one completes the proof of the theorem following the line of the proof of Theorem 5.8.

7.3. We give some examples of λ of moderate growth in the direction P.

(1) Let (π, H) be a representation of G on a Banach space such that $H_K = V$. Let $\lambda \in H'_\infty$ be such that $\dim U(\underline{\bar{n}}) \cdot \lambda < \infty$. Then clearly λ satisfies 7.1 (2), (3).

(2) Let (π,H) be a representation of G on a Hilbert space such that $H_K = V$. Let $\pi^*(g) = \pi(g^{-1})^*$. Then (π^*,H) is a representation of G, the conjugate dual representation to (π,H). If $w \in \check{H}_\infty$ (H_∞ relative to π^*) then $v \mapsto \langle w,v \rangle$ is of moderate growth in the direction P.

The first example is the one that motivated the theorem. It answers a question posed by Piatetski-Shapiro. Let me state the question (which was originally posed with a specific example in mind, however the generalization is obvious). Let (π,H) be an admissible finitely generated Banach representation of G. Let χ be a unitary character of N. Let $\lambda \in H'$ be such that $\lambda(\pi(n)v) = \chi(n)\lambda(v)$, $n \in N$, $v \in H_\infty$.

Then does $\lambda(\pi(\exp tH)v)$ have an asymptotic expansion of the form

$$\lambda(\pi(\exp tH)v) \sim \sum_{i=1}^{r} e^{\xi_i t} \sum_{n=0}^{\infty} e^{nt} P_{i,n}(t;v), \quad t \to -\infty$$

with $t \mapsto P_{i,n}(t;v)$ a polynomial of fixed degree (independent of v) in t and ξ_i independent of v ?

Clearly Theorem 7.2 answers this question in the affirmative if one replaces P by $\overline{P}$. Furthermore the ξ_i in the expansion are just the weights of H on $V/\underline{n}V$.

Let us give the reason for wanting such an expansion. Let λ be as above. Let $f_{\lambda,v}(e^t) = \lambda(\pi(\exp tH)v)$. Then

$$(3) \quad f_{\lambda,v}(x) \sim \sum_{i=1}^{r} x^{\xi_i} \sum_{n=0}^{\infty} x^n P_{i,n}(\log x;v) \quad \text{as} \quad x \to 0.$$

One checks easily that

$$|f_{\lambda,v}(x)| \le C_{n,v} x^{-n} \quad \text{for} \quad x \ge 1 \quad \text{all} \quad n = 1,2,\ldots .$$

Since $|f_{\lambda,v}(x)| \le C_v x^{\mu}$ for all $0 < x$, we find that if we form

$$M(\nu:v) = \int_0^\infty f_{\lambda,v}(x) x^{\nu} \frac{dx}{x},$$

Then the integral defining $M(\nu:v)$ converges absolutely for $\operatorname{Re}\nu \geq \mu_0$. Clearly $M(\nu:v) = \int_0^1 f_{\lambda,v}(x)x^\nu \frac{dx}{x}$ + an entire function of ν on $\mathbb{C}$. Using (3) it is a simple matter to see that $\nu \mapsto M(\nu:v)$ has a meromorphic extension to all of $\mathbb{C}$ with the only possible poles at the points $-(\xi_i+n)$ with order at most (degree $P_{i,n}$) + 1.

One can construct examples where the expansions definitely do not converge to the function.

We also note that in example (2) above if we replace v by $\check{H}_K$ and P by $\overline{P}$ we get as a special case the expansion in Theorem 5.8 without the continuity in λ asserted in Theorem 5.8.

8. Poisson integral formulas

8.1. In this section we derive a general integral representation of a class of functions on G that has played a role in many contexts. Let $A(G)$ denote the space of all $f \in C^\infty(G)$ with the following properties.

(1) f is right K-finite.

(2) f is $Z(\underline{g})$-finite.

(3) There exists d (depending on f) such that if $g \in U(\underline{g})$

$$(g \cdot f(x)) \le C\|x\|^d, \quad x \in G.$$

For example $A(G) \cap C^\infty(\Gamma\backslash G)$ for $\Gamma \subset G$ a discrete subgroup so that $\Gamma\backslash G$ has finite volume is precisely the space of automorphic forms on $\Gamma\backslash G$.

We assume that G is linear.

8.2. Theorem. If $f \in A(G)$ then there exists (σ, W) a finite dimensional representation of P_0, $\lambda \in (H^\sigma_\infty)'$ and $v \in H^\sigma_K$ such that $\lambda(\pi_\sigma(g)v) = f(g)$, $g \in G$.

Proof. Set $V_f = \text{Span}\{U(\underline{g}) \cdot K \cdot f\}$. Here $x \cdot \phi(y) = \phi(yx)$, $x,y \in G$, ϕ a function on G, and $U(\underline{g})$ acts as left invariant differential operators (i.e., by differentiation on the right). Then (1), (2) imply that V_f is an admissible finitely generated $(\underline{g},K)$-module. We define $\delta(u) = u(1)$ for $u \in V_f$. Then $\delta \in V_f^*$. We assert that $\delta \in V_f'$ (see 6.1). Indeed, let $u \in V_f$; then $u = \sum k_i \cdot x_j \cdot f$, $k_i \in K$, $x_j \in U(\underline{g}_{\mathbb{C}})$. Set $f_{\delta,u}(g) = \sum k_i \cdot x_j f(g)$. We assert that

$$x \cdot f_{\delta,u}(k) = \delta(k \cdot x \cdot u).$$

Indeed

$$\begin{aligned} x \cdot f_{\delta,u}(k) &= \sum x \cdot k_i \cdot x_j f(k) \\ &= (k \cdot x \cdot \sum k_i \cdot x_j \cdot f)(1) \\ &= \delta(k \cdot x \cdot u). \end{aligned}$$

By assumption 8.1 (3) $|f_{\delta,u}(g)| \le C_{\delta,u}\|g\|^d$ with d depending only on f. Thus $\delta \in V_f'$.

Let σ be a finite dimensional representation of P_0 such that there is a surjective $(\underline{g},K)$-module homomorphism $T : H_K^\sigma \to V_f$. Let X be the closure of $\mathrm{Ker}\, T$ in H_∞^σ. Then T induces a surjective isomorphism on $T : H_\infty^\sigma/X$ onto $\overline{\overline{V}}_f$ (Proposition 6.5). Now by definition of $\overline{\overline{V}}_f$, λ extends to a continuous functional on $\overline{\overline{V}}_f$. Hence $\mu = \lambda \circ T$ is a continuous functional on H_∞^σ. Let $v \in H_K^\sigma$ be such that $T(v) = f$. Then $\mu(\pi_\sigma(g)v) = f(g)$, $g \in G$. Q.E.D.

8.3. We note that the converse of 8.3 is also true. That is, if σ is a finite dimensional representation of P_0 and if $\lambda \in (H_\infty^\sigma)'$, $v \in H_K^\sigma$ then

$$(g \mapsto \lambda(\pi_\sigma(g)v)) \in A(G).$$

Indeed, $|\lambda(\pi_\sigma(g)w)| \le \|g\|^{d_\lambda}\nu(w)$, $w \in H_\infty$, with ν a continuous semi-norm on H_∞. Thus if $f(g) = \lambda(\pi_\sigma(g)v)$ then $|x\cdot f(g)| = |\lambda(\pi_\sigma(g)\pi_\sigma(x)v)| \le \|g\|^{d_\lambda}\nu(\pi_\sigma(x)v)$.

8.4. We apply this to an interesting special case. Let $f \in A(G)$ and suppose that

(i) $f(gk) = f(g)$, $k \in K$.

(ii) There exists $\chi : U(\underline{g})^K \to \mathbb{C}$

such that $u \cdot f = \chi(u)f$, $u \in U(\underline{g})^K$.

Then V_f (or in the proof of Theorem 8.2) is a quotient of Y^ν (see 4.2) (here $\chi = q_\nu$ as in 4.2). We choose ν such that $\mathrm{Re}(\nu,\alpha) \ge 0$, $\alpha \in \Phi(P_0,A_0)$. Then by Kostant's Theorem ([9]), $Y^\nu \to H_K^\nu$ given by $g \cdot \overline{1} \mapsto g \cdot 1_\nu (1_\nu(nak) = a^{\nu+\rho}\, n \in N_0, a \in A_0, k \in K)$ is a $(\underline{g},K)$-module isomorphism. Thus we have

Theorem. If $f \in C^\infty(G/K) \cap A(G)$ and $u \cdot f = q_\nu(u)f$, $u \in U(\underline{g})^K$, and if $\mathrm{Re}(\nu,\alpha) \ge 0$, $\alpha \in \Phi(P_0,A_0)$ (this can always be arranged) then there exists $\lambda \in (H_\infty^\nu)'$ so that

$$f(g) = \lambda(\pi_\nu(g)1_\nu).$$

We state this in a somewhat more suggestive way. We note that $H_\infty^\nu = C^\infty(K/M)$, $(H_\infty^\nu)' = \mathcal{D}(K/M)$ the distributions on $C^\infty(K/M)$.

$$1_\nu(g) = a(g)^{\nu+\rho}.$$

Set $P_\nu(gK, k\cdot M) = a(kg)^{\nu+\rho}$. Then the theorem says that if f is as in the statement of the theorem then there exists $T \in \mathcal{D}(K/M)$ such that

(1) $$f(g) = \int_{K/M} P_\nu(gK,b)\,dT(b).$$

In [8] it was shown that if $f \in C^\infty(G/K)$ and $u \cdot f = q_\nu(u)f$, $u \in U(\underline{g}_{\mathbb{C}})^K$, then f has a representation of the form (1) with T an analytic functional (hyperfunction) on K/M. Thus the condition of moderate growth (8.1(3)) is a necessary and sufficient condition for the hyperfunction boundary value of [8] to be a distribution.

<u>8.5</u>. We conclude this paper with another example which was the original motivation for the definition of V'.

We assume that G is quasi-split and a Lie subgroup of $G_{\mathbb{C}}$, a complex Lie group with Lie algebra $\underline{g}_{\mathbb{C}}$. Quasi-split means M_0 is a Cartan subgroup of G. Let $\underline{h}$ be the complexified Lie algebra of M_0 and let $\underline{n}^+ = (\underline{n}_0)_{\mathbb{C}}$. Then $\underline{h} \oplus \underline{n}^+$ is a Borel subalgebra of $\underline{g}_{\mathbb{C}}$. Let Φ^+ be the roots of $\underline{h}$ on $\underline{n}^+$. Let $\Delta \subset \Phi^+$ be the set of simple roots in Φ^+. Then

(1) $$\bigoplus_{\alpha\in\Delta} (\underline{n}^+)_\alpha \oplus [\underline{n}^+,\underline{n}^+] = \underline{n}^+.$$

Let $\chi : N_0 \to S^1$ be a unitary character. Let $d\chi : \underline{n}_0 \to i\mathbb{R}$ be the differential of χ. Then χ is called generic if $d\chi|_{(\underline{n}^+)_\alpha} \neq 0$ for all $\alpha \in \Delta$. Fix χ a generic character of N_0.

If $V \in \mathcal{H}$ we define $Wh(V) = \{\lambda \in V^* | \lambda(X\cdot v) = d\chi(X)\lambda(v),\ v \in V,\ X \in \underline{n}_0\}$.

8.6. Theorem. (Notation as in 8.5). Let $\xi \in E({}^0M_0)$, $\nu \in (\underline{a}_0)^*_{\mathbb{C}}$; then

$$\dim Wh(H_K^{\xi,\nu}) \cap (H_K^{\xi,\nu})' \leq 1.$$

Proof. If $\lambda \in Wh(H_K^{\xi,\nu}) \cap (H_K^{\xi,\nu})'$ then since $(H_K^{\xi,\nu})^= = H_\infty^{\xi,\nu}$ (6.7), λ extends to a continuous functional on $H_\infty^{\xi,\nu}$. Now it follows from [10] (see also [3] for a simple proof using Bruhat theory) that

$$\dim Wh(H_K^{\xi,\nu}) \cap ((H_\infty^{\xi,\nu})'|_{H_K^{\xi,\nu}}) \leq 1.$$

8.7. Kostant [10] (Casselman, Zuckerman proved the result for the case of $GL(n,\mathbb{R})$ which was done a bit earlier) has shown that $\dim Wh(H_K^{\xi,\nu}) = |W(A_0)|$ ($W(A_0)$ the Weyl group of (G,A_0), the so called "small Weyl group").

Goodman-Wallach [3] have shown that each $\lambda \in Wh(H_K^{\xi,\nu})$ extends to an analytic functional on $H_\omega^{\xi,\nu}$ (we showed more but this will suffice in this discussion). Thus if $v \in H_K^{\xi,\nu}$ and $\lambda \in Wh(H_K^{\xi,\nu})$ then

$$g \mapsto \lambda(\pi_{\xi,\nu}(g)v)$$

defines an analytic function $f_{\lambda,v}$ on G such that $f_{\lambda,v}(ng) = \chi(n)f_{\lambda,v}(g)$.

The above theorem says that up to scalar multiple there is a unique $\lambda \in Wh(H_K^{\xi,\nu})$ such that $f_{\lambda,v} \in A(G)$ for $v \in H_K^{\xi,\nu}$.

This result had been conjectured to us by Piatetski-Shapiro.

We close with one more result about Whittaker vectors. Let $G_{max} = \{g \in G_{\mathbb{C}} | Ad(g)\underline{g} \subset \underline{g}\}$.

8.8. Theorem. Let $V \in \mathcal{H}$ be irreducible; then

(1) $\dim Wh(V) \cap V' \leq 1$.

(2) If $G = G_{max}$ and if (π,H) is a Banach representation of G with $H_K = V$ then $H'_\infty|_V \cap Wh(V) = V' \cap Wh(V)$.

<u>Proof</u>. Let $\xi \in E(M_0)$, $\nu \in (\underline{a}_0)^*_{\mathbb{C}}$ be such that there is $T : H_K^{\xi,\nu} \to V$ a surjective $(\underline{g},K)$-module homomorphism. Then $V' \circ T \subset (H_K^{\xi,\nu})'$ and $Wh(V) \circ T \subset Wh(H_K^{\xi,\nu})$. Thus

$$(V' \cap Wh(V)) \circ T \subset Wh(H_K^{\xi,\nu}) \cap (H_K^{\xi,\nu})'.$$

Thus $\dim Wh(V) \cap V' \leq 1$ by Theorem 8.6. This proves (1).

To prove (2) it is enough to show that $\bar{V}'|_V \cap Wh(V) = V' \cap Wh(V)$. (Indeed, $\bar{V}'|_V \subset H'_\infty|_V$ for any (π,H) a Banach representation with $H_K = V$.)

For this we use a result of Kostant [10] (which depends on a theorem of Vogan) which implies that if $\xi' \in E({}^0M_0)$, $\nu' \in (\underline{a}_0)^*_{\mathbb{C}}$ then there is a <u>unique</u> irreducible subquotient, $V_{\xi',\nu'}$, of $H_K^{\xi',\nu'}$ with the property that $Wh(V_{\xi',\nu'}) \neq 0$. We also need the basic theorem of Kostant [10] that if $0 \overset{\alpha}{\to} A \overset{\beta}{\to} B \to C \to 0$ is exact in H then $0 \to Wh(C) \overset{\alpha^*}{\to} Wh(B) \overset{\beta^*}{\to} Wh(A) \to 0$ is exact. Using the above two (hard) results the rest of the proof of (2) is now easy. Indeed, let ξ',ν' be such that there is an injective $(\underline{g},K)$-homomorphism $S : V \to H_K^{\xi',\nu'}$. Since (2) is clear if $Wh(V) = 0$ we assume $Wh(V) \neq 0$. Then $S(V) = V_{\xi',\nu'}$. By the above remarks, $Wh(H_K^{\xi',\nu'}) \cap (H_\infty^{\xi',\nu'})'|_{H_K^{\xi',\nu'}}$ is concentrated in $V_{\xi',\nu'}$. Thus since

$$(H_\infty^{\xi',\nu'})'\Big|_{V_{\xi',\nu'}=S(V)} = \bar{V}'$$

the second assertion follows.

<u>Notes</u>. 1. If we knew $\bar{V} = \bar{\bar{V}}$ always then the condition $G = G_{max}$ is unnecessary to the theorem.

2. Theorem 8.8 (1) which does not need the big machinery of the proof of (2) is a generalization of Shalika's multiplicity 1 result.

References

[1] W. Casselman and D. Miličic, Asymptotic behavior of matrix coefficients of admissible representations, Duke Math. J., 49 (1982), 869-930.

[2] J. Dixmier, "Enveloping algebras," North-Holland, Amsterdam, 1977.

[3] R. Goodman and N. R. Wallach, Whittaker vectors and conical vectors, J. Func. Anal., 39 (1980), 199-279.

[4] Harish-Chandra, Representations of semi-simple Lie groups, I., Trans. Amer. Math. Soc., 75 (1953), 185-243.

[5] ______, Representations of semi-simple Lie groups, II., Trans. Amer. Math. Soc., 76 (1954), 26-65.

[6] ______, Spherical functions on a semi-simple Lie group I, Amer. J. Math., 80 (1958), 241-310.

[7] S. Helgason, A duality for symmetric spaces with applications to group representations, Advances in Math., 5 (1970), 1-154.

[8] M. Kashiwara, A. Koroata, K. Minemura, K. Okamoto, T. Oshima and M. Tanaka, Eigenfunctions of invariant differential operators on a symmetric space, Ann. of Math., 107 (1978), 1-39.

[9] B. Kostant, On the existence and irreducibility of certain series of representations, Bull. Amer. Math. Soc., 75 (1969), 627-642.

[10] ______, On Whittaker vectors and representation theory, Invent. Math., 48 (1978), 101-184.

[11] B. Kostant and S. Rallis, On representations associated with symmetric spaces, Bull. Amer. Math. Soc., 75 (1969), 884-888.

[12] R. P. Langlands, On the classification of irreducible representations of real algebraic groups, preprint, Institute for Advanced Study, 1973.

[13] J. Lepowsky, Algebraic results on representations of semi-simple Lie groups, Trans. Amer. Math. Soc., 176 (1973), 1-44.

[14] J. Lepowsky and N. R. Wallach, Finite and infinite dimensional representations of linear semi-simple Lie groups, Trans. Amer. Math. Soc., 184 (1973), 223-246.

[15] C. Rader, Thesis, University of Washington, 1971.

[16] G. Schiffmann, Integrales d'entrelacement et fonctions de Whittaker, Bull. Soc. Math. France, 99 (1971), 3-72.

[17] J. T. Stafford and N. R. Wallach, The restriction of admissible modules to parabolic subalgebras, Trans. Amer. Math. Soc., 272 (1982), 330-350.

[18] F. Treves, "Topological Vector Spaces, Distributions and Kernels," Academic Press, N.Y. 1975.

[19] P. C. Trombi, The tempered spectrum of a real semi-simple Lie group, Amer. J. Math., 99 (1977), 57-75.

[20] N. R. Wallach, "Harmonic analysis on homogeneous spaces," Marcel Dekker, New York, 1973.

[21] ______, On the Enright-Varadarajan modules, a construction of the discrete series, Ann. Sci. Éc. Norm. Sup., (4) 9 (1976), 81-102.

[22] ______, Representations of semi-simple Lie groups, Proc. Canad. Math. Soc. Math. Cong., 1977, 154-245.

[23] ______, "Representation theory and harmonic analysis on real reductive groups," to appear.

[24] G. Warner, "Harmonic analysis on semi-simple Lie groups I," Grund. Math. Wiss., 188, Springer, 1972.

[25] ______, "Harmonic analysis on semi-simple Lie groups, II," Grund. Math. Wiss., 189, Springer, 1972.

[26] J. Dixmier and P. Malliavin, Factorisations de fonctions et de vecteurs indéfiniment différentiables, Bull. Sci. Math., 102 (1978), 307-330.

Department of Mathematics
Rutgers University
New Brunswick, NJ 08903